CLIMATE
ABANDONED

We're on the Endangered Species List

Edited by Jill Cody, M.P.A.

Writing Endeavors Press
Carmel CA 93923

www.climateabandoned.com

Ordering Information:
Special discounts are available on quantity purchases by corporations, associations, and others. For details, contact the publisher at the website above.

Publisher's Cataloging-in-Publication Data

Names: Cody, Jill, editor.
Title: Climate abandoned: we're on the endangered species list / Edited by Jill Cody, M.P.A.
Series: Abandoned.
Description: Includes bibliographical references. | Carmel, CA: Writing Endeavors Press, 2019.
Identifiers: ISBN 978-0-9977962-3-0 (pbk.) | 978-0-9977962-4-7 (ebook)
Subjects: LCSH Global warming. | Global warming--Prevention. | Climatic changes. | Climate change mitigation. | Climatic changes--Economic aspects. | Climatic changes--Government policy. | Sustainability. | Global temperature changes. | Environmental policy. | BISAC SCIENCE / Global Warming & Climate Change | POLITICAL SCIENCE / Public Policy / Environmental Policy
Classification: LCC TD171.75 .C54 2019 | DDC 363.738/746--dc23

Cover Design by Killer Covers
Interior Design by BookStarter

Printed in the United States of America

Pale Blue Dot: A Vision of the Human Future in Space

Look again at that dot. That's here. That's home. That's us. On it everyone you love, everyone you know, everyone you ever heard of, every human being who ever was, lived out their lives. The aggregate of our joy and suffering, thousands of confident religions, ideologies, and economic doctrines, every hunter and forager, every hero and coward, every creator and destroyer of civilization, every king and peasant, every young couple in love, every mother and father, hopeful child, inventor and explorer, every teacher of morals, every corrupt politician, every "superstar," every "supreme leader," every saint and sinner in the history of our species lived there--on a mote of dust suspended in a sunbeam.

– **Carl Edward Sagan**, an American astronomer, cosmologist, astrophysicist, astrobiologist, author, science popularizer, and science communicator in astronomy and other natural sciences.

For my beautiful granddaughter, Molly, with hope that she will have a bright and happy future in an ever-challenging world.

How to Read This Book

I invite you to read this book in any order you like! The urgent voices contained within these pages come from a variety of topics, perspectives, and experiences. Feel free to pick and choose to read those chapters that resonate with you first. I bet that after you do so, your appetite for other chapters will grow. Before you know it, you will be more knowledgeable than 90% of others regarding our civilization's climate crisis. That last statement may or may not be an exaggeration on my part but, sadly, it is probably more accurate than not.

Every chapter was written by individuals (www.climateabandoned.com) who possess expertise in their respective subject areas and, nearly all of them, are Leadership Corp Team members of former Vice President Al Gore's non-profit, "Climate Reality Project (www.climaterealityproject.org). To become a member, they each (including myself) were required to apply, be pre-screened, and selected to be personally trained by the former Vice President. In return, they committed to performing at least ten "An Inconvenient Truth" slideshow presentations in their communities.

Chapter authors were identified through a "Call for Authors" post I submitted to the Climate Reality Project via its Hub. The Hub binds authorized presenters together from around the world. The topics discussed in the book are ones who volunteered to write a chapter on their specific interest or expertise. Each chapter author did their best to relate the science accurately and to credit appropriately. Please be in touch if you have any concerns or comments.

Disappointedly, some topics are not covered in this book, such as how the climate crisis pertains to world population, livestock production, and food security, the reason why is an authorized presenter did not respond the "Call" to write on those topics. Maybe there is a second book on the horizon?

What makes this book different is the addition, at the end of each chapter, of an organization development tool called "Keep/Stop/Start." These are suggestions made by the chapter author to enable the reader to think about what they can *keep* doing, *stop* doing, and *start* doing. As Peter Fiekowsky, President and Founder of the Foundation for Climate Restoration stated:

> I say that only action will change the physical world. Knowledge certainly changes our perception of the physical world, but it doesn't change the world. Climate is a field which is ultimately physics. In contrast, other popular topics, such as civil rights, are mainly about perception—if everyone perceives that civil rights are being honored, then the job is complete. Knowledge can change action—but debate can displace action.

The debate is over. Action is what changes the physical world. We need to keep the climate crisis top-of-mind consistently, we need to *stop* being silent, and we must *start* focusing on our gut-level climate change issues toward addressing them, in addition to our other life challenges. The "Keep/Stop/Start" suggestions at the end of each chapter will jump start your personal strategies!

Constructive outrage! All of us who spent time and toil to write our respective chapters did so out of constructive outrage. We are outraged by the lack of urgency, regard, action, and the intentional obfuscation regarding the climate crisis by our national leaders and corporate multi-billionaires to make however hard or costly, the right decision to preserve a livable future for civilization. They appear to have no sense of self-preservation. They have *abandoned* humanity. Their own self-interest seems to be their only concern. Snowballs tossed on the Senate floor is an insult and an homage to the fossil fuel barons.

We know this book will be troublesome to read. It will take courage on your part to brave through the information. No one wants to be the bearer of bad news and, maybe all chapter authors won't agree on some details, but they all do agree that it is a message that must be said and heard. These *urgent* voices demonstrate the courage to face this potentially civilization destroying subject head-on. We know you can meet it head on too but in doing so, take a breath, take a walk, hug a loved one on either two or four legs. After reading the "Keep/Start/Stop" section at the end of every chapter, you will feel inspired, empowered, and less victimized by a seemingly overwhelming subject.

We need you to add your voice to ours. Read what topics speak to you, then become an *urgent* voice that cannot be ignored. Let's move ourselves off the Endangered Species List!

Preface

It's Too Hot for Coffee!

"If we don't take action, the collapse of our civilizations and the extinction of much of the natural world is on the horizon." – **David Attenborough**, BBC Planet Earth

It is two minutes to midnight. According to the Bulletin of the Atomic Scientists, a non-profit organization comprised of 15 Nobel laureates and the keepers of the internationally recognized Doomsday Clock, the scientists decided in January 2019, *not* to move the clock's hands. Every January, these esteemed scientists determine, based on nuclear and other weapons of mass destruction, climate change, and disruptive technologies where to place the hands towards the moment of apocalyptic destruction of planet Earth.

The Bulletin of Atomic Scientists was founded in 1945 by a former Manhattan Project scientist after the nuclear bombings of Hiroshima and Nagasaki, and the creation of the Doomsday Clock followed in 1947 (https://thebulletin.org/doomsday-clock/doomsday-dashboard/). Its purpose was to fill an urgent need the organization saw for educating scientists and the American public about the relationship between the world of science and the world of national and international affairs.

Looking on the bright side, I suppose it is encouraging that the Nobel laureates did not move the clock's hands any closer to midnight this year. They did so even after discussing their concerns about the Trump Administration's abandonment of the Iran nuclear deal, announcing the withdrawal from the Immediate-range Nuclear Forces Treaty (that subsequently occurred), the increase of global carbon dioxide emissions, and the retreat from the 2015 Paris Climate Accord.

However, if the Bulletin of Atomic Scientists met just one month later, they may have made a very different decision. On February 16, 2019, a New York Times opinion piece written by David Wallace-Wells, deputy editor of New York Magazine, said that in regards to climate change, it is time to panic. He stated in, "Time to Panic - The planet is getting warmer in catastrophic ways. And fear may be the only thing that saves us", that the age of climate panic is here (https://www.nytimes.com/2019/02/16/opinion/sunday/fear-panic-climate-change-warming.html). Tick tock.

By now, we all should be aware of the impending challenges ahead. Yet, we are not acting and solving problems at the World War II preparation level that is required. Maybe if we realized, at a gut level, that we won't be able to have our morning cup-of-coffee anymore then could we see the daily constructive outrage and activism that's needed?

"Coffee is not ready to adapt to climate change without help," said Doug Welsh, Peet's Coffee vice president and roast-master. Climate scientists are now saying that few coffee-growing regions will be spared the effects of climate change. It is literally becoming too hot for coffee.

As Starbucks founder, Howard Schultz, stated, "Make no mistake, climate change is going to play a bigger role in affecting the quality and integrity of coffee." The future loss of coffee is such a serious business issue to Starbucks that in 2013, the company bought a 600-acre plot in the province of Alajuela, Costa Rica that is now a field laboratory and testing facility to analyze the threats of climate change to Starbucks' business.

The threats are real. Coffee plants need to be grown in colder regions. As global temperatures have been rising, farmers have been moving their growing areas higher up mountains, searching for a cooler temperature.

Columbia, one of the world's best coffee growing nation, is already aware of how climate change will disrupt their small family farms with a wide range of climate-related weather such as floods, mudslides, erosion, droughts, and invasive insects.

Columbian leaders see the livelihoods of their 300,000 coffee producers endangered. When asked in August 2018, 90% of Columbia's coffee farmers said they have already seen changes in the average temperature and in the flowering and fruiting cycles of their coffee crops. The farmers saw the changing climate as an existential threat. Some of us who love coffee may see it that way too. Scientists are now working fast and furiously with 30 plant varieties in 20 different countries to attempt to find those varieties that may thrive in a warmer climate and in foreign soils.

Also, if you like strawberries, better eat them now. Savior now the almonds, walnuts, freestone peaches, cherries, apricots, kiwis, nectarines, and wine grapes. Yes, grapes for wine too! California, with its year-round sunshine and extent of fertile agricultural land, harvests a bounty of fruits, nuts, and wine to the tune of over $50 billion annually. A team from the University of California asked the question, what impact will climate change have on this agricultural and fiscal bounty (https://www.motherjones.com/food/2018/03/strawberries-al-monds-climate-change-drought-snowpack-california-yields/)?

What they found, with just a few degrees of temperature change, the land to grow these highly valued and loved crops will be too warm and agricultural yield will be significantly reduced. Also, daytime and nighttime heat waves are expected to increase in frequency and intensity. When temperatures rise, pests, crop diseases, and weeds become a significant problem affecting the quality of the harvest.

There is an old song, the chorus of which said, "Yes, we have no bananas." It was a funny song, but what will happen to bananas is not funny at all. Bananas are a vital crop grown in the tropics and subtropics. They're shipped all over the world as a nutritional food source to millions of people. In a recent study from Biodiversity International, it concluded that with rising planetary temperatures about 50% more of the land will become more suitable for growing banana crops. However, with that same rise in temperature, the banana crop growing cycle will be shortened. This may look like a positive, but with temperatures rising the banana crop's water demand will increase and, as with strawberries, pests and diseases (such as the black leaf streak disease) will also advance.

The effects of the climate crisis on our food will be dramatic like a slow-moving thriller building up to a shocking result. We face losing via temperature increases that favor pests and diseases and by ocean acidification: anchovies, apples, avocados, bananas, beans, beer, blue crab, cherries, chicken, chickpeas, chocolate, cod, coffee, corn, cranberries, honey, lobsters, maple syrup, oysters, peaches, peanuts, potatoes, pumpkins, rice, sardines, scallops, shrimp, soybeans, strawberries, turkey, and wine. See anything you love on that list? If we abandon our responsibilities to Mother Earth, we abandon our style of living too.

What a struggle it is, in a book meant to explain various climate crisis impacts, to be hopeful and optimistic when David Wallace-Wells basically states that now is the time to panic and to be fearful. To do so, then maybe the right and urgent decisions will be made to save our and our offspring's lives. Many of us following the crisis, have been worried and slightly panicky for many years, but it wasn't acceptable within the climate science world to say much about it for fear we would scare people into paralysis (https://www.breakthroughonline.org.au/pub-lications). That thinking has been flipped on its head, and now the message across the globe is ... panic!

Panic to action, not panic to paralysis!

The most critical and urgent action is to seek out, pressure, and support elected officials and those in positions of power to take this threat seriously. This is no longer an issue that can be kicked to the next generation. We are the pivotal generation.

"What's really breathtaking is how ill-prepared we are for such changes." – Bill McKibben, *"This is How Human Extinction Could Play Out,"* – Rolling Stone

Introduction

"Man stupid, fix Earth." – Koko, the gorilla

The old saying goes, "Outta the mouths of babes ..." This bottom line and to-the-point quote above is outta the mouth (or hands) of a bright, child-like primate. Koko, who passed away on June 19, 2018, having been taught a modified version of the American Sign Language, was a female western lowland gorilla. When Koko spoke, we all listened. When Koko painted, we all admired.

The actual quote, "Man stupid, fix Earth," is up for debate. Isn't everything? However, having lived for many years near where Koko lived, I followed her life, her sayings, and her paintings. I remember her sobbing and signing how sad she was when her beloved kitten, "All Ball," was run over by a car. For the rest of her life, after that tragic event, she talked about her "All Ball." Her caregivers got her another kitten, but the relationship was never the same. I know she could understand intangible concepts so, for me, the quote above is not off the mark.

Fixing Earth, as Yale's Program on Climate Change Communication puts it, requires lifting the "Spiral of Silence" on the topic (http://climatecommunication.yale.edu/publications/climate-spiral-silence-america/). In their survey, they learned that most Americans feel that climate change is a personally important issue, but that it was not a frequent topic of discussion nor did they hear much about it in their daily lives.

I would add that the topic may not be discussed because Americans can't trust what they've heard, as you will read in the Selling Doubt chapter. Over the past three decades, there has been a multi-million-dollar campaign to paralyze the average citizen by creating doubt in their mind. Creating uncertainly is easy to do. It doesn't have to prove anything. Understanding science is hard because its purpose is to prove a hypothesis painstakingly. America's Achilles Heel is the "quick fix." A speedy generality or falsehood feeds the "quick fix" mentality, and then America can move on.

Unlike the American media, foreign media is attempting to sound the alarm but, sadly, no other country can play the role the United States could. It was the United Kingdom's Daily Mail that reported the findings of University of California researchers last year who said that there is now a 1 in 20 chance that there could be a catastrophic climate event that could wipe out civilization by 2100. You think 2100 is far away? Do the math. And, the ride towards that year could be ruinous for millions before then.

To abate the *spiral*, Congress recently introduced a resolution titled, "The Green New Deal." The resolution was introduced in response to the realization that humanity has about a decade to get carbon emissions under control before catastrophic climate change impacts become unavoidable.

The most recent October 2018 UN Intergovernmental Panel on Climate Change (IPCC) stated that to avoid catastrophic climate breakdown requires "rapid, far-reaching and unprecedented changes in all aspects of society."

The Green New Deal resolution calls for a 10-year national mobilization of primary goals such as "meeting 100-percent of the power demand in the United States through clean, renewable, and zero-emission energy sources (https://en.wikipedia.org/wiki/Green_New_Deal). At least it appears the silence may have been broken and the debate is just beginning. However, the resolution is not legislation, and legislation of this type was needed years ago. Let's hope the debate doesn't take long. In a democracy, a debate is necessary and healthy. Yet, when it comes to the climate crisis, we need to act immediately. We must "fix Earth." Is "man stupid?" Well, we have abandoned our existential responsibilities far too long.

~ ~ ~

Abandonment: "To withdraw one's support or help from, especially in spite of duty, allegiance, or responsibility; desert." – The Free Dictionary

~ ~ ~

Politicians and others often profess to love our country, but words are not enough. Where does their *allegiance* really lie? We must be realistic when it comes to the problems we face, and put our sense of *duty* and *responsibility* into action in addressing the multitude of challenges and crises that have overtaken our society. It is for these reasons I wrote my first book, *America Abandoned: The Secret Velvet Coup That Cost Us Our Democracy*, that led to my producing and hosting a radio program, "Be Bold America!," on KSQD 90.7FM.

My purpose in editing and contributing to my second book, *Climate Abandoned: We're on The Endangered Species List*, is to increase awareness of the breadth and depth of our country's abandonment of humanity and all living things that the planet miraculously created. This book is for concerned readers who have observed some severe weather patterns such as the numerous and extreme wildfires or the extraordinarily powerful hurricanes we've seen in recent years and how intense they have become — and to offer practical suggestions on how to control the negative impact of abandonment on our lives and our environment with an organization development tool called a "Keep/Stop/Start", included at the end of each chapter. Only through awareness can change genuinely begin.

With great appreciation, this book, *Climate Abandoned: We're on the Endangered Species List*, would not have been possible if not for the dedicated work, experience, contribution, and passion of each contributing chapter author. They stuck with this book project through thick and thin during the two years it took to finish. They generously offered their time, research, and knowledge about the climate crisis to become an urgent voice calling to you.

Are you listening?

Contents

Note: This anthology will help you understand the serious dangers in each of the following areas.

What Did You Do Once You Knew?

It's 3:23 in the morning and I'm awake

Because my great-great-grandchildren won't let me sleep.

My great-great-grandchildren ask me in dreams

"What did you do while the Planet was plundered?

What did you do when the Earth was unravelling?

Surely you did something when the Seasons started failing?

As the mammals, reptiles, birds were all dying?

Did you fill the streets with protest when Democracy was stolen?

What did you do once you knew?

— **Drew Dellinger**

Chapter 1

Climate Crisis & The Greenhouse Effect

by

Mathieu Thuillier, M.P.A.

"Few people would dare traveling by plane if the risk of a crash was 1.8 percent." – **Johan Rockström,** Professor of environmental Studies, University of Stockholm

In most Westernized Judeo-Christian homes, children are raised to believe that once a year on Christmas Eve, Santa Claus gathers up his gang of eight or nine flying reindeer and departs from his home at the North Pole to deliver presents to families across the world. While this legend has changed over time, as folklore tends to do, it has stood the test of time and is repeated again and again to each successive generation. In the coming years, however, we may have to amend this holiday story, primarily because old Saint Nick won't have a home anymore. That's right, folks: Due to the effects of climate change, the North Pole as the icy area we know it may very well cease to exist.

In 2015, for the first time, global warming reached 1 degree Celsius (1.8 degrees Fahrenheit) above pre-industrial levels. What does that single degree mean for us and the planet? Believe it or not, it changes everything. The variation may sound quite small and insignificant, but the consequences are already dire. Just that one degree on the planetary scale is all it takes to make Arctic sea ice a thing of the past. Indeed, over the last few decades, the extent of this ice has declined by 13.3 percent per decade. In 1980, it covered almost eight million square kilometers at the annual minimum, making it roughly the size of the United States sans the states west of the Rocky Mountains. By 2016, it had shrunk to 4.7 million square kilometers. That one degree of separation in global temperature means a lot, and it has led to the collapse of the Arctic.

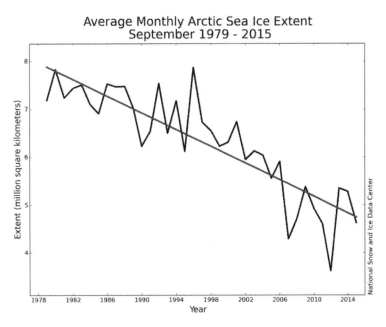

(Source: National snow and ice data center)

Peter Wadhams, a leading climate scientist who has made over fifty expeditions in the region, declared to *The Guardian* in 2016, "Next year, or the year after that, I think it will be free of ice in summer, and by that- I mean that the central Arctic will be ice-free. You will be able to cross over the North Pole by ship." Many thought this prediction was a bit off, but the facts have almost proved him right. In another not-so-great first, during summer 2017, a Russian tanker sailed along the northern coast of Russia, from Norway to South Korea, without an icebreaker escort. No, it did not navigate directly through the North Pole, but the perspective of an ice-free Arctic is dangerously closing in on us, and that would be a first in the last 100,000 years.

One thing is for sure: The next generation will inhabit a very different Earth than we do now. If things do not improve, the white hat over the North Pole will soon disappear, to be replaced by a gigantic, deep blue sea. Santa and his elves will either have to relocate to Antarctica or learn to swim.

Tinkering with the Machine

Why, though, is the temperature rising? The culprit, something dubbed the greenhouse effect, was identified some time ago, but it is, in fact, less of an enemy and more of a blessing for us. Indeed, based on the distance between the sun and our planet, the average global temperature should be around zero degrees Fahrenheit instead of the mild fifty-nine degrees Fahrenheit we enjoy nowadays.

The greenhouse effect is a natural consequence of the existence of a dense atmosphere around our planet. As energy produced by the sun reaches the atmosphere, some of it is reflected directly into space. Therefore, only part of the solar radiation reaches the Earth's surface and heats it. That heat then redirects toward space, in the form of long-wave radiation. Unfortunately, part of this long-wave radiation is trapped in our atmosphere by the so-called greenhouse gases, which mainly consist of water vapor, carbon dioxide, methane, nitrous oxide, and ozone. The extra energy trapped in the atmosphere warms our planet enough to sustain life. We call it a greenhouse effect because it works very similarly to a gardener's greenhouse. Light and heat come in, but only a small minority of it radiates back out, thus creating a warmer interior that enables things to grow.

The concentration of greenhouse gases in the atmosphere, especially carbon dioxide, has varied throughout the history of our planet, as has the amount of radiation trapped in the atmosphere and this has a direct impact on the temperatures we experience.

After water vapor, carbon dioxide is the most important of greenhouse gases, intimately coupled with the global temperature. This relationship has understood for over a century. In 1896, a Swedish scientist by the name of Svante Arrhénius calculated that "any doubling of the percentage of carbon dioxide in the air would raise the temperature of the Earth's surface by four degrees." As a Swede, he assumed this would be a blessing, as it would allow northern countries to develop flourishing agriculture. A century later, scientists agree that doubling the concentration of atmospheric CO_2 would increase global temperature between 1.5 and 4.5 degrees Celsius and result in dire environmental consequences worldwide. It is safe to say that modern-day scientists are a bit less enthusiastic than Arrhénius was.

The variation in the concentration of carbon dioxide in the past had natural causes, a notion particularly adamant climate change deniers love to proclaim. Volcanic activity does explain some of the changes, as eruptions inject notable quantities of carbon dioxide into the atmosphere, along with other gases. On the flipside, the weathering of silicate rocks—the long dissolution of rocks under the effect of the carbonic acid found in rain droplets—traps vast quantities of carbon dioxide in our oceans and rivers. Moreover, oceans exchange massive amounts of CO_2 with the atmosphere, but the extent of the inflow of carbon stored in the ocean varies over time; among other factors, determined by the changes in water circulation patterns. These are only a few of the natural processes that contribute to the regulation of carbon dioxide in the atmosphere.

Thanks to the Industrial Revolution, humankind has had significant influence over the natural carbon cycle, so much so that our role has begun to overshadow the natural variations. The main process through which carbon is released in the atmosphere is the burning of fossil fuels. Coal, oil, and natural and shale gas have formed over the course of millions of years, through the slow transformation of decaying plants and animals under extreme heat and pressure; mostly, they are concentrates of very energetic carbon. Now, in just three centuries, we've attempted to burn the entirety of this concentrated energy source, and we're close to succeeding. Nowadays, more than 80 percent of the energy consumed by industrial nations comes from fossil fuels. As these fuels are combusted, carbon that was stored deep within the Earth is released into the atmosphere, and it remains active for at least 100 years. This means the carbon released by Rudolf Diesel in his conception of the first compression ignition engine in 1892 only just dissipated from the atmosphere around the time the Mall of America in Bloomington, MI was beginning to open its doors twenty-five years ago.

We are capable of closely monitoring the concentration of carbon dioxide in the air, recently by direct measures and by analyzing ice cores before 1950. During the first decades of industrialization, CO_2 level in the atmosphere was about 280 parts per million (ppm). In other words, in an isolated million molecules of air, 280 were CO_2 molecules. In 2017, the concentration reached 410ppm, rising at a pace of 2 to 3ppm per year.[1] Thus, CO_2 atmospheric concentration has increased by 40 percent since 1750, to reach levels unprecedented in at least 800,000 years.

Of course, limiting the conversation to carbon dioxide is a mistake. Among the other greenhouse gases that are especially worth keeping an eye on is methane (CH_4). It does not have the same lasting impact in the atmosphere as carbon dioxide, as the molecules don't last much longer than twelve years up there, but the radiative effect is senseless. One methane molecule harbors warming power twenty-three times bigger than one molecule of carbon dioxide. Not only that, but CH_4 concentration has risen very fast in the atmosphere, up 150 percent since 1750, according to the IPCC. The rise in the future will be even faster and more furious, and there are many reasons why.

Methane can be the result of a biological process, the action of microbes comfortably living inside the intestines of cows, other ruminants, and termites. A single cow produces about 120kg of methane per year. Now, while flatulent cattle may be an endless source of jokes, they are a significant nuisance, especially when one considers the prediction that meat consumption will double over the first half of the twenty-first century and milk consumption almost doing the same. In other words, the human population is demanding more and more mooers. We should probably stop asking, "Got milk?" because the answer is that we do.

Under elevated temperatures and pressure, the breakup of organic matter also produces methane. Usually, this byproduct is trapped somewhere deep in the soil and does not contribute to global warming, at least not until some oil company decides to extract it for industrial or domestic use.

Another problem occurs when it is caught under ice. Indeed, colossal quantities of methane are trapped in Arctic regions under the *permafrost*, hard-as-a-rock soil that is frozen year-round. Furthermore, there is about twice as much carbon stuck under the permafrost than there is in the atmosphere right now. All this methane, produced by the slow decomposition of plants and trees, is only stopped from reaching the atmosphere by the formerly impenetrable barrier created by a thick, icy crust of soil. As temperatures rise, the ice in this barrier thaws that then weakens it. In these regions, temperatures are known to warm three times faster, on the global scale, and as the climate grows hotter the thawing of the permafrost frees immense quantities of methane. This escaped methane in the atmosphere, in turn, stimulates the warming of the planet, which thaws even more permafrost, which frees even more methane, which accelerates the warming, and so forth. The point is this: We are at the precipice of a vicious cycle.

---

[1] SCRIPPS Institution of Oceanography

Humanity is about to be confronted by a Goliath of a mounting problem. While we struggle to reduce our emissions, nature undermines these efforts, injecting extreme quantities of greenhouse gases into the atmosphere, without any possibility of regulation and this is the fear of many-a-scientist: We are creating a monster in the form of an uncontrollable, climate-warming machine. A recent study estimates that the thawing of permafrost may contribute up to 0.5 degree Celsius (0.9 degrees Fahrenheit) warming over this century alone.

Realistically, human activity is not all that bad for the climate. Ironically enough, pollution has thus far rescued us from some of the worst effects of global warming. That said, while spewing enormous quantities of carbon dioxide into the atmosphere by burning fossil fuels, we have simultaneously injected vast amounts of other pollutants in the air. The most common are sulfate aerosols, a byproduct of the burning of coal. While greenhouse gases absorb a lot of energy without reflecting any, sulfate aerosols do the opposite. They don't absorb energy but can reflect into space a significant part of the sun's radiation. They even interact within clouds to make them reflect even more sunlight than they otherwise would. In a nutshell, sulfate aerosols counteract the effect of greenhouse gases, cooling our planet. Scientists believe these aerosols mask a severe impact of rising greenhouse gas concentrations, and without this pollution, our world would even be warmer.

If that is the case, why not inject as much sulfur in the air as possible, in the hopes of entirely offsetting our carbon emissions? Surely, we can dump a load of sulfuric acid high in the atmosphere every once in a while, and go on with our carbonated lives. This may sound absurd, but this thought has gained quite a bit of traction over the last years as an easy, rather inexpensive solution to global warming. One tiny problem is that sulfate aerosols lead very short lives in the atmosphere. After a few days, months, or years, they return to the ground, often in the form of acid rain, which is damaging to the environment as well as to human health.

Moreover, once we start this process, there would be no turning back, because if we failed to replenish the high atmosphere with sulfuric acid, the temperature would ultimately catch up, rising five to ten times faster than it would otherwise. Which would you rather do, endeavor on a problematic decision-making process to find feasible measures we can take to mitigate climate change and develop sustainable energy sources or enjoy temporary easement from global warming while we try to invent acid rain-proof umbrellas? Scientists are still working out the kinks, and there are a lot of them to work out!

While aerosols do counteract the effects of our greenhouse gas emissions, the tendency is to curb these emissions as much as possible. Most developed countries, the U.S. being one successful example, have developed technology and legislation to diminish aerosol emissions. Why curb aerosols? In addition to acid rain, sulfate aerosols are known to cause lung irritation and smog that lingers over many cities. As China and India get richer, they will likely tackle this problem and increase their efforts to combat atmospheric pollution. Thus, emissions of sulfate aerosols are expected to drop by half throughout this century and will directly benefit populations affected by this pollution. Conversely, it will suppress one of the main dampeners of global warming.

A Resilient Nature

The mean global temperature has not always been fifty degrees Fahrenheit. Frankly, it has varied a lot throughout history, much colder and/or warmer on the surface of Earth in the past. Studies of our past prove that nature can thrive in a hotter environment. Why, then, are scientists so alarmed that temperatures are rising now? The problem is not the level of the temperature, as the pace of change.

Indeed, nature has proven to be very resilient, with a capacity to adapt to almost any conditions, settling in the most hostile of environments. However, it takes time to adapt and to colonize new territories. At the end of the last Ice Age, some 20,000 years ago, the mean temperature of the planet rose by 5 degrees. As the climate warmed and ice started to recede, animals and plants migrated north. It's worth noting that that change occurred over several thousand or tens of thousands of years. This time, we are facing warming of a comparable magnitude, only it will happen on fast-forward, in just a few decades. In 2013, Stanford climate scientists estimated, based on our knowledge of the past climate, that the pace of change could be ten times faster than any change recorded in the past sixty-five million years.

Because they depend on a narrow window of temperature for survival, about half of the species of the planet, both animal and plant, are on the move.[2] They tend to move northward, so they benefit from similar climatic conditions, but not all can keep up with the pace of our changing climate. Living things that lack legs or wings, such as trees, are at a true disadvantage. For our purposes here, we'll ignore the fictional *Lord of the Rings* Ents. In real life, trees can't move, but they can migrate. Indeed, where they sprout matters, and their distribution can change as old trees die and new ones grow. That is what is happening in the tree world right now. For example, studies show that in the States, climate change has pushed trees an average of twenty miles north and twenty-five miles west over the last thirty years. If you think that's bad, brace yourself: We've only just crossed the threshold of the doom and gloom.

Remember polar bears, the Great Barrier Reef, and small island nations? If you don't, you ought to read up on them because they will soon disappear. The polar bear population is shrinking as Arctic ice melts. There are now about 30,000 polar bears left, and scientists are very concerned about the number that have starved to death. Because of warming waters, from 2015 to 2016, the Great Barrier Reef experienced one of the worst bleaching events in its history. Scientists lament this as a terminal stage for the extremely delicate ecosystem.[3] On some of the Pacific atolls, as the soil is infiltrated by saltwater and the coast recedes, authorities are already planning for the seemingly inevitable resettlement of the population.

Every degree matters and the impact of every single degree gained or lost is dramatic and catastrophic. Keenly aware of this, at the end of 2015, the leaders of Planet Earth gathered in Paris, France and agreed that something must be done to stop global warming "well below 2 degrees Celsius above pre-industrial levels and pursuing efforts to limit the temperature increase to 1.5 degree Celsius."

While the agreement had just entered into force, the publication of a new study in *Nature Climate Change* in July 2017 dampened the enthusiasm of the international community. According to researchers at the Universities of Washington and California, the perspective of a limitation of the global warming to 1.5 degrees is unrealistic, and our chances of limiting the rise of temperatures to 2 degrees is a mere 5 percent. Though technological advances are expected to reduce our emissions considerably, these will not suffice. The sad truth is that even the Paris Accord, which was not legally binding and has recently been shunned by President Trump, set unreachable goals, as admirable as they were.

Planet Earth seems to be on a path of warming by three degrees Celsius. The risk of a 6-degree warm is estimated at 1.8 percent, and they may look like low figures, but not according to Johan Rockström, Professor of Environmental Sciences at Stockholm University, who says, "Few people would dare traveling by plane if the risk of a crash was 1.8 percent." The difference between a three-degree warm-up and a six-degree one is enormous. For South Asia, for example, the latter would mean exposing 4 percent of the population to heatwaves, with heat and humidity that would prevent the human body from being able to cool itself. Such intense heatwaves can prove fatal even to fit people sitting in the shade, in less than six hours. In the Gulf region, temperatures could rise over sixty degrees Celsius, and the same phenomenon of deadly heatwaves would occur before the end of the century.[4] In Europe, the death toll attributed to global warming would rise to 152,000 a year, according to a study published in 2017 in The Lancet Planetary Health. At the same time, experts estimate that these events would be improbable if the terms of the Paris agreement were respected.

---

[2] *Science*, August 11, 2011
[3] *The Guardian*, April 9, 2017
[4] Based on studies conducted by MIT researchers, published in *Nature Climate Change and Science Advances*

But a Quite Fragile Mankind

The skeptics are right: In the past, the climate has been much warmer than it is today, yet they have lived to tell about it. Our species will probably survive any change, but rather than asking what we can endure, shouldn't we be focused on what is desirable? This is precisely the question Johan Rockström, one of the leading scientists in sustainability science, asks.

To answer this question, we must travel back in time. The last 800,000 years were marked by a variation between two weather states: long periods of glaciations (with an average temperature about four degrees lower than it is today) and somewhat short warm interglacial periods, with an average temperature as high as today. To an inhabitant of Stockholm, these four degrees constitute "the difference between having two kilometers of ice above our heads and the warm, lush environmental conditions that we are so used to," explains Rockström. Even minimal variations in temperature heap enormous consequences upon our planet.

During our journey, humankind has suffered the frequent, abrupt variations of climate, and we did not always thrive. We were hunters and gatherers, scattered all over this third rock from the sun. At some point, about 75,000 years ago, during a particularly harsh glacial period, there were about 15,000 humans left on Earth, roughing it out on the Ethiopian highlands. "We were essentially extinct," explains Rockström. Our species had to wait for a warm and incredibly stable period, the Holocene, to start its impressive development. Only when the seasons became predictable did humans start to develop agriculture, and from there, to build the wonders we know. During the Holocene, temperature flattened out, plus or minus one degree. As Rockström sees it, this means only one thing: "The Holocene is the only stable state of the Earth that we know can support the modern world as we know it."

So, why imperil the stable climate that has proven to be such an ally in our development as a species? Why would we willingly or foolishly jump into an unknown, more violent world? These questions are only rhetorical since most modern geologists agree that we have left the Holocene and have crossed over to a new geological period, the Anthropocene. This is not, by any means, an innocent moniker; *Anthropocene* means the era of humankind, and it is named as such because scientists reckon that our species, for the first time, is the driving force of nature and climate, rather than the other way around. Now, <u>we</u> define what kind of climate our children and grandchildren will have to survive.

The Risk of Catastrophic Change

We may be the driving force, but that doesn't necessarily put us in the driver seat to decide the outcome. Pushed too far, the climatic machine could prove uncontrollable. Why? Because climate is not a linear system, which means every ton of carbon dioxide injected into the atmosphere will not have the same impact on our climate. Up to this day, polar ice caps have contributed to the cooling of the Earth. White ice reflects into space a larger part of incoming solar radiation than dark surfaces. Thus, an essential portion of solar energy exits the system without impacting it. If that white ice is replaced with dark blue seawater, incoming radiation will contribute to the warming of the oceans and, eventually, of the atmosphere. Therefore, the planet will heat up faster if the ice disappears. You probably realize you feel much warmer on hot days when you wear a black t-shirt as compared to a white one, and that is based on a simple concept: White reflects light, while dark absorbs it. Similarly, if permafrost thaws, vast quantities of greenhouse gas will be introduced into the atmosphere. Even if humanity has managed to embrace an entirely carbon-free lifestyle by then, the system will still continue to heat up.

Countless mechanisms impact the system, and most work as stabilizers. When the climate system is pushed a little in a specific direction, it tends to return to its initial state or at least to dampen the change. Planet Earth has na inertia all its own, and if we push the system too hard and too fast, we risk crossing a threshold and tipping everything off balance, putting us in a new state. In our case, the sudden injection of massive quantities of greenhouse gases into the atmosphere puts us at risk for forever departing from the stable, mild climate that has characterized the last 10,000 years. In exchange, we could enter a whole new era, forced to endure and adapt to a wild, unstable, much warmer climate. At least for a while, humans will suffer if this balance is upset.

Therefore, it's crucial for us to respond to this problem now. If we cross that threshold in a few years, it won't matter if we stop emitting greenhouse gases. By then, it may be too little, too late, with the climate stuck in a perpetual warmer state. There is a point of no return, and we are on course to reach it. This is the theory developed by Dr. Rockström and his colleagues at the Stockholm Resilience Center. His team has identified nine essential mechanisms and cycles that rule nature. For most, they have defined boundaries beyond which we will enter unchartered territory. Beyond these tipping points, it is not certain that nature can compensate and return to its former equilibrium. One major issue for climate research is to determine where the tipping points are, but as of now, we know too little to quantify it. At the same time, we know enough to understand that the risk is real.

According to Stockholm Resilience Center researchers, we have already crossed the boundary when it comes to climate change, and we are already in the risk zone. To avoid reaching the point of no return, we need to act now, without delay. Every year of hesitation puts us at higher risk.

We have entered the Anthropocene, the era of humankind. We are now the defining force on this planet, and too few of us take this responsibility seriously. Indeed, we have gained power over nature, yet we refuse to harness it for the greater good. Our actions will define the future of our planet and our species. We choose to live as we please, without any care about the consequences of our actions. We behave like rebellious teenagers who would smoke cigarettes, eat fast food, and play videogames all day as if there will be no adverse effects. The only difference is that we know the risks.

Though the amount of available information about the crisis is ever on the rise, we have collectively made only a single lackluster concession so far: We will reduce our emissions, as long as it does not change anything in our lives. We have chosen to rely on technological advances to clean up our mess and to reduce our greenhouse gas emissions. For now, this seems to be working to stabilize global emissions, but we cannot wait for technology to save us. We must question our way of life. Is there something we can do to go a bit further? Definitely!

By now, we are sure that some of the riches of nature will be gone when our grandchildren are born, but we can save the rest. By now, we know for a fact that coming generations will live in a much more hostile environment, with increased frequency and intensity of hurricanes, tropical storms, and extreme heatwaves, but we can do something to keep these catastrophes and their severity somewhat at bay. We can do a lot to minimize the extent of the damage, and we can save millions of lives. We can preserve the wellbeing of billions of individuals on the planet, but only if we act now.

Climate Crisis "Keep/Stop/Start" Actions

Below for your consideration, are several author-suggested thought, behavior, and action items. For a further explanation of the Keep/Stop/Start organizational tool, please see How to Read This Book.

KEEP:

1) *Keep reading and gathering information about the climate crisis.* The more we understand it, the easier it is to take action. There are seemingly infinite dimensions to the issue: economics, society, science, business, culture, education, etc. Determine which area is most interesting and relevant and useful to you, then pursue learning about it. There are many ways we can curb our habits, educate our peers, and reduce damage, but they all require continued research and persistent action.

2) *Keep enjoying the natural world.* As noted in this chapter and many other parts of this book, the environment is rapidly changing as the climate warms. In the future, glaciers will melt, rivers will dry up, forests and vegetation will shift, animals will migrate, and the environment will look dramatically different than it does today. Although we should continue to do everything we can to combat these changes, we should also enjoy the natural world as it exists now. The Earth our grandchildren will inherit will look much different than the one we occupy now so we should take advantage of what it has to offer while we still can.

STOP:

1) *Stop thinking it is a done deal, that the crisis is beyond our control.* Although we are quickly approaching the tipping point, we have not entirely passed the point of no return. There are still many things we can do to mitigate and adapt to climate change. However, if we hold on to the mindset that the battle is over, then we have already lost. We must continue the fight with a sense of urgency and insistence to prevent as much harm as possible, but it all starts with our perception of our capacity to do so.

2) *Stop allowing misinformation and ignorance to persevere.* When it comes to a polarizing subject like climate change, there are many arguments both for and against it that take advantage of statistical manipulation, biased opinions, and scientific misunderstanding. There is little to be gained from engaging in discussions based on incorrect information, assumptions, and myths; thus, we must correct the falsehoods and present the truth. That said, we don't need to use hyperboles and exaggerations to illustrate the necessity for action either. The climate crisis is a daunting enough predicament on its own, and we don't need to embellish it with scare tactics and overstatements. Honest conversation to inform the uninformed about the scientific evidence, the historical context and the repercussions we are likely to face will inspire people to act far more than spewing doom and gloom and accusations.

START:

1) *Start thinking about what you can do to reduce your individual emissions.* Is it necessary to spend your holidays on the other side of the world? Is there a cleaner or sturdier alternative to the products you want to buy? And, most of all, do you need those products at all? When considering how to reduce emissions, most people automatically think about cutting down on energy use and transportation habits, but many less invasive methods will not demand such drastic changes to your daily life. Altering the content and source of your diet, paying closer attention to the types of materials you use and purchase, and being mindful of the businesses you invest in and the politicians you support can all have a significant effect on your own emissions, and they don't require an overhaul of your day-to-day life.

2) *Start investing in clean energy.* Investing will benefit your wallet as well as the planet. Renewable energy like wind, solar, hydro, geothermal, and biofuel are now economically competitive with fossil fuels. Regardless of what our current president says, markets are likely to favor these resources in the future. Investing now will allow for individual and collective financial prosperity and will work wonders for transitioning our economy to become more sustainable and reliable, in addition to curbing greenhouse gas emissions and reducing climatic change.

~ ~ ~

Mathieu Thuillier, M.P.A. obtained his master degree in political science from Lille (France) and is a specialist of environmental policies and nuclear energy. He studied journalism at the Sorbonne, Paris, before working as a newspaper and radio journalist for ten years. Eight years ago, Mathieu moved to Sweden to become a teacher and climate activist, and he currently resides in Stockholm. He joined the Climate Reality Project two years ago and completed training in Toronto, Canada. Read more at: www.climateabandoned.com.

Chapter 2

Climate Crisis & Biodiversity in a Warming World

by

Rituraj Phukan

(To my lovely wife Rosee and our beautiful child Rene Raj)

"Biodiversity starts in the distant past and it points toward the future." – **Frans Lanting**

I was ecstatic when we finally rolled into Churchill, located on the western shore of Canada's Hudson Bay, after all the talk in the dining car about whether we'd need to be airlifted from Thompson. I had chosen the train journey to experience the diverse biomes, from prairie grasslands to the milieu of boreal conifers and muskeg, and ultimately, the marine and Arctic tundra, and I did not initially understand why there was a five-hour scheduled stop at Thompson. We came to find out that the climate change-impacted train runs on an unstable track undermined by thawing permafrost and thus often progressively slows during its northward trundle. But lucky us! Our delay was eventually truncated to only five minutes. Although we may have been fortunate that our travel time was much shorter than expected, Churchill isn't really lucky at all, for it is on the frontlines of a battle, the battle of climate change.

A biodiversity-rich ecotone[1] that was once the historic center of the North American fur trade, Churchill was where early European colonialists established the foundation for industrial exploitation of wildlife, and it now advertises itself as the polar bear capital of the world. I was there to take part in an Earthwatch expedition dubbed Climate Change at the Arctic's Edge, at the Churchill Northern Studies Center (CNSC), located about fourteen miles from town.

We had just passed the last of the ubiquitous polar bear warning signs at the outskirts of town when someone screamed.

"Polar bear!"

The van stopped. An adult bear had crossed our railway track and was now less than 200 meters from the road, heading in our direction. She briefly stopped, stood up on her hind legs, and sniffed the air. She was more stunning than I had ever imagined: huge, glistening white, and full of grace. Then she turned and briskly walked toward the station. Already, in less than half an hour after we reached Churchill, my polar bear dreams had come true, but for reasons I wish weren't the case.

Lead CNSC scientist Dr. LeeAnn Fishback, in the driver seat, informed us that intrusions into Churchill have increased in recent years because of climate change. Due to the early meltdown and late formation of ice in Hudson Bay, the bears had been forced to spend more time on land, and in town, to find food. She slowed to point out the Polar Bear Holding Facility, a former military holding station and the only bear jail in the world, where intruding bears are often kept in solitary confinement, and fed only snow and water until they can be released far from town. It was heartbreaking to envision a bear behind bars solely because it was doing what it needed to do for survival when it, like other species affected by climate change, was not the least bit responsible for what was happening to its habitat.

Earth has enjoyed a relatively stable climate for the last 10,000 years, and life on the planet has evolved to exist within specific temperature ranges. But unprecedented and abrupt climate change in recent decades has

---

[1] Ecotone is a transition area where two communities meet and integrate. https://en.wikipedia.org/wiki/Ecotone

adversely impacted ecosystems and biodiversity around the world, and most species, like the polar bear, have been unable to adapt to their varied environments.[2] To make matters worse, fragmentation and corresponding isolation of habitats have made large-scale natural migration nearly impossible.

To estimate the effect of climate change on species, scientists use what is known as a *climatic or bioclimatic envelope,* or a range of temperatures, rainfall, and other climate-related parameters, in which a species currently exists. Applying the Goldilocks Principle, conditions need to be just right, or species will tend to move toward the edges of their climatic envelopes. The polar bear, for example, wants to migrate to cooler, moister environments, usually uphill or toward the North or South Poles. In many cases, however, such migration proves impossible because of unfavorable environmental parameters, geographical or human-made barriers, and competition from species that already inhabit the new area.[3] Melting ice is one of the principal factors that prevent the polar bear from going where it needs to go.

Climate change is a universal problem, but the Arctic region is warming at double the average global rate. [4]Moreover, physical changes such as sea ice loss, permafrost thaw, coastal erosion, ocean acidification, and sea level rise in the Arctic have had a profound and undeniable ecosystem impact.[5] The Arctic Council has documented ecosystem changes across the entire region and noted that these changes are accelerating in response to physical climate changes3. The Council's biodiversity working group, the Conservation of Arctic Flora and Fauna,[6] reports there has been disappearance or dramatic modification of habitats, ecosystems, and populations; shifts in their geographic ranges; changes in the timing of ecological events like flowering, migration, and breeding; and outbreaks of pests and disease affecting plants and animals. The Council's 2016 Arctic Resilience Report describes nineteen regime shifts, defined as large and persistent changes in the structure and function of social-ecological systems, that have already made an enormous impact on native wildlife, climate stability, and the people who depend on both. Arctic air temperatures for the past five years (2014-18) have exceeded all previous records, confirming declining snow cover, melting of the Greenland Ice Sheet and lake ice and expansion of harmful toxic algal blooms.[7]

Changes in the Arctic environment have sweeping consequences for the rest of the world and may also foreshadow what could happen elsewhere on the planet. Understanding how those changes, including warming temperatures and melting sea ice, have generated marine and terrestrial biodiversity losses in the Arctic is only the first step. Looking at those losses on a planet-wide basis, and why they're so important, is critical to our collective responses and reactions to overall climate change.

Marine Biodiversity

Polar bears have emerged as the face of something far more horrible than Coca-Cola; they are now the poster children for the mal-effects of climate change. Photographs of emaciated or stranded bears are standard in media reports about Arctic warming[8], and for a good reason: What happens to these furry creatures should matter to all of us. First, they are both historically and currently integral to the cultures and economies of Arctic communities, and indigenous people in particular. Second, as apex predators, they are critical players in their ecosystem, and their survival – or failure to survive – may portend what could happen to many other species. And finally, their

[2] Shah, Anup. Climate Change Affects Biodiversity.
[3] Bellard, Céline, Paul Leadley, Wilfried Thuiller, Franck Courchamp. Impacts of climate change on the future of biodiversity.
[4] Arctic Council (2016). Arctic Resilience Report. M. Carson and G. Peterson (eds). Stockholm Environment Institute and Stockholm Resilience Centre, Stockholm. http://www.arctic-council.org/arr.
[5] Meltofte, H. (ed.) 2013. Arctic Biodiversity Assessment. Status and trends in Arctic biodiversity. Conservation of Arctic Flora and Fauna, Akureyri.
[6] CAFF. 2017. State of the Arctic Marine Biodiversity: Key Findings and Advice for Monitoring. Conservation of Arctic Flora and Fauna International Secretariat, Akureyri, Iceland. ISBN: 978-9935-431-62-2
[7] Osborne, E., J. Richter-Menge, and M. Jeffries, Eds., 2018: Arctic Report Card 2018, https://www.arctic.noaa.gov/Report-Card.
[8] WWF Global. Global warming impacts in the Arctic and Antarctic

beauty and grace are unparalleled, and as such, they symbolically remind us that Earth, and all her flora and fauna, are gifts to, and not the property of, humankind.

Polar bears rely mainly on seals for sustenance, and seasonal melting of sea ice, which serves as a hunting platform for the bears, forces the bears to go ashore and fast for the warmer months. Typically, bears would fast for about four months. Now, faster melting of sea ice means longer fasts – or the need to forage where humans live. One-third of the global population of polar bears have already felt the effects of these sea-ice changes. Studies have shown that for each week of early ice breakup, bears come ashore ten kilograms lighter and, therefore, in poorer condition.[9] Females are also giving birth to fewer cubs;[10] in fact, sightings of polar bear mothers with triplets have become rarer in the past decade.

Warming temperatures are also impacting polar bears adversely. An increase in wildfires and spring rainfall work to collapse their maternity dens, as well as the dens of the ringed seal, their prey, so their habitats become unstable and their food becomes scarcer. Although polar bears still inhabit their historic range in nineteen Arctic subpopulations, eight of these are in decline, primarily because of climate warming. With the continuing loss of sea ice, along with additional stressors like low genetic diversity, human habitation, industrial activities, toxic substances in their food web, and reduced populations of potential prey, polar bears are likely to completely disappear from the southern Arctic ranges. This will happen within thirty to forty years, and some studies predict that two-thirds of the polar bear population will be completely wiped out by mid-century.[11]

The expansion of harmful toxic algal blooms, coinciding with warming oceans, threatens food sources in the Arctic region with deaths reported in birds, seals, walruses, and whales.[12] The Arctic Report Card 2018 adds another dimension to the mounting challenges to Arctic wildlife. Microplastic contamination is on the rise, posing a threat to seabirds and marine life that can ingest debris. There has been a 20-fold increase in marine litter in some areas of the Arctic and doubling of microplastic concentrations in others over the course of a decade.[13]

Arctic marine species are also at high risk or on the move, especially southern Arctic species, as global temperatures rise and traditional food sources diminish. Other changes in sea ice, such as age and salinity, also affect marine life. For example, multi-year sea ice is being replaced by first-year sea ice, which causes shifts in ice algal communities, with cascading effects on the ice-associated ecosystem.[14] Rapidly changing conditions are also a potentially serious threat to species that rely heavily on sea ice. Walruses, for example, like to rest on sea ice located directly over prime feeding areas, but often times they are now stranded at coastal sites because of late-season ice formation. Also, walrus calves run the risk of being stampeded when the ice eventually forms, and the population moves out to the food location. Changing ice conditions are also impacting Narwhals, who wind up becoming entrapped beneath the unfamiliar ice patterns and wind up dying from starvation, suffocation, or predation.[15]

When species have to travel farther than ever before to find food, they expend far more energy than should be necessary. They also have to compete with, or otherwise interact with, new species they've never before encountered, which can seriously disrupt the Arctic's delicate marine ecosystems.[16]

---

[9] Galicia, M. P., Thiemann, G. W., Dyck, M. G., Ferguson, S. H. and Higdon, J. W. (2016), Dietary habits of polar bears in Foxe Basin, Canada: possible evidence of a trophic regime shift mediated by a new top predator. Ecol Evol, 6: 6005–6018. doi:10.1002/ece3.2173

[10] http://www.telegraph.co.uk/news/earth/environment/climatechange/8311137/Polar-bears-having-fewer-cubs-due-to-global-warming.html

[11] Lunn, N. J., Servanty, S., Regehr, E. V., Converse, S. J., Richardson, E. and Stirling, I. (2016), Demography of an apex predator at the edge of its range: impacts of changing sea ice on polar bears in Hudson Bay. Ecol Appl, 26: 1302–1320. doi:10.1890/15-1256

[12] Anderson, D.M., M. L. Richlen, K. A. Lefebvre, 2018: Harmful Algal Blooms in the Arctic [in Arctic Report Card 2018], https://www.arctic.noaa.gov/Report-Card.

[13] Peeken, I., M. Bergmann, G. Gerdts, C. Katlein, T. Krumpen, S. Primpke, M. Tekman, 2018: Microplastics in the Marine Realms of the Arctic with Special Emphasis on Sea Ice [in Arctic Report Card 2018], https://www.arctic.noaa.gov/Report-Card.

[14] CAFF. 2017. State of the Arctic Marine Biodiversity: Key Findings and Advice for Monitoring. Conservation of Arctic Flora and Fauna International Secretariat, Akureyri, Iceland. ISBN: 978-9935-431-62-2

[15] Meltofte, H. (ed.) 2013. Arctic Biodiversity Assessment. Status and trends in Arctic biodiversity. Conservation of Arctic Flora and Fauna, Akureyri

[16] M. Duarte, Carlos. Impacts of Global Warming on Polar Ecosystems

Humpback, fin, and minke whales, as well as harp and other seals and walruses, are all poised to expand their commutes and disperse further north in search of food. Gray whale calls have been recorded far north of their present range in the western Beaufort Sea. And the Atlantic killer whales, swimming through huge swaths of open water created by melting ice, will likely threaten the existence of resident Arctic marine life such as the bowhead, narwhal, and beluga whales.[17]

Similarly, the expanding distribution of Atlantic cod is increasing predation pressure on the polar cod, which is an important, nutrient-rich prey fish that other fish, seabirds, and marine mammals rely upon.[18]

Arctic seabirds, including black guillemots and ivory gulls, have also had to adjust their hunting grounds dramatically as a result of rising temperatures and melting sea ice, and the impact these environmental changes have had on the fish that the birds feed upon, both in the Arctic and beyond. Ducks that typically breed on the Siberian tundra and winter at sea have shorter migration pathways now, because of shorter periods of winter ice cover, and this can have an important impact on their breeding patterns. The European population of the Atlantic puffin has declined by 50 percent over 3 generations, due to changes in fish stocks attributed to climate change. Seabirds in the northern Hudson Bay have had to adjust their hunting habits because of the northward expansion of the capelin, a forage fish in the smelt family. And Caspian tern chicks and fledglings might hold the record for most drastic nesting change: observed at the Cape Krusenstern National Monument in Alaska, above the Arctic Circle, they were nearly 1,000 miles farther north than any previously recorded jump of nesting range for any species.

At the lower end of the food chain, the distribution and availability of benthic resource species, which are those flora and fauna that live in the bottom sediment of water bodies, is also believed to be changing due to climate change. The boreal copepod, a microscopic crustacean, is expanding north from the Atlantic and replacing its more nutritious Arctic relatives. And indigenous communities have observed that walrus stomachs contain fewer clams than they did historically.[19]

The changes in the polar ecosystems are not confined to the region surrounding the North Pole. The Antarctic ecosystems are equally fragile, threatening the long-term survival of marine species in the southern polar region.

I experienced the impact of warming on these ecosystems first hand during the 2013 International Antarctic Expedition led by Robert Swan, the first man to walk to both Poles, when I learned that changes in the relative area of water covered by sea ice in the Antarctic affect the entire Antarctic food web, down to the smallest species.

Krill is the primary food source for many marine species, including the Antarctic toothfish[20], and the krill population has declined substantially around the West Antarctic Peninsula as a result of climate change, as well as the rise in ultraviolet radiation thanks to the depletion of the ozone layer and a diminished supply of algae beneath sea ice. The Adélie and Chinstrap penguin population decline has also been attributed to the decrease in sea ice and the corresponding krill population. Emperor penguins, which breed and raise their offspring almost exclusively on sea ice, are also highly vulnerable in the warming Antarctic.[21]

Some species are actually benefiting from the reduction in sea ice. Gentoo penguins, which eat a much more flexible diet, have increased in recent decades,[22] and studies have shown that less ice has meant more elephant seal pups in the breeding colonies on Macquarie Island near Antarctica. These seals once colonized areas thousands of miles south, during an earlier period of warming about 7,500 years ago, and they only returned north when the

[17] Phukan, Rituraj. Global Bio-diversity and Climate Change, Wild Sojourns, May- June 2016
[18] M. Duarte, Carlos. Impacts of Global Warming on Polar Ecosystems
[19] CAFF. 2017. State of the Arctic Marine Biodiversity: Key Findings and Advice for Monitoring. Conservation of Arctic Flora and Fauna International Secretariat, Akureyri, Iceland. ISBN: 978-9935-431-62-2
[20] Ingels J, Vanreusel A, Brandt A, et al. Possible effects of global environmental changes on Antarctic benthos: a synthesis across five major taxa. Ecology and Evolution. 2012;2(2):453-485. doi:10.1002/ece3.96.
[21] Forcada, Jaume. The Impact of Climate Change on Antarctic Megafauna. British Antarctic Survey, Natural Environment Research Council, Cambridge, United Kingdom.
[22] Phukan, Rituraj. Global Bio-diversity and Climate Change, Wild Sojourns, May- June 2016

open waters froze after a few thousand years. Incidentally, some scientists are attaching tiny sensors to the heads of these deep-diving seals so they can learn more about Antarctic warming from warm waters below the surface.[23]

Another species that has been directly impacted by sea level rise and salinization is the sea turtle, for whom there has been a reported decline of nearly 50 percent between 1970 and 2012.[24] Researchers have also documented a change in the ratio of male-to-female sea turtles, as warmer nest temperatures yield more female babies, and disruption of migration and feeding patterns, due to changes in ocean currents. Likewise, another study revealed that ocean temperatures have impacted the male: female ratio of northern elephant seal pups, which could have disastrous effects on the viability of seal populations in the long term.[25]

Salinization of coastal areas could wipe out entire populations of endangered birds, and animals and African penguin populations in the South Africa Western Cape has declined by nearly 70 percent during the last decade[26].

While the Arctic and Antarctic ecosystems are critically at risk from climate change, tropical coral reefs are suffering at what might be the most frightening pace of all. Rich biomes teeming with millions of marine species of flora and fauna, these naturally occurring communities are dying. The year 2016 was particularly bad for coral reefs around the world, and warming ocean waters, combined with an unusually long El Niño, left behind a trail of destruction.[27] The Great Barrier Reef, the largest living structure on Earth, was devastated, with the most severe coral-bleaching event on record.

Coral bleaching occurs when algae is expelled from coral polyp tissues, and this happens as a result of rising ocean temperatures. In addition to the Great Barrier Reef, coral bleaching has been observed in waters surrounding Hawaii, Fiji, and New Caledonia. There have only been three mass bleaching events recorded, with all of them occurring since 1998, and the problem is that repeated and sustained bleaching events are fatal for coral and the organisms that rely on the reef biome, and the nutrients provided by its algae, for survival[28]. In the Indian Ocean, nearly 70 percent of coral reefs appear to have died, and it is predicted that up to three-quarters of all existing reefs will disappear by 2050. Scientists predict annual coral bleaching by 2050 under business-as-usual scenarios – meaning no significant climate change mitigation — and 90 percent of the world's corals could be lost.[29]

We often hear about tragedies taking place in other parts of the world, and we listen with sympathy but without real understanding. For those of us who don't live in the Arctic, Antarctic, or tropical regions, we may find it hard to relate to some of these startling ecosystem disruptions. But climate change hit all of us hard when a grieving killer whale made headlines around the world in the summer of 2018 as she carried her dead calf for seventeen days on a thousand-mile journey. Known as Tahlequah, she is a member of the Pacific Northwest's endangered southern resident killer whale clan who has called the Puget Sound home for as long as we know. These orcas have been revered by indigenous people and have awed tourists, but with a population cut in half in the last fifty or so years (primarily through capture for the marine entertainment industry), their viability is becoming increasingly doubtful, especially as the females have greater and greater difficulty carrying healthy babies to term and infant mortality rates rise.

Lack of food takes the lion's share of the blame for the killer whale predicament, although toxic contaminants, and disturbance from marine noise and vessel traffic, are also responsible. Unlike many other killer whales,

[23] Norkko A, Thrush SF, Cummings VJ, Gibbs MM, Andrew NL, Norkko J, Schwarz AM. Trophic structure of coastal Antarctic food webs associated with changes in sea ice and food supply. Ecology. 2007 Nov;88(11):2810-20. PubMed PMID: 18051650.

[24] Climate Change and Biodiversity Loss. School of Public Health. Center for Health and the Global Environment.

[25] Bellard, Céline, Paul Leadley, Wilfried Thuiller, Franck Courchamp. Impacts of climate change on the future of biodiversity.

[26] Birdlife International and National Audubon Society (2015), The Messenger: What Birds Tell Us About Threats from Climate Change and Solutions for Nature and People. Cambridge, UK And New York, USA: Birdlife International and National Audubon Society.

[27] Phukan, Rituraj. Global Bio-diversity and Climate Change, Wild Sojourns, May- June 2016

[28] Green, Rhys E., Mike Harley, Lera Miles, Jörn Scharlemann, Andrew Watkinson and Olly Watts. Global Climate Change and Biodiversity. The RSPB, UK Headquarters, The Lodge, Sandy, Bedfordshire SG19 2DL

[29] Skibba, Ramin. Bringing the Plight of Coral Reefs to Our Screens, Inside Science. July 19, 2017. https://www.insidescience.org/news/bringing-plight-coral-reefs-our-screens

Tahlequah and her clan primarily rely on the cold-water chinook salmon for food, and the salmon population itself is dwindling at alarming rates. Overfishing is one cause for the decline in the salmon population, and the introduction of Atlantic farms to Pacific waters has also been blamed. But climate change is also wreaking critical havoc on this population through rising water temperatures, acidification, chemical contaminants, an increasingly hospitable habitat for pests and diseases that afflict salmon, and even artificial light.[30]

The chinook salmon, which ironically jumps up ladders and over rocks as it travels upstream into freshwater to spawn, supports a vast marine ecosystem. But it's also key to terrestrial and aviary ecosystems, the cultural and dietary lifestyles of Native people, the economy of non-Native people. The demise of the chinook salmon could be catastrophic.

And finally, although we tend to focus on marine mammals, fish, seabirds, and beautiful coral reefs when we think about our oceans, we can't forget plankton, the tiny creatures that serve as a foundation for the entire oceanic food chain by providing oxygen and sequestering carbon dioxide.[31] In recent years, there has been a catastrophic collapse of phytoplankton and zooplankton, due to warmer surface temperatures, ocean acidification, pollution, and the surprisingly rapid decline of whales, who have the distinct honor and responsibility of fertilizing phytoplankton.[32] One Blue whale, for example, defecates three tons of nitrogen and iron-rich feces a day to feed phytoplankton which, in turn, pass along the nutrients to zooplankton, fish, and nearly everything else that lives in the sea. Without the whale, such as the North Atlantic right whale on the brink of extinction, plankton can't survive. At the same time, whales can't survive without zooplankton[33]. So, it's a vicious cycle with no end in sight.

The ocean provides about 50 percent of the oxygen we breathe, and along with the sun, it is responsible for the global circulation system that transports water from the land to the sea to the atmosphere and back to the land again. Recent reports say that the number of ocean dead zones with zero oxygen have quadrupled in size, and those with very low oxygen sites near coasts have multiplied tenfold since the 1950s. Most sea creatures cannot survive in these zones, which means mass extinction of marine biodiversity is likely in our future. The collapse of oceanic biodiversity will inevitably have disastrous consequences for humanity. [34]

Terrestrial Biodiversity

Just a month before I arrived in Churchill, a grizzly was seen for the first time ever at the CNSC, just a week after another rare grizzly sighting in Manitoba at Wapusk National Park. The grizzlies of Alaska and Canada have extended their northward range with global warming, bringing them in contact with the polar bears who are forced to spend more time on land. There was no evidence of grizzly bears in the area before 1996, not even in the trapping data from centuries of Hudson Bay Company operations, but recent records published in *Canadian Field Naturalist* show increasing frequency of sightings.[35]

When grizzlies and polar bears meet, they must compete for food. But they might also mate. Although European zoos have experimented with grizzly and polar bear hybrids, none were discovered in the wild until 2006. The hybrids are known by various portmanteau names (pizzly, grolar bear, polizzly, etc.), but Canadian wildlife officials prefer nanulak, from the Inuit *nanuk*, for "polar bear" and *aklak*, for the "grizzly." Unfortunately,

[30] https://www.eopugetsound.org/magazine/ssec2018/marine-survival-4
[31] Forcada, Jaume. The Impact of Climate Change on Antarctic Megafauna. British Antarctic Survey, Natural Environment Research Council, Cambridge, United Kingdom.
[32] Ingels J, Vanreusel A, Brandt A, et al. Possible effects of global environmental changes on Antarctic benthos: a synthesis across five major taxa. Ecology and Evolution. 2012;2(2):453-485. doi:10.1002/ece3.96.
[33] Norkko A, Thrush SF, Cummings VJ, Gibbs MM, Andrew NL, Norkko J, Schwarz AM. Trophic structure of coastal Antarctic food webs associated with changes in sea ice and food supply. Ecology. 2007 Nov;88(11):2810-20. PubMed PMID: 18051650.
[34] Cool It. The Climate Issue. National Geographic.
[35] Lunn, N. J., Servanty, S., Regehr, E. V., Converse, S. J., Richardson, E. and Stirling, I. (2016), Demography of an apex predator at the edge of its range: impacts of changing sea ice on polar bears in Hudson Bay. Ecol Appl, 26: 1302–1320. doi:10.1890/15-1256

hybridization has emerged as another threat to the polar bear population according to University of Alberta bear biologist Andrew Derocher, who believes the polar bear population may lose out genetically.[36]

The decline of polar bears will also affect the Arctic fox, as they are mutually beneficial. The Arctic fox is regarded as a prime indicator of the overall health of the tundra, due to its sensitivity to the habitat and the fact that the animal is unlikely to survive in any other biome. The small creature is highly adapted to its habitat, with thick fur, specialized heat-retaining circulatory system in its feet, and the ability to lower metabolic rate, factors that would be detrimental in warmer environments.[37]

I had read about the abundance of the Arctic fox in the Churchill area, so I was surprised to see several red foxes in the vicinity of the CNSC. Red fox expansion into the Arctic, I learned, correlates with climate warming and threatens the Arctic fox because the red foxes are superior hunters and can prey on northern species. Even worse news for the Arctic fox: milder and shorter winters are likely to cause population decline in the lemmings and voles that they eat. Consequently, the International Union of Conservation of Nature (IUCN) predicts existential threat to the Arctic fox from persistent global warming and its consequences.

While at the CNSC, I took part in ongoing citizen science research to document the current status of ecosystems, benchmark future changes in the tundra wetlands, and participate in an outdoor *mesocosm* experiment, a tool that allows scientists to study a natural environment under controlled conditions. The goal of this experiment was to identify and document the effects of warming and increased nutrient loads on the growth rate and survival of tadpoles of the boreal chorus and wood frogs. It was fascinating to discover the micro-ecosystems of the Hudson Bay lowlands, which contain the highest density of ephemeral wetlands in the world and rank as the third-largest wetland region. These lowlands contain the single-largest peatland system on Earth, which is essential because peatland is particularly rich in carbon. That means the Hudson Bay lowlands are potentially the most carbon-dense terrestrial eco-region on Earth.

I also gleaned insight on the International Tundra Experiment (ITEX), a long-term international collaboration of researchers examining the responses of Arctic and alpine plants and ecosystems to climate change at around twenty different Arctic sites. During these studies, it became evident that fragile tundra and wetland ecosystems are experiencing rapid change, with unknown consequences for the biodiversity of the Arctic and subarctic areas.

The caribou, or reindeer, provides another interesting example of how warming affects biodiversity in northern climates. Ecologists from the James Hutton Institute, Norwegian Institute for Nature Research and Norwegian University of Life Sciences, have found that reindeer in Svalbard are literally shrinking in response to climate change. Warmer summer temperatures lead to more productive pastures, which, in turn, leads to a doubling of the reindeer population over the last twenty years. This, of course, leads to more competition for food. Warmer winters mean more rain, which freezes on the snow, locking out access to the food beneath[38]. In response, females abort their calves or give birth to much lighter young, as evidenced by the Svalbard study that documented a 12 percent shrinkage in size of adult reindeer over 16-years.[39]

The continuous decline of caribou populations, particularly declining calf and adult survival is linked to changing climate conditions.[40]

As with marine diversity, terrestrial diversity is at risk around the world and not just in Arctic regions. Some believe that the natural world is in the midst of a mass extinction, with more than 24,000 species at risk; one recent study claimed that the total number of all wild animals is likely to decrease by two-thirds by the year 2020,

[36] Atwood, T. C., Marcot, B. G., Douglas, D. C., Amstrup, S. C., Rode, K. D., Durner, G. M. and Bromaghin, J. F. (2016), Forecasting the relative influence of environmental and anthropogenic stressors on polar bears. Ecosphere, 7: n/a, e01370. doi:10.1002/ecs2.1370
[37] WWF Global. Global warming impacts in the Arctic and Antarctic
[38] Arctic Council (2016). Arctic Resilience Report. M. Carson and G. Peterson (eds). Stockholm Environment Institute and Stockholm Resilience Centre, Stockholm. http://www.arctic-council.org/arr.
[39] Meltofte, H. (ed.) 2013. Arctic Biodiversity Assessment. Status and trends in Arctic biodiversity. Conservation of Arctic Flora and Fauna, Akureyri.
[40] Russel, D.E., A. Gunn, S. Kutz, 2018: Migratory Tundra Caribou and Wild Reindeer [in Arctic Report Card 2018], https://www.arctic.noaa.gov/Report-Card.

just a few short years from now, in comparison to their numbers in 1970. Another suggests the climate crisis is expected to cause the extinction of about one-quarter of all land species by 2050.

The recent IUCN Red List, which is recognized as a comprehensive global approach to evaluating conservation status for plants and animals, now includes the giraffe as another animal facing extinction and lists the eastern gorilla and whale shark as moving ever closer to their demise as well. Smaller terrestrial species are also in trouble: up to 50 percent of all amphibian species, which have occupied Earth for over 250 million years and survived the extinction of the dinosaurs, could be lost in this century. [41]

There is no denying that these casualties are due to climate change, habitat loss, over-exploitation, and other anthropogenic factors. Since 1992, when the landmark Convention on Biological Diversity was ratified, an additional 450 million hectares of natural habitat has been destroyed. Almost half of the world's 800 eco-regions, places with distinctive animal and plant communities, are classified as very high risk, with 41 deemed to be in crisis, losing habitat so rapidly that there is little left to protect.[42]

India is one of the seventeen countries considered to be mega-diverse, or possessing the largest number of animals and plant species. I live in Assam, which lies in the transition zone between the eastern Himalaya and Indo-Burma biodiversity hotspots. I help young people make the connection between biodiversity, water, and climate change; the Kids for Tigers program teaches students how conservation of the tiger in its primary forest habitat protects their water sources and helps to mitigate climate change.

For example, changes in precipitation patterns can alter the nature of remaining tiger habitats in India. A deficit in soil moisture due to less rainfall will increase tree mortality in the deciduous forest habitat. This will trigger a shift toward open tropical dry forests, which are considered to be less productive for tigers. Rising sea levels and salinization also impact coastal mangroves and wetlands, which are expected to decimate the Sundarbans habitat of the tiger.[43] The result of these changes means that apex predators like the tiger will experience a lack of suitable alternative habitats.

In Bhutan, tigers have been fortunate and able to extend their range upslope with the expanding tree line. However, this has meant displacing the already endangered snow leopards. Further impacting the snow leopards: the disappearance of alpine pastures that sustain the Himalayan tahr, which the snow leopard preys upon, because of the rapidly rising temperatures in the Himalayas — thrice as fast as the global average.[44]

Assam also harbors a large population of the Indian one-horned rhinoceros, which thrives in the grasslands sustained by abundant rainfall during the annual monsoons. Changes in the annual monsoon patterns can deplete these grasslands as well the floodplains of India and Nepal, resulting in extreme flood or drought. The rhinos then displaced will not only be faced with food availability issues but also will be at higher risk of being poached.

As secretary general of Green Guard Nature Organization, a civil society group engaged in human-elephant conflict mitigation at the grassroots for nearly two decades, I have seen how a change in rainfall patterns and the availability of fodder is severely impacting Asiatic elephants. Conflict situations have increased as the elephants have been forced to stray out of protected areas. Moreover, Asian elephants have a low adaptive capacity, due to a limited dispersal ability as most of their habitat is fragmented, as well as slow reproductive rates and only moderate amounts of genetic variation within the species. All elephants are very sensitive to high temperatures, and extreme heat leaves them susceptible to disease. Elephants need high quantities of freshwater, and accessibility to it has a direct influence on their daily activities, reproduction, and migration[45].

[41] Shah, Anup. Climate Change Affects Biodiversity.
[42] Green, Rhys E., Mike Harley, Lera Miles, Jörn Scharlemann, Andrew Watkinson and Olly Watts. Global Climate Change and Biodiversity. The RSPB, UK Headquarters, The Lodge, Sandy, Bedfordshire SG19 2DL
[43] Climate Change Endangering Asia's wildlife. Sanctuary Asia.
[44] Phukan, Rituraj. Global Bio-diversity and Climate Change, Wild Sojourns, May- June 2016
[45] The Heinz Centre. 2012. Climate-Change Vulnerability and Adaptation Strategies for Africa's Charismatic Megafauna. Washington, DC, 56 PP.

Precipitation changes directly impact the habitat of other charismatic mega-fauna around the world. In Malaysia and Indonesia, the existence of orangutans is threatened by warmer temperatures, rainfall decrease, drought, and forest fires, all of which adversely affect the availability of tropical fruits, the mainstay of their diet, as well as their breeding success. The rare Sumatran rhino is also now exposed to a new risk, as dry seasons lengthen, and forests and peatland become prone to fires. In Panama, changes in precipitation have led to the drastic decline of plant species growing in moist conditions. Amphibian populations like golden toads have declined dramatically in neotropical montane forests.[46]

On the African continent, precipitation changes are likely to render several habitats unsuitable for certain species. African elephants consume over 200 liters of water a day, and any changes in rainfall result in elephants straying out of protected areas, increasing their vulnerability to conflict and poachers. The number of mammal species in sub-Saharan Africa could decline by up to 40 percent, and one study offers the grim prognosis that 66 percent of animal species in South Africa's Kruger National Park will go extinct.[47]

Birdlife International and the National Audubon Society published a milestone report, "The Messengers: What Birds Tell Us About Threats from Climate Change and Solutions for Nature and People," during COP21, the United Nations Climate Conference in Paris in December of 2015. The report holds that birds are the planet messengers, literally the "canaries in the coalmine." Our feathered friends confirm that climate change is a posing extreme danger, with documented distribution shifts Pole-ward. And, migrating to higher ground to escape warming temperatures, disrupted interactions with predators, competitors and prey, mismatches in the timing of migration, breeding and food supply and population declines resulting from these and other effects.[48]

This pioneering report revealed ongoing and impending climate and sea level rise impact on bird populations worldwide, as the possible wipeout of the great frigate bird, which breeds in low-elevation Pacific islands. It also outlined extinction threats to common birds like the limited-range Allen's hummingbird in North America, the lilac-breasted roller in Africa, the golden bowerbird, a Queensland endemic, and the helmeted hornbill in Asia.

The State of the World's Birds, a five-year compilation of population data released in April 2018 has revealed a biodiversity crisis, with logging, invasive species, hunting, and trapping and climate change and severe weather as the top five risks to bird populations worldwide. According to this definitive study of global bird populations, one in eight bird species is threatened with global extinction, including abundant and popular birds like turtle doves, puffins, and snowy owls.[49]

Many bird species are already struggling to adapt to the pace of these changes. Rising temperatures are driving species' distributions towards the poles and towards higher ground. Migratory and breeding cycles are changing, leading to disrupted relationships with prey, predators, and competitors, driving population declines.[50]

Nearly one-quarter of bird species so far studied have already been negatively affected by climate change, with declines in abundance and range size being the most common impacts. A recent review of the scientific literature shows that 24% of the 570-bird species studied globally have already been negatively affected by climate change to date, while only 13% have responded positively. The research indicates that even the relatively modest temperature increase experienced to date has had a considerable impact on global biological diversity.

The first Global Flyways Summit in April 2018 addressed the critical declines in many migratory bird populations all eight flyways of the world and listed climate change as one of the major threats to migratory birds, which are recognized as flagships and indicators for nature conservation.

[46] Green, Rhys E., Mike Harley, Lera Miles, Jörn Scharlemann, Andrew Watkinson and Olly Watts. Global Climate Change and Biodiversity. The RSPB, UK Headquarters, The Lodge, Sandy, Bedfordshire SG19 2DL

[47] The Heinz Centre. 2012. Climate-Change Vulnerability and Adaptation Strategies for Africa's Charismatic Megafauna. Washington, DC, 56 PP.

[48] Birdlife International and National Audubon Society (2015), The Messenger: What Birds Tell Us About Threats from Climate Change And Solutions For Nature And People. Cambridge, UK And New York, USA: Birdlife International and National Audubon Society.

[49] BirdLife International (2018) State of the world's birds: taking the pulse of the planet. Cambridge, UK: BirdLife International.

[50] BirdLife International (2018) State of the world's birds: taking the pulse of the planet. Cambridge, UK: BirdLife International.

One of the most worrisome consequences of warming and change in climatic envelopes is the documented decline of pollinators. Although some species may benefit from warming that leads to range extension in areas freed of ice cover, several bee populations are on the decline. Today, more than 30,000 bee species exist worldwide, and they are vital to the health of the planet[51] as the most important group of pollinators for farming and wild plants and the fertility of most flowering plants, including nearly all fruits and vegetables. Bees are declining due to a variety of factors, including human development, pesticides, disease, and changing climate.[52]

The climate crisis also affects pollination by disrupting synchronized timing of flower blooming and bee pollination. Flowers are blooming earlier in the growing season due to rising temperatures before many bees have a chance to pollinate. In India, a brutal combination of climate change and human interference has wiped out entire colonies of giant honeybees, which many plants and trees depend on for survival.

As mentioned, a few species do benefit from warming. These include trumpeter swans, which almost went extinct in the late 1800s but recovered after northward expansion of their summer range. Wild boar populations have multiplied; as warmer winters increase the survivability of older and younger boars, there is increased food availability across wider expanses, and they have enjoyed better reproduction success. Rat snakes flourish in the thermally advantageous environment, resulting in population increase and range expansion.[53] Starfish thrive in increased temperatures and carbon dioxide levels, as they eat more and grow faster in warmer temperatures. Cephalopods like cuttlefish, octopus, and squid have multiplied tremendously in the last few decades, and scientists attribute their success to extreme adaptability; they are able to modify the rate at which they grow or the size at which they mature and reproduce in response to warming ocean temperatures.[54]

In the 1980s, the population of Brown Argus butterflies was diminishing, but due to climate change, these animals are on the rise again in southern England and are expanding in the northern parts of the country. Another species that does well in warmer spring conditions are the long-tailed tit, a bird whose numbers have increased across Europe.[55]

But for the most part, biodiversity is on the wane. Scientists have described numerous examples of trophic cascades in marine and terrestrial ecosystems, which demonstrates the consequences of taking out one species, e.g., a mammal or a predator, from the food chain[56]. When too many species are removed, the foundation of a healthy ecosystem collapses. Although biodiversity varies from habitat to habitat, an abundance of different species in any habitat provides the resilience and strength necessary for a system, as a whole, to survive and thrive, despite the inevitable changes.

A recent study by a group of scientists from the University of East Anglia, the James Cook University and the World Wildlife Fund (WWF) compared the potential impact of warming on 80,000 plant and animal species in 35 biodiverse regions around the world based on different warming models. With a temperature rise above 3° C, the projected loss is more than 60% of plant species and almost 50% of animal species in the Amazon. Even with an average global temperature rise limited to 2°C, as called for by the Paris Agreement to limit catastrophic change, many of the 35 Priority Places classified by WWF are projected to lose a significant proportion of their species as the climate becomes unsuitable for them and almost 25% of the species in Priority Places are at risk of local extinction.[57]

---

[51] Palmer, Brian. Would a World Without Bees Be a World Without Us? Assessing our chances of survival without the prodigious pollinator. May 18, 2015
[52] Walsh, Bryan. The Plight of the Honeybee. TIME magazine. Aug. 19, 2013
[53] Green, Rhys E., Mike Harley, Lera Miles, Jörn Scharlemann, Andrew Watkinson and Olly Watts. Global Climate Change and Biodiversity. The RSPB, UK Headquarters, The Lodge, Sandy, Bedfordshire SG19 2DL
[54] Bellard, Céline, Paul Leadley, Wilfried Thuiller, Franck Courchamp. Impacts of climate change on the future of biodiversity.
[55] Moreno Di Marco, James E.M. Watson, Oscar Venter, Hugh P. Possingham, Global Biodiversity Targets Require Both Sufficiency and Efficiency, Conservation Letters, 2016
[56] Phukan, Rituraj. Global Bio-diversity and Climate Change, Wild Sojourns, May- June 2016
[57] Jeffries, Barney., Jeffries, Evan and Elliott, Katherine. Wildlife in a Warming World- The effects of climate change on biodiversity in WWF's Priority Places. WWF-UK

In the long run, the projected extinction of species will disrupt the web of life and create unforeseen challenges, and extinctions are impacts that cannot be reversed. While the earth has seen waves of extinction in the distant past, none has been driven by humanity. New research underscores the influence that humankind has had in transforming the planet in relatively recent years; scientists have found that over 50 percent of the world's land area is now dominated by human activity, with 9 percent of this change happening in the last 25 years alone. [58]

The October 2018 Special Report on Global Warming of 1.5° C by the Intergovernmental Panel on Climate Change (IPCC) has also highlighted the comparative biodiversity impacts of 1.5° C and 2° C or more, rise in global temperatures. For instance, at 2° C, around 8% of vertebrates will lose at least half of their present range, which is two times the projections for a 1.5° C rise. Similarly, 16% of plants would lose at least half their range at 2° C, again twice the projections for a 1.5° C rise. The IPCC report also predicted that insect populations losing out half their range at 2° C will be three times the 6% loss at 1.5°C. Coral reefs would decline by 70-90 percent with global warming of 1.5°C, whereas over 99 percent would be lost with 2° C[59].

The Living Planet Report 2018 shows population sizes of wildlife *decreased by 60% globally between 1970 and 2014.* The Living Planet Index, produced for WWF by the Zoological Society of London, uses data on 16,704 populations of mammals, birds, fish, reptiles, and amphibians, representing more than 4,000 species, to track the decline of wildlife. **Plummeting numbers of mammals, reptiles, amphibians, birds, and fish around the world are an urgent sign that nature needs life support.[60]**

There is a strong association between rapid climate warming and declines of bird and mammal populations globally, showing that population declines have been greatest in areas that have experienced most rapid warming. If the rate of climate warming continues to increase then we can expect greater bird and mammal population declines, these losses will be greatest at locations which experience the most rapid climate warming. These findings echo aspects of previous global studies which suggest that future climate change will lead to large range contractions and increased species extinction risk[61].

The recent heat waves down under have killed hundreds of thousand bats, with an estimated 23,000 spectacled flying fox deaths in Queensland during November 2018, equating to almost one-third of the species in Australia. The heat waves also claimed the lives of wild horses and fishes in Australia and ancient trees in New Zealand. The Earth's warming may have exacerbated conditions leading to starfish die-offs along the Pacific coast, according to one report published earlier this year. Another recent study has linked warmer forests to dramatic declines in arthropod insectivore populations with broad implications for the web of life. There has been a 98% decline in insect mass over 35 years in the forests of Puerto Rico and 86% decrease in Monarch butterfly numbers in California.

For many of us, these scientific reports and projections are essential but abstract predictions about some time in the future. We know they will somehow influence what our world will look like for our children and grandchildren, or what foods might—or might not—be available for them. We recognize that local economies in one part of the world or another could suffer from the loss of a particular species. But still, the reports are abstract, unless they hit us close to home, or unless we're able to find empathy and compassion in our hearts for our animal brothers and sisters.

Negotiators at the United Nations (UN) Biodiversity Conference of the Parties (COP14) held at Egypt in December 2018 adopted the Sharm El Sheikh Declaration inviting the UN General Assembly to convene a Summit

---

[58] Moreno, Ibid

[59] IPCC, 2018: Global warming of 1.5°C. An IPCC special report on the impacts of global warming of 1.5°C above pre-industrial levels and related global greenhouse gas emission pathways, in the context of strengthening the global response to the threat of climate change, sustainable development, and efforts to eradicate poverty [V. Masson-Delmotte, P. Zhai, H. O. Pörtner, D. Roberts, J. Skea, P.R. Shukla, A. Pirani, W. Moufouma-Okia, C. Péan, R. Pidcock, S. Connors, J. B. R. Matthews, Y. Chen, X. Zhou, M. I. Gomis, E. Lonnoy, T. Maycock, M. Tignor, T. Waterfield (eds.)]. In Press.

[60] WWF. 2018. Living Planet Report - 2018: Aiming Higher. Grooten, M. and Almond, R.E.A.(Eds). WWF, Gland, Switzerland.

[61] E. B. Spooner, Fiona & G. Pearson, Richard & Freeman, Robin. (2018). Rapid warming is associated with population decline among terrestrial birds and mammals globally. Global Change Biology. 24. 10.1111/gcb.14361.

on Biodiversity for heads of State by 2020 for an international agreement on reversing the global destruction of nature and biodiversity loss threatening all forms of life on Earth. COP14 called upon the UN General Assembly to designate 2021 to 2030 as the UN Decade of Ecosystem Restoration.[62]

The urgency of the crisis has led to talk for a transformational 'New Deal for Nature and People' in 2020, when the next UN Biodiversity Conference is scheduled to be held at China, for an effective strategy to halt the collapse of life on Earth. The year 2020 marks the deadline for nations to agree on new global targets for the protection and management of forests, rivers, oceans, pollinators and other wildlife.

Nature-based Solutions provide a low-cost and low-risk mitigation and adaptation potential with the added benefits of food and water security and biodiversity conservation. It is, therefore, no surprise that Nature-based Solutions is one of the six core action portfolios prioritized by the Secretary-General, which are recognized as having high potential to curb greenhouse gas emissions and increased global action on adaptation and resilience, for the UN Climate Summit in September 2019.

Surprisingly, biodiversity was hardly ever discussed during the twenty-fourth session of the Conference of the Parties (COP24) of the United Nations Framework Convention on Climate Change (UNFCCC), held in Katowice, Poland, from 2 to 14 December 2018. The only exception was the adoption of 'The Ministerial Katowice Declaration on Forests for the Climate' introduced by the Polish presidency, which stated that there is no future without counteracting climate change, and there is no future without forests.

In the months since COP24, students have been out on the streets demanding urgent action to avoid catastrophic climate change. What had started as one brave girl's silent protest outside the Swedish Parliament has now grown into a clamorous uprising by thousands of students across the globe. For Greta Thunberg, the 15-year old who started the 'school strike for the climate' movement in August 2018, the futility of going to school when the future was uncertain prompted action. The tardy progress of climate negotiations over two decades and the IPCC 1.5 C report has exasperated the youth, and rightly so.

We are the first generation that has a clear picture of the value of nature and the grave situation we are facing. We may also be the last generation that can do something about it. We all have a role to play in reversing the loss of nature – but time is running out. Between now and 2020 we have a unique opportunity to influence the shape of global agreements and targets on biodiversity, climate, and sustainable development – for a positive future for nature and people.[63]

The polar bear wandering through Churchill, not far from the polar bear jail, was one individual creature who impacted me profoundly. Tahlequah, the grieving orca mother, affected many in the Pacific Northwest. And images of bats dying from dehydration in record Australian temperatures have horrified many around the world. The scientific reports inform us, but the individual stories are what calls us to action.

So, what *can* each of us do to mitigate these losses?

Climate Crisis "Keep/Stop/Start" Actions

Below for your consideration, are several author-suggested action items. For a further explanation of the Keep/Stop/Start organizational tool, please see How to Read This Book.

KEEP:

Keep learning about the planet, the natural systems that harbor life, and the diversity of our fellow inhabitants. We are born on a beautiful planet and we shall return to Mother Earth after we die, just like all the other creatures

---

[62] UN Biodiversity Conference 2018, Sharm El-Sheikh, Egypt Announcement: Sharm El-Sheikh to Beijing Action Agenda for Nature and People.
[63] WWF. 2018. Living Planet Report - 2018: Aiming Higher. Grooten, M. and Almond, R.E.A.(Eds). WWF, Gland, Switzerland

that have ever lived on the Earth. In this way, we are part of the Earth, and we will remain so for eternity. We must KEEP that in mind all through our conscious lives.

In his book, Love Letter to The Earth, Zen Master Thích Nhat Hanh invites us to understand that WE ARE THE EARTH. He writes, "We often call our planet Mother Earth. Seeing the Earth as our mother helps us to realize her true nature. The Earth is not a person, yet she is indeed a mother who has given birth to millions of different species, including the human species."

The modern environmental movement has led to a better understanding of the natural world and respect for other life forms. The lessons of nature are best learned in the great outdoors, and I find every walk in the forest inspirational; as I immerse myself in my surroundings, I see the relevance of biomimicry for the past and future of mankind. Reconnecting with the wilderness has manifested in the anxiety and commitment of the younger generations towards the planet. We have to definitely KEEP doing that.

Where there is a political will, there is a way! We must KEEP that in mind, KEEP fighting for the planet, KEEP up the pressure to protect our forests and biodiversity and KEEP teaching the younger generations about the magical planet Earth.

STOP:

Stop believing that nature can heal itself or that a miracle solution will emerge from somewhere! We, humans, are responsible for the mess, and we must start the healing process. One of the most important lessons I learned was from the first man to walk to both the poles, Robert Swan who says that "The Greatest Threat to Our Planet Is The Belief That Someone Else Will Save It."

We must take responsibility for conservation of the natural habitats that harbors biodiversity, for the survival of mankind depends on the survival of these ecosystems. We are part of the web of life; STOP believing that humankind can survive as a standalone species on the planet. Urban life has alienated people from the natural world. We will have to START understanding how life works, all over again.

One of my favorites quotes, attributed to the American poet Gary Snyder, is "Nature is not a place to visit. It is home." This takes us back to oft-repeated idiom that humans must learn to live in harmony in nature. The idea of the co-evolution of humanity with nature is not new, and in societies who closely interact with nature, humans are conceptualized as part of it. The Industrial Age brought with it the Cornucopian belief that nature is an endless basket of resources, as well as a convenient dump for human waste. We must STOP treating Mother Earth as a dump and promote a holistic understanding of the integral relationship between humanity and nature as part of the one-planet, larger system.

Humans are not just a fundamental part of nature and utterly dependent on it but also a powerful force of nature. In the age of the Anthropocene, human beings are sufficiently numerous, and, with our actions magnified by technology, we have become a powerful biophysical force at a global scale. We have the choice of forceful action to arrest biodiversity collapse before we drown in an unknown cascade.

START:

1) *Start believing that we are the problem*; human population growth and increased consumption is disrupting the web of life. Almost everything we do has an environmental impact, including what we choose to eat, so START managing your contribution to global warming. Besides significant emissions of Greenhouse gases, the production of meat and other animal products places a heavy burden on the environment. The vast amount of fodder required for meat production is a significant contributor to deforestation, habitat loss and species extinction.

2) *Start making conscious efforts* and set viable personal goals to decrease your carbon footprint. Most of us are slaves to wasteful habits and lifestyles and we can make a positive global impact towards slowing down the sixth extinction by embracing simple lifestyle changes. Leo Tolstoy had famously said, "Everyone thinks of changing the world, but no one thinks of changing himself."

3) *Start believing in yourself.* You can be the leader to help break stereotypes and help change the world for the better. It was Mahatma Gandhi who said, "We must become the change we wish to see in the world."

~ ~ ~

Rituraj Phukan is an environmentalist, explorer, and writer based in Assam, a biodiversity-rich, climate-impacted province in the far east of India. He is secretary general of Green Guard Nature Organization, a grassroots civil society group that works with fringe forest communities to explore and establish sustainable solutions for management of human-wildlife conflicts.

Chapter 3

Climate Crisis & Ice

by

Peter M. J. Hess, Ph.D.

(To Viviane, Michael, and Robert Hess, for sustaining me with enduring hope in what may become an Earth without ice.)

"Only by regaining this intimacy with the natural world we can begin to understand the ramifications of what it is to lose so much of Earth's ice, species, and biosphere." – **Dahr Jamail,** The End of Ice

I. Introduction

I was not anticipating a crevasse fall, although I should have paid heed to the warning signs. On summer solstice 2009, my climb of northern California's 14,179-foot (4,321 m) Mt. Shasta collided with an unexpected late-June snow storm. With the wind blowing about 75 mph (120 kph) on the summit plateau, my partners and I abandoned the goal of ascending the final pyramid and beat a hasty retreat down the steep and icy north ridge between the Hotlum and Bolam Glaciers. Periodic bursts of thunder accompanied the beginnings of a blizzard.

In whiteout conditions the telltale longitudinal surface depressions beneath hidden crevasses tend to disappear. Plunge-stepping down the slope, my right leg punched without warning through a snow bridge into the inky void of a concealed crevasse. Unwittingly we had strayed into the crevasse field of the western lobe of the Hotlum Glacier. Flailing wildly, I felt nothing below me at first. But fortunately, the slot was not wide, and when I kicked forward the front fangs of my right crampon struck the down-slope wall of the crevasse. With a well-placed ice-axe stroke I extricated myself and sat down trembling. The intense ground-level whiteout lifted partially, revealing the deep bergschrund crevasse (between the glacier and the headwall) looming above and to the right of me. In our descent I had missed it by only a few feet! I took a few moments to re-center myself[1].

Ice has an ambivalent import in mountaineering: a glacier can offer a convenient ramp to bypass thousands of feet of crumbly volcanic rock, while at the same time it can conceal dangerous crevasses. Having climbed on more than twenty glaciers in North America, Mexico, and Europe, I delight in their beauty and the technical challenges of navigating and camping on them. Glacier travel is an avocation I have hoped to pass on to my sons, but with growing sadness I recognize that they might not have the same range of opportunity.

Glaciers on the Cascade volcanoes are retreating and disappearing at a rapid pace, as are glaciers in most other regions of our planet[2]. And for billions of people on Earth, ice serves a far more significant role than that of merely being a mountaineer's paradise: glaciers offer a reliable supply of stored water for drinking, cooking, washing and irrigation. The loss of Andean and Himalayan glaciers to melting will put billions of people at risk of significant loss of water availability.[3]

II. The Role of Ice in the Biosphere

That June evening in 2009, as we descended Mt. Shasta to make a snowy bivouac at timberline, I reflected on the role ice has played in the evolution of life on Earth. For the first two billion years after coalescing out of the

---

[1] For an essay on this climb in light of theology, see Peter M. J. Hess, "Doubt, faith and crevasses on my mind," in *God and Nature Magazine*, (Summer 2015), https://godandnature.asa3.org/essay-doubt-faith-and-crevasses-on-my-mind-by-peter-m-j-hess.html

[2] See for example the rapid melting of glaciers in Canada's Yukon Territory, https://www.theguardian.com/world/2018/oct/30/canada-glaciers-yukon-shrinking

[3] D. B. Jones, et al., "Mountain rock glaciers contain globally significant water stores," in *Sci.Rep.* 2018; 8: 2834, https://www.ncbi.nlm.nih.gov/pmc/articles/PMC5809490/

solar nebula, the young planet's surface was too hot for water to condense into liquid form. Vapor accumulated in the atmosphere through volcanic out-gassing and the bombardment of the planet by ice-rich comets. When Earth had cooled enough – beginning about 2.4 billion years ago – water formed on the surface as lakes, rivers, and oceans, and then it began to freeze.

As the planet cooled further, it may more than once have been covered almost completely with ice, for long periods at a time. According to this "snowball Earth theory,"[4] ice would have extended from the poles nearly to the equator, severely constricting the range of living organisms. Nevertheless, life was still possible because tidal wave action would have kept leads open in the thinner ice of equatorial regions, permitting penetration of enough light to allow photosynthesis to continue.[5]

The last such global glaciation ended about 625 Mya when carbon dioxide had built up in the atmosphere to a level sufficient to raise temperatures and melt the ice back toward the polar regions. In the next half billion years the planet became considerably warmer – 'a hothouse earth' – and marine life expanded and diversified. The Cambrian Explosion took place across forty million years, during which period most of the major phyla of animals first appear in the fossil record.[6] The genera proliferated, organisms gained a significant foothold on land, and eventually the age of the dinosaurs rolled around.

The Chicxulub asteroid event at the Cretaceous-Tertiary boundary (65 Mya) was a major life-shaping event[7]. A bollide 10-15 kilometers in diameter struck what is now the Yucatan Peninsula, sending out enormous waves of crustal ejecta and clogging the atmosphere with 4.5 million tons of rock, soil, dust, and smoke. This impact created the equivalent of a "nuclear winter" that hastened the extinction of most dinosaurs. However, within a short time the atmosphere warmed considerably, facilitating the ascendancy of mammals, although this rise in temperature did not take place as fast as the current temperature increase resulting from human industrial activity.

The recent climatic cycle of ice ages punctuated by warmer inter-glacial periods began at the end of the Pliocene Epoch, 2.6 Mya, with warm and cold phases alternating over tens of thousands of years. During this time *Homo sapiens* emerged, and since the end of the last ice age 12,000 years ago, humans have migrated around the globe into almost every ecological habitat.

In what might be called "the blink of a geological eye," modern humans discovered how to extract and use fossil fuels in the form of coal, oil, and natural gas. We have been so effective in discovering applications for these fuels in areas as diverse as transportation, agriculture, manufacturing, building, and medicine that our population has exploded sevenfold in two centuries, with few signs of stopping before we reach ten billion.[8] The burning of fossil fuels has warmed Earth to the point that most of the ice on the planet may disappear for millennia, if not forever. The loss of ice and rise in temperature will have enormous consequences for all life on earth. Arctic research scientist Peter Wadhams explains that

Human beings are truly carrying out a global experiment involving an unprecedented level of interference with the natural system.[9]

In the section that follows we will look at some of the key issues raised by Wadhams in his sobering book, *A Farewell to Ice: A Report from the Artic.*

---

[4] Joseph Kirschvink, (1992), "Late Proterozoic low-latitude glaciation: the snowball earth," in J. W. Schopf and C. Klein, eds., *The Proterozoic Biosphere – a Multidisciplinary Study.* Cambridge: Cambridge University Press, pp. 51-52.

[5] Peter Wadhams, *A Farewell to Ice: a Report from the Artic* (Oxford University Press, 2017), 23-27.

[6] Douglas Erwin and James Valentine, *The Cambrian Explosion: The Construction of Animal Biodiversity* (W. H. Freeman, 2013). See also the Cambrian Period web page of the University of California Museum of Paleontology, http://www.ucmp.berkeley.edu/cambrian/cambrian.php.

[7] Walter Alvarez, *T-Rex and the Crater of Doom* (Princeton University Press, 2013).

[8] "Population Crisis: World to hold 10 billion people by 2050 leading to global famine fears," https://www.express.co.uk/news/science/820099/POPULATION-CRISIS-10-BILLION-global-famine

[9] Peter Wadhams, *A Farewell to Ice: A Report from the Artic* Oxford University Press, 2017.

III. Feedback Loops, the Loss of Ice, and Climate Change

Ice exerts a profoundly important stabilizing influence on the global climate. A "snowball earth" would have had very few storms (but almost no livable land surface), whereas warm, ice-free oceans pump huge amounts of energy into the atmosphere, making a diverse biosphere possible. This thermal energy from the oceans helps intensify hurricanes, direct the jet stream, and create long-term precipitation patterns.[10] Permanent polar and Greenland ice acts as a brake or a moderating influence on the atmosphere, and as this ice disappears the global climate will become less stable, threatening food production, transportation, industry, and indeed, every aspect of energy-intensive civilization as we have come to know it.

A. The Future of Arctic Sea Ice: The Death Spiral

The Arctic is one of the main drivers of our global climate, governing the Atlantic thermohaline circulation system including the Gulf Stream that affords Europe in a milder climate that that of other high-latitude regions on the planet. Once Artic ice began to melt a few decades ago, various factors came into play that serve as feedback loops. If enough of these self-reinforcing loops combine to increase the retention of solar radiation and begin melting methane hydrates in both terrestrial and marine permafrost, the Artic will enter a "death spiral." Methane is a greenhouse gas twenty-three times more potent than carbon dioxide and it persists in the atmosphere far longer. The release of methane from its frozen state on the sea floor will force more atmospheric and oceanic warming, which in turn will cause the release of more methane bubbles, in a spiral that may end only with the release of most of the sequestered methane.

In the not-too-distant future the Arctic Ocean will become ice-free for five months of the year (July-November), opening the region to new commercial shipping routes. With only about 6% of Earth's surface area, the Arctic holds an estimated 22 percent of Earth's oil and natural gas reserves and is thus an incredibly rich resource area. Oil companies are jockeying for position to be ready to drill in this harsh environment as soon as the permanent ice is gone[11]. Ironically, the burning of these reserves will reinforce the melting of the ice cap that is making their extraction possible in the first place. An additional severe risk is that the consequences of oil spills in cold Arctic waters would be catastrophic both for the polar marine environment and for the terrestrial mammals and indigenous peoples who depend on that ecosystem for their livelihood. Wadhams contends that the Arctic region has already entered a death spiral, in the form of a series of eight reinforcing feedback loops.[12]

Ice-Albedo Feedback

The earth's "albedo" (from Latin "white) refers to the fraction of incoming solar radiation that is reflected straight back into space. For a given orbital distance from the sun, the higher a planet's albedo (i.e., the lighter it is) the colder it will be; the darker it is the more it will be warmed by retained solar radiation. On Earth the albedo of open water is 0.1; that of fresh snow on ice is 0.9. Ridges or angular surfaces in arctic ice increase albedo, as do holes melted by airborne particles of ash or soot contaminating the ice. As ice melts and exposes more water, the air temperature warms, melting the remaining ice faster, creating a self-reinforcing feedback loop. Wadhams warns that the ice-albedo feedback is the biggest threat to our continued existence on earth.[13]

---

[10] Dhar Jamail, "When the Ice Melts: the catastrophe of vanishing glaciers," *The Guardian*, https://www.theguardian.com/news/2019/jan/08/when-the-ice-melts-the-catastrophe-of-vanishing-glaciers?fbclid=IwAR1HU-Bm3CawHMCELcMTnoFL9F_8mH39MJRA_u05vPkNMeFbeyJ86i4UlZg

[11] Henry Fountain and Steve Eder, "In the Blink of an Eye, a Hunt for Oil Threatens Pristine Alaska," *The New York Times* (December 3, 2018), https://www.nytimes.com/2018/12/03/us/oil-drilling-arctic-national-wildlife-refuge.html

[12] Wadhams, *A Farewell to Ice*, Chapter 7, "The Future of Artic Sea Ice – the Death Spiral," 82-103.

[13] Wadhams, *A Farewell to Ice*, 105-107.

Snowline Retreat Feedback

Warmer air over the Artic causes snowline to retreat north as faster spring snowmelt warms the coastal tundra. In 2012 the mid-season snow extent was six million square kilometers less than it was in the late twentieth century. Wadhams estimates that snow line retreat and sea ice melt are adding equivalent amounts to overall global warming[14].

Water Vapor Feedback

The water vapor loop is temperature dependent: for every degree Celsius rise in temperature, seven percent more water evaporates and enters the atmosphere as water vapor, forcing more polar heating over the entire arctic basin. This added heat leads to more melting of ice and more exposure of sea water to the evaporative effect of the sun, and thus to more atmospheric water vapor.[15]

Ice Sheet Melt Feedback and Sea Level Rise

Two types of sea level rise – steric and eustatic – affect the elevation of the global oceans. (1) Steric sea level rise results from water in the oceans absorbing atmospheric heat from the burning of fossil fuels and expanding in the ocean basins. Heat transferred from the atmosphere at first warmed only the top level of the ocean, but now it extends downward into the deep ocean. As warming oceans expand, the water level rises.

(2) Eustatic sea level rise is the result of glaciers melting on continental mountain ranges and on the Greenland and Antarctic ice caps. This melting ice delivers cubic kilometers of water into the oceans, raising sea level and eating away at land-based glaciers, causing them to flow into the sea even faster. The situation with Greenland is particularly worrisome. Since the Greenland ice sheet sits at both a high latitude and an altitude of 2-3 km, it used to remain frozen solid all year long, except for some melting around the coastal edges in summer. But in the last decade or so warming air temperatures have resulted in significant surface melting, causing extensive pools of water to form over much of the glacial surface. Until recently climatologists had assumed that this water would refreeze quickly enough after the short Greenland summer that it would take several thousand years for the ice sheet to disappear. [16]

But then came the discovery of the phenomenon of "moulins": huge drain holes following glacial fractures that penetrate straight through the ice cap to where the glacier sits on rock, from whence the melt water finds its way through channels to the sea. A new study published in *Geophysical Research Letters*, a journal of the American Geophysical Union, shows that meltwater lakes forming on the ice surface can drain through moulins alarmingly fast, prompting further glacial disintegration:

> The massive ice flow speedup caused by all this water reaching the bed literally pulls the ice at the surface apart over a wide area, temporarily creating pervasive cracks, which seed new moulins where these cracks intersect meltwater flowing on the surface.[17]

The surface water surging downward through these vertical conduits penetrates through mile-deep ice, efficiently funneling the majority of summer meltwater from the glacial surface to its base. Lubrication of the ice sheet facilitates its slide off the Greenland rock surface, causing outlet glaciers to calve icebergs at a faster rate. Additionally, as the ice sheet shrinks down the rate of melting increases in the warmer air of its new lower

[14] Wadhams, *A Farewell to Ice*, 107.
[15] Wadhams, *A Farewell to Ice*, 109.
[16] A Farewell to Ice, 109-115.
[17] "Glacial moulin formation triggered by rapid lake drainage." American Geophysical Union, January 18, 2018, https://phys.org/news/2018-01-glacial-moulin-formation-triggered-rapid.html#jCp.

elevation. The alarming conclusion of this study is that the Greenland ice sheet is now losing 300 cubic kilometers of water equivalent per year.[18]

Arctic River Feedback

As snowline retreats south in Alaska, Canada, and Siberia, the Arctic albedo decreases, warming the coastal lands. Water flowing north through the tundra absorbs this increased heat and dumps it into the Arctic Ocean, melting sea ice and further decreasing Earth's albedo.[19]

Black Carbon Feedback

Airborne particulate matter from the burning of fossil fuels and from forest fires is transported by wind currents to the Artic region and deposited on snow. Black carbon absorbs solar radiation, melting down into holes in the snowpack. This feedback loop is dependent on pollution levels and wind dispersal[20].

Ocean Acidification Feedback

As the ice melts, the surface water in the oceans is exposed to carbon dioxide. As CO2 reacts with water it forms carbonic acid (CO2 + H2O -> H2CO3), rendering the ocean water more acidic. Marine organisms such as planktonic foraminifera lose their ability to make strong shells out of calcium carbonate (CaCO3), and thus are more vulnerable to predators. This leads to the dual problem that (1) the food chain is disrupted from the very bottom, and (2) tiny marine organisms no longer fall 4,000 meters to the ocean floor where they used to sequester carbon dioxide permanently. In an increasingly acidic ocean these organisms die before their shells harden, releasing carbon dioxide that would otherwise have been removed from the oceanic-atmospheric system[21].

Arctic Methane: a catastrophe in the making

Potentially the most catastrophic feedback loop of all is the release of Artic methane hydrates. Methane per molecule is at least 23 times more effective as a heat-trapping gas than is carbon dioxide. On the seabed lie frozen sediments that are extensions of the tundra permafrost on land, deposited in previous ice ages when sea level was lower. Embedded within this marine permafrost are methane hydrates or clathrates, a mixture of methane gas compressed together with water ice and held stable by the immense pressure of the water column above it. The amount of methane frozen in hydrates and clathrates is estimated to be 10,400 gigatons, more than thirteen times the total carbon in the atmosphere[22].

As atmospheric warming forces the rapid northward retreat of Arctic sea ice from relatively shallow continental shelves (only 50-100 meters deep), this leaves the open sea exposed to katabatic winds blowing downslope off Siberia. These winds push surface water down to the bottom, warming the seafloor for the first time in several tens of thousands of years, and methane hydrates are beginning to thaw out after being frozen in the sediment for millennia. They rise to the surface through the water column in huge bubble plumes, a process called ebullition, and breaking the surface they enter the atmosphere.

The slow, gradual release of gigatons of stored methane would be bad enough for long-term climatic stability. But what worries some scientists even more is the sudden, catastrophic release of methane hydrates. Natalia Shakhova of Tomsk Polytechnic University in Russia studies the East Siberian Arctic Shelf (ESAS), the largest undersea shelf in the world at two million square kilometers, and also one of the shallowest.[23] Shakhova believes

[18] Wadhams, *A Farewell to Ice*, 112.
[19] Wadhams, *A Farewell to Ice,* 115.
[20] Wadhams, *A Farewell to Ice*, 115-116.
[21] S. Uthicke, et al., "High risk of extinction of benthic foraminifera in this century due to ocean acidification," Nature: *Scientific Reports* volume 3, Article number: 1769 (2013), https://www.nature.com/articles/srep01769.
[22] Wadhams, *A Farewell to Ice*, 121-1246.
[23] Jamail, *The End of Ice*, 197-198.

the ESAS contains 10-15% of Earth's methane hydrates, and that releases of these may be abrupt like the rupturing of a sealed over-pressurized pipeline.

Methane releases could be triggered at any moment by an event such as an earthquake or the sudden thawing of permafrost. Methane emission rates could change by orders of magnitude within a matter of minutes, and a fifty-gigaton (50 billion tons) release of methane hydrates is "highly possible at any time." [24]Such an influx of methane into the atmosphere would be the equivalent of a thousand gigatons of carbon dioxide (two-thirds of the total of 1,475 gigatons of CO_2 introduced by humans since 1850). Shakhova states that this would "cause an approximately twelve times increase of modern atmospheric methane burden with consequent catastrophic greenhouse warming."[25]

B. Continental Ice: What's happening in Antarctica?

If the loss of Arctic sea ice will have a tremendous destabilizing effect on the global climate, what about the Antarctic? Wadhams notes that as far as climate is concerned "the Arctic is a driver and the Antarctic can be thought of as a passive trailer in the global warming road race to oblivion."[26] But although losing Arctic ice will have no net effect on ocean level rise – since that sea ice is already floating – it is quite a different story with respect to large land masses such as Greenland and Antarctica, where glaciers are in retreat. Greenland alone is pouring cubic kilometers of water into the global oceans.

Antarctica is the most isolated of the continents, surrounded by the largest body of water on Earth, the Great Southern Ocean. The extent of Antarctic sea-ice coverage fluctuates annually, unlike the inexorable shrinking of Arctic ice. In winter Antarctic ice grows by millions of square kilometers, retreating in the summer. Scientists are greatly concerned is that as ice shelves break off on the margins of the continent, they no longer buttress massive glaciers on the landmass, which can then flow into the Great Southern Ocean. In 2002 the 200-meter-thick Larsen B Ice Shelf broke away from the Antarctic Peninsula, slowly disintegrating into a vast flotilla of icebergs moving northward into shipping lanes.[27]

Other glaciers likewise are showing signs of disintegration. A recent NASA-led study of the Thwaites Glacier in West Antarctica made the disturbing discovery of a gigantic cavity at the bottom of the glacier, two-thirds the area of Manhattan and almost 1,000 feet (300 meters) tall.[28] The cavern is big enough to have contained fourteen billion tons of ice, most of which has melted out over the last three years. Moreover, since many Antarctic glaciers extend for miles beyond their grounding lines – floating out over the open ocean – it is ominous that the Thwaites Glacier is not securely anchored to bedrock. When a glacier loses ice weight it can float over the rocks where it used to stick, allowing the grounding line to retreat inland, exposing more of its underside to sea water, and accelerating its melt rate. Thwaites holds enough ice to raise the world ocean 65 centimeters – a little over two feet.

Antarctic sea ice holds neighboring land-based glaciers in check. If the Ross and Filchner-Ronne ice shelves collapse, this will permit glaciers in the interior – such as those in the Transantarctic Mountains – to flow unimpeded off the continent and debouch directly into the ocean.[29] While a total melt-down might yet be several centuries off, the loss of all Antarctic glaciers would raise sea global level by 68 meters, or 225 feet. Our descendants will no doubt regard such total inundation of all coastal cities in the world as a civilization-altering catastrophe.

[24] Natalia Shakhova, et al., "Anomalies of methane in the atmosphere over the East Siberian Shelf: is there any sign of methane leakage from shallow shelf hydrates?, *Geophysical Research Abstracts,* 2008.
[25] Natalia Shakhova, et al., "Current rates and mechanisms of sub-sea permafrost degradation in the East Siberian Arctic Shelf," *Nature Communications* 8 (June 22, 2017).
[26] Wadhams, *A Farewell to Ice,* 170.
[27] Wadhams, *A Farewell to Ice,* 157.
[28] "Huge Cavity in Antarctic Glacier Signals Rapid Decay," Jet Propulsion Laboratory, California Institute of Technology (January 30, 2019), https://www.jpl.nasa.gov/news/news.php?feature=7322.
[29] Wadhams, *A Farewell to Ice,* 168.

IV. CONSEQUENCES of Loss of Ice

The permanent loss of Arctic, Greenland, and Antarctic ice will have dire consequences for life on earth in at least three areas. (1) Rising sea level will erase coral atoll nations and inundate coastal areas everywhere on the planet. (2) Ocean acidification from dissolved CO_2 may irreparably damage the marine and terrestrial food chains. (3) Loss of ice as a climate stabilizer may make much of the world subject to caprices of wild storms, floods, and extended droughts, threatening the agricultural stability that made civilization possible. As Peter Wadhams says:

> The *indirect* effects of Arctic sea ice retreat are overwhelmingly negative for the planet as a whole, and negative to such an extent that Artic sea ice retreat has to be seen as an unmitigated disaster for the Earth.[30]

Let us briefly sketch these three areas.

1. Sea Level Rise threatens civilization (Wadhams, 118-20)

The most immediately obvious threat of melting polar and Greenland ice is the global sea level rise that will erase entire coral atoll nations that sit at only a few meters of elevation. How can this be addressed? Wealthy nations with coastal cities like the U.S., the U.K. and the Netherlands might have the engineering expertise and practical experience to erect extensive sea walls and tidal barriers, but they will find it very expensive. In fact, future generations everywhere will need to make challenging decisions: will Italy and the world community deem the architectural heritage of Venice to be worth saving at all costs, or will the city be abandoned in the end to rising Adriatic waters?

Poorer countries such as Bangladesh – where twenty million people live within two meters of current sea level – will find holding back the sea an impossible task.[31] As communities and entire regions are swallowed by the sea, tens of millions of people will be forced from their homes and will become climate refugees, seeking asylum in neighboring countries who themselves may be agriculturally hard-pressed to support their own people. And sooner or later all but landlocked nations will struggle with the destruction of every coastal city and its infrastructure.

2. Ocean acidification and loss of sea ice threatens the food chain

The consequences of ocean acidification as more ice-free surface water is exposed to carbon dioxide are dire[32]. Ninety percent of coral reefs worldwide show significant dieback from bleaching, and this spells disaster-in-the-making for these marine nurseries and nesting grounds.

Humans are as dependent on our physical environment as are all the species we are extinguishing around us, such as "sea butterflies," a taxonomic suborder of small pelagic sea snails, named *Thecosomata*.[33] Thecosomes include some of the world's most abundant gastropod species, about the size of a lentil, that form the sole food source of their relatives, the *Gymnosomata*. Both are also consumed by sea birds, whales, seals, and commercially important fish. Thus, thecosomes sit at the base of the marine food chain that leads up through a wide variety of pelagic and anadromous fish to penguins, pinnipeds (seals) and cetaceans, and on land to and polar and grizzly bears.

---

[30] Wadhams, *Farewell to Ice*, 104.
[31] Warren Cornwall, "As sea levels rise, Bangladeshi islanders must decide between keeping the water out—or letting it in," *The Guardian*, (March 1, 2018), https://www.sciencemag.org/news/2018/03/sea-levels-rise-bangladeshi-islanders-must-decide-between-keeping-water-out-or-letting?r3f_986=https://www.google.com/
[32] Jamail *The End of Ice*, 72-79.
[33] Hannah Waters, "Amazing Sea Butterflies Are the Ocean's Canary in the Coal Mine," *Smithsonian Magazine* (May 14, 2013), https://www.smithsonianmag.com/science-nature/amazing-sea-butterflies-are-the-oceans-canary-in-the-coal-mine-61813612/

Thecosomes are vulnerable to increased levels of atmospheric carbon dioxide, because dissolved CO_2 creates carbonic acid (H_2CO_3). Increased oceanic acidity means that sea butterflies can no longer form shells, threatening their very survival. If we continue with "business as usual," thecosomes will become regionally extinct as soon as 2050, with dire consequences for the entire marine food chain on which billions of humans increasingly depend.

A recent study of polar bear physiology reinforces our understanding of the challenges polar bears face as sea ice shrinks and becomes increasingly fragmented. The bears' primary prey of seals becomes less available, and the bears have to become more active and cover more distance to feed, expending more energy than they recoup through eating the meager amount of seal blubber available. [34]This winter fifty-two starving polar bears have invaded the Russian Belushya Guba work settlement of 2,000 people in the remote Novaya Zemlya archipelago. Deprived of their normal diet of seal meat, the bears have taken to foraging in garbage dumps, threatening local residents. Here we see a direct result of climate disruption, as the crash of one species' evolutionary diet forces them to make a drastic change that affects both the bears and the humans living in their environment.[35]

3. Agricultural failure from extreme weather and loss of water

Except for hunter-gatherers, human communities depend on stable agriculture. This in turn depends on predictable weather and reliable water sources. The loss of ice will lead to climate changes that will make local and regional weather unpredictable. Hurricanes, typhoons, tornadoes and the wobbling of the less stable polar vortex can adversely affect crops and orchards and commercial forests. Extended droughts, dust bowls, and wildfires likewise will exercise an adverse impact on agriculture.

Of direct relevance to communities in and around the Andes mountain range and the Himalayan massif is climate-change-induced loss of ice through the melting of glaciers. Since the South American dry season can last up to six months, life in some parts of the Andes is almost exclusively dependent on glacial water. As glaciers shrink away in a warming climate, Peruvians, Bolivians, and Chileans may have to abandon high Andean communities.[36]

Similarly, most of the 5,500 glaciers in the Mt. Everest region are expected to retreat drastically or even disappear completely by the end of the 21st century. This will carry significant implications:

> Farming and hydropower generation downstream of the Himalayan peaks is likely to be greatly affected. Over one billion people in Asia depend on rivers fed by glaciers for their food and livelihoods. While increased glacier melt initially increases water flows, ongoing retreat leads to reduced meltwater from the glaciers during the warmer months[37].

There are many rivers arising in the Central Asian massif flowing out of glacial valleys. Among these are the Indus, Ganges, Brahmaputra, Irawaddy, Yangtze, and Yellow Rivers. As glaciers that used to supply a consistent flow of water begin to retreat, a billion downstream users will be put at risk, directly or indirectly.

V. Strategies for addressing Greenhouse Gases
1. Reduction of GHG emissions to zero
2. Addressing Overpopulation and Overconsumption
3. Sequestration of Carbon

[34] Anthony Pagano, "Polar bears starving as arctic ice vanishes," *LiveScience* (Feb. 1, 2018) https://www.livescience.com/61619-polar-bears-starving.html
[35] Brandon Spector, "52 Polar Bears 'Invade' a Russian Town to Eat Garbage Instead of Starve to Death" *LiveScience* (February 11, 2019), https://www.livescience.com/64741-polar-bears-are-taking-back-russia.html
[36] Mathias Vuille, "Melting South American Glaciers Could Leave Locals High and Dry," *Inverse* (December 7, 2018), https://www.inverse.com/article/51526-andes-glacier-melt-threatens-local-communities.
[37] John Vidal, "Most glaciers in Mount Everest area will disappear with climate change," *The Guardian* (May 27, 2015) https://www.theguardian.com/environment/2015/may/27/most-glaciers-in-mount-everest-area-will-disappear-with-climate-change-study.

Conclusions

The relentless melting away of Earth's permanent ice is a tremendous loss that hits humans on many levels: climatological, environmental, aesthetic, and personal. As a mountaineer my relationship with ice has never been more personal and exhilarating than it was during my crevasse fall on Mt. Shasta, when the Hotlum Glacier held me for a brief moment. Aesthetically the loss is as painful to bear as the personal: three decades ago I attended a lecture in Berkeley by renowned photographer and wilderness adventurer Galen Rowell, entitled "Antarctica: The Shining Continent." Galen held us enraptured with his description of the majesty of ice and his photographs of spectacular glaciers and Antarctic mountains, material later included in *Poles Apart: Parallel Visions of the Arctic and Antarctic.*[38] It is heart-breaking to realize that this majestic beauty and its attendant ecosystems are melting away before our eyes.

But the stakes are far greater than merely aesthetics and personal enjoyment: they involve nothing less than the stability of the biosphere and the long-term survival of human civilization. Three decades ago we had ample warning from environmental scientists that we needed to reduce overconsumption and human overpopulation, to curtail carbon dioxide emissions, and to begin atmospheric carbon drawdown. But we largely squandered this grace period, and we now contemplate the approaching "death spiral" described by Peter Wadhams:

> We are not far from the moment when the feedbacks themselves will be driving the change – that is, we will not need to insert more CO_2 to the atmosphere at all, but will get the warming anyway. This is a stage called *runaway warming*, which is possibly what led to the transformation of Venus into a hot dry, world... We are fast approaching the stage when climate change will be playing the tune for us while we stand by and watch helplessly, with our reductions in CO_2 emissions having no effect.[39]

With everything we now know about the critical, life-giving role played by permanent terrestrial ice and the consequences of its irretrievable loss, will humans at last be motived to think through and implement strategies to address this issue?

Climate Crisis "Keep/Stop/Start" Actions

Below for your consideration, are several author-suggested thought, behavior, and action items. Action nourishes inspiration! For a further explanation of the Keep/Stop/Start organizational tool, please see How to Read This Book.

KEEP:

Keep informing yourself about the crucial role played by polar and Greenland ice caps in stabilizing Earth's climate. As GHGs build up in the atmosphere, temperatures in the polar regions are warming faster, pushed by feedback loops. For millions of years, the presence of ice on our planet has served like a flywheel, serving as a stabilizing factor for ocean level, ocean water pH, and climate. The loss of permanent ice will be a catastrophe unparalleled in human history, and understanding this issue must become a core element of discussions in areas as diverse as education, science, technology, politics, religion, and culture.

---

[38] Galen Rowell, *Poles Apart: Parallel Visions of the Arctic and Antarctic* (Berkeley: University of California Press, 1997).
[39] Wadhams, *Farewell to Ice,* 108-109.

STOP:

Stop allowing climate change deniers to distract you from the reality of ice loss and the seriousness of its consequences. There are excellent resources for countering climate change skepticism. In addition to the resources in this volume, one of the most helpful and elegantly laid-out websites is "Skeptical Science: getting skeptical about global warming skepticism." See https://skepticalscience.com/.

START:

Start learning about the range of strategies available to us to arrest the production of greenhouse gases and begin carbon sequestration. *Climate Abandoned* and numerous other resources address carbon drawdown and sequestration. An excellent resource is Paul Hawken, ed., *Drawdown: The Most Comprehensive Plan Ever Proposed to Reverse Global Warming*[40]. As scientists and informed citizens, as parents and friends, it is important for us all to know about and use resources that can help us combat climate fatalism fostering nothing but inaction and despair. Hope, resilience, and informed action must become the driving forces of our lives.

~ ~ ~

Peter M.J. Hess, Ph.D. is a Roman Catholic theologian who specializes in issues at the interface of science and religion. He has served as international director for the Center for Theology and the Natural Sciences (CTNS) in Berkeley, California and director of Outreach to Religious Communities with the National Center for Science Education (NCSE) in Oakland, promoting dialogue in the areas particularly of evolutionary biology and climate change. Peter Hess earned an MA in philosophy and theology from Oxford University and a PhD in historical theology from the Graduate Theological Union in Berkeley. He is co-author of *Catholicism and Science* (Greenwood Press, 2008) and writes on the religious and ethical implications of a growing human population in light of climate change. Peter lives in Berkeley, California. Read more at: www.climateabandoned.com.

---

[40] Paul Hawken, ed., *Drawdown: The Most Comprehensive Plan Ever Proposed to Reverse Global Warming* (Penguin Books, 2017).

Chapter 4

Climate Crisis & Oceans
Part 1: Protect oceans, the heart of Earth's climate system

by
Julie Packard, M.A.

On Thursday [9-13-18] and Friday [9-14-18], leaders from around the world will gather in San Francisco for the Global Climate Action Summit. They're here to learn, share ideas and strengthen their commitments to fighting climate change, the greatest challenge facing humanity.

What makes this gathering unusual is that, for the first time, ocean issues are a top priority. The attention is overdue.

For too long – the heart of Earth's climate system – has been ignored in climate conversations. To solve the climate crisis, we must address the health of the largest ecosystem on our planet, and our first line of defense against the impacts of climate change.

And we must act quickly.

Sea levels are rising, placing tens of millions of coastal residents in harms way. Intensifying storms are costing human lives and causing billions of dollars in damage to homes and businesses. Around the world, people depend on the ocean for so much: food, jobs, transportation and stabilizing our climate. Now is the time to recognize that human health is directly tied to ocean health.

Climate change is already disrupting fundamental ocean processes that sustain life on Earth. Warming water is choking tropical corals and stunting kelp forest growth along the California coast. And carbon pollution is making seawater more acidic, dissolving the shells of plankton that are the foundation of ocean food webs.

The good news is that the ocean is resilient. It can recover.

At the summit, ocean leaders will issue a call to governments, industry and concerned citizens outlining what it will take to protect our living ocean. We must reverse the destruction of coastal habitats, create more global marine protected areas and improve the sustainability of global fisheries and aquaculture. We must help coastal communities prepare for, and adapt to the growing impacts of extreme weather and sea level rise, and invest in science, which is the bedrock of sound decision-making.

Finally, and most importantly, we must redouble our commitment to dramatically reduce greenhouse gas emissions.

By embracing this vision, we can assure the ocean gets the attention, and the protection, it deserves. The lives of Earth's 7.5 billion people depend on it.

As a California native, I couldn't be prouder of my home state. California is advancing ambitious climate policies and moving toward zero emissions. Cities along our beautiful coast are factoring climate change into their land-use planning, building resilience for the challenges ahead.

California has the first-in-the-nation, statewide network of marine protected areas. Innovators the private sector are turning their creativity to climate solutions. Philanthropists are investing in science, and in science-based approaches that put us on a path toward sustainability.

At a moment when federal leadership on climate change has receded, the Global Climate Action Summit is proof that states and cities aren't waiting for Washington, D.C. to lead.

We cannot wait any longer. We must all act as if our lives depended on it. Because they do.

~ ~ ~

Julie Packard, M.A. is the executive director and vice chair of the Monterey Bay Aquarium's Board of Trustees. Julie Packard is executive director of the Monterey Bay Aquarium, which she helped found in the late 1970s. She is an international leader in the field of ocean conservation, and a leading voice for science-based policy reform in support of a healthy ocean. Under her leadership, the Aquarium has pioneered innovative exhibits and education initiatives and has evolved into one of the nation's leading ocean conservation organizations. Julie is a leader in ocean conservation worldwide, and brings a lifelong passion for the natural world. Julie was a member of the Pew Oceans Commission and serves on the Joint Oceans Commission Initiative, where she works to implement comprehensive reform of U.S. Ocean policy. *(Op-Ed reprinted with permission by the Monterey Bay Aquarium)*

Chapter 5

Climate Crisis & Oceans
Part II: Earth's Heartbeat is Getting Weak

by
Jill Cody. M.P.A. and Richard Nolthenius, Ph.D.

"The ocean is the heart of the climate system." – **Julie Packard**

To painfully extend the analogy made in the previous chapter, our Earth's heart is nearing cardiac arrest, and we are witnessing fibrillations; the twitching of its muscle fiber, rapid irregular contractions, and lack of synchronism between heartbeat and pulse.

One can see the pulse of the ocean in its wave action. As a recent study[1] by the University of California Santa Cruz Institute of Marine Sciences discovered, waves are crashing onto our coasts and beaches with more force than previously seen. The ocean waves have become larger and stronger due to ocean warming and the increase in wave energy.

The study's result was published in "Nature Communications",[2] January 14, 2019, and it stated that *"Climate change is modifying oceans in different ways, including ocean-atmosphere circulation and water warming. The effects of climate change will particularly be present on the coast, where humans and oceans meet. Surface gravity waves generated by winds have far-reaching implications for coastal areas."*

More than 600 million people live less than 10 meters away from the ocean's coastline. The impact on life, property, and the political stability of 600 million people cannot be overstated. The property, some very expensive, will be pulsed away forcing the inhabitants to move, maybe losing their livelihood, impacting economies, agriculture, forcing a political crisis, and losing historical and vacation sites. States, such as California, are now being forced to face whether or not to save the beaches that millions of people enjoy or close them off and protect the housing with boulders and concrete walls to fortify them from the climate-induced increased wave action. This is California's Sophie's Choice.

We see the ocean is nearing its version of a cardiac arrest in four other alarming ways.

Decreasing Oxygen Levels

If the oxygen were sucked out of the room in which you're sitting, it would attract your horrified attention for fear of suffocating. However, oxygen in the ocean is dangerously declining and is now suffocating sea life. We don't pay much attention because it is happening slowly, but it is as devastating to life as if we suddenly lost all the air in the room. Oxygen levels actually have been dropping three times faster than the computer models predicted.

There are, in fact, places where the ocean's heart has died. They are called "Dead Zones." They're regions where the oxygen is entirely gone, and most forms of ocean life can no longer survive. Rising global temperature impacts the oxygen's solubility in the water and restricts its movement into deeper water while, concurrently, sea life has become more stressed due to the warmth and increased acidic levels. This spells trouble for us too.

---

[1] Riguero et al. 2019
[2] https://www.nature.com/articles/s41467-018-08066-0

With the lack of oxygen, fish suffocate. Even if you don't care about the fish themselves, worldwide, 35 million people financially depend on the fishing industry and an untold number of people are nourished by it. "Over the past 50 years, the ocean suffered a loss of about 85 billion tons (77 billion metric tons) of oxygen, affecting an accumulated area approximately the size of the European Union. Globally, the amount of zero-oxygen ocean water has quadrupled, while the area occupied by low-oxygen zones has increased by 10 times," as a Live Science article related.[3] Lowering ocean oxygen kills fish while favoring the propagation of stinging jellies.

Ocean Acidification

Think back to your high school chemistry class. Remember the Litmus paper you dunked into a glass of water? If the paper turned "red" the water was acidic.[4] That is exactly what is happening to our oceans today due to the interaction of the carbon, from fossil fuels, in our atmosphere coming in contact with the ocean saltwater. When this contact is made, the ocean absorbs the carbon and, thus, becomes more acidic. Acid eats away at hard-body-animals such as coral, bivalves (clams, oysters, mussels, scallops), and plankton. The acidity of the oceans has increased 30% since pre-industrial days, due to human CO_2 emissions.

The National Oceanic and Atmospheric Administration (NOAA) stated that climate change is "the greatest global threat to coral reef ecosystems.[5] Coral reefs are alive. As the ocean temperature rises, the symbiotic algae leave the corals, bleaching them, and then the rising acidity further inhibits the ability of the fragile calcium carbonate corals to survive. The weakened ecosystems then suffer infectious diseases.

Bivalves, such as the mussels and oysters in your linguini, are in trouble too. Not only does the acid eat away at their hard shells (aragonite calcium carbonate), but it also weakens that stringy stuff that allows them to stick themselves onto rocks (or maybe your boat's haul). The vexing problem is that these string things are non-calcified structures, so the acid is weakening them by 40%, these little creatures are being attacked by acid in both the calcified and the non-calcified parts of their bodies.[6]

"It's hard to imagine that organisms you can't see make a difference, and we all struggle with that. You look out in the ocean and you see water, but that water is just teeming with millions and millions of tiny organisms, and when you sum them up, they have a huge impact" as Mike Behrenfeld, botany and plant pathology professor at Oregon State University observed. These microscopic creatures are vigorously coupled with the climate and life on the planet. As the ocean warms, it contains less oxygen and phytoplankton suffers, and dangerous anaerobes do better. Since 1950, phytoplankton levels have decreased globally by roughly 5-8%. You can't see them, but our food chain depends on them.

Heart Warming

Water, water everywhere ... and it's hot! Unlike us, the Earth's heart is 70% of its body. We couldn't lose 70% of our body and still live so we can't expect Earth (or us) to live if 70% of it suffocates to death. Just as on land, ocean warmth is shifting marine ecosystems at an unprecedented scale. Coupled with the depletion of oxygen, acidification, and heat we could very well experience the same result as the planet experienced 252 million years ago known as the "Great Dying." Due to the oceans becoming too hot, it was the worst of our planetary extinctions, and killed nearly all sea life on Earth. This is a risk for us if we, collectively, don't act to change this direction immediately.

As a matter of fact, researchers with the University of Santa Barbara who published their findings in Science, used "hindcasting" methods to create a picture of the effects of climate change, overfishing, and other variables

---

[3] https://www.livescience.com/61338-ocean-losing-oxygen.html.
[4] https://www.youtube.com/watch?v=j3HhPxRSiA0
[5] https://oceanservice.noaa.gov/facts/coralreef-climate.html
[6] https://www.nature.com/articles/nclimate1846

over an 80-year span from 1930-2010. They examine fish populations from 38 regions and studied 235 fish populations comprised of 124 species that represented about third of the global fishing catch over that 80-period. They discovered that fish catches have diminished significantly and probably will fall further due to warming ocean.[7]

The millions of tons of CO_2 that human activity belches into the air daily have been warming the planet since the Industrial Age began. This climate warming forcing is equivalent to that of four Hiroshima bombs per second, and 93% of that warming is deposited into the oceans where it remains. There is nowhere for it to go, except ultimately back into the atmosphere and then out into space, but only if we drop atmospheric CO_2 levels so that the atmosphere will try to cool – heat only flows from hot to cold, never the other way. Not to be too Biblical, but the ocean giveth by absorbing our heat insult, and the ocean will taketh away by pumping that heat right back into the atmosphere if we try to cool it by lowering atmospheric CO_2. And so far, we're only adding more heat.

The ocean temperature was the hottest ever in 2018. In the Advances of Atmospheric Sciences[8] the signatory scientists stated: *"Increases in ocean heat are incontrovertible proof that the Earth is warming. The long-term trend of ocean heat is a major concern both in the scientific community and for the public at large."* On August 3, 2018, off of the Scripps Institution of Oceanography pier in San Diego, California, the ocean temperature reached a 102-year all-time high of 78.8 degrees. The Scripps researchers said the rise in ocean temperature was similar to the high land temperatures experienced that same summer. They also noted that the warming cannot be blamed on an El Nino year. The last El Nino pattern ended in 2016, but the heat just stayed.

The ocean is warming but it also has "hotspots." These are areas of the ocean that experience heatwaves consisting of temperature extremes for five days or more. Dan Smale at the Plymouth, UK, Marine Biological Association explained, "You have heat-wave-induced wildfires that take out huge areas of forest, but this is happening underwater as well." Kelp forests, seagrass, and coral reefs are being lost to these ocean hotspots and they are foundational to ocean life. Even though El Nino events are a natural ocean cycle, the temperature rise in the ocean is in addition to these cycles and make the heatwave worse. As Rutgers University, who was not part of the team, stated, "This [research] makes it clear that heatwaves are hitting the ocean all over the world ... The ocean, in effect, is spiking a fever. These events are likely to become more extreme and more common in the future unless we can reduce greenhouse gas emissions." What the Earth's heart is trying to say is humankind, each and every one of us, must come together quickly as we did during World War II to fight the enemy ... our greenhouse gas emissions.[9]

Ocean Current

In keeping with the *heart* analogy, slowing ocean currents may be seen as a lack of synchronicity in the ocean's heart; unquestionably the lack of blood flow in its arteries. Two researchers, using different methodologies, reported the same stunning findings and they were that the climate crisis is weakening the ocean's massive circulation system.[10] Both studies found that Greenland had dumped vast quantities of freshwater into the North Atlantic Ocean diluting the salinity of the current which resulted in a 17% decrease in the flow of the **Atlantic Meridional Overturning Current** (AMOC). The scientists disagree on when this ocean current began slowing, and it would be nice to have alignment, but the end result in which they do agree - is scary. A weakened ocean current can drastically change weather patterns and cripple ocean eco-systems. The global ocean thermohaline current is now weak enough to be in a "bi-stable" regime, and close to shut down. If it shuts down, it will be centuries before it could start again, even if the thermal gradients were favorable for the existence of the re-

[7] https://www.carbonbrief.org/ocean-warming-has-caused-sustainable-fish-stocks-to-drop-by-4-since-1930s
[8] https://link.springer.com/content/pdf/10.1007%2Fs00376-019-8276-x.pdf
[9] https://www.theguardian.com/environment/2019/mar/04/heatwaves-sweeping-oceans-like-wildfires-scientists-reveal
[10] http://blogs.discovermagazine.com/d-brief/2018/04/11/ocean-current-climate-change-amoc/#.XHnQJnjx5YU

emerged current. The AMOC and global current did, in fact, shut down at the height of the last interglacial warm period. Paleontologist Dr. Peter Ward warns that a severe shut down can lead to loss of oxygen transport to the ocean bottom, followed by anaerobic hydrogen sulfide producing bacteria to flourish, generating enough deadly hydrogen sulfide to cause a mass extinction. This is the likely cause of most of our 5 past mass extinctions.

The ocean is everything to human survival. And our ocean is now hot, suffocating, and sluggish. As Michael Brune, Sierra Club executive director, says: *"The world's oceans are the canaries in the coal mine when it comes to the climate crisis. The writing has been on the wall for years."*

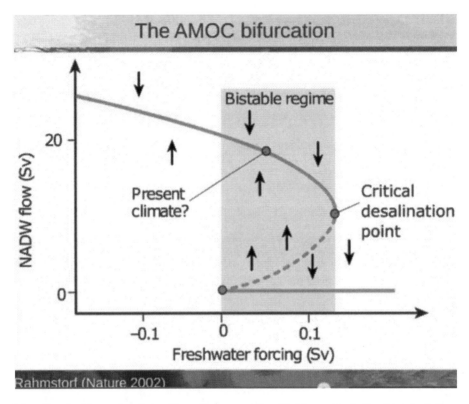

As freshwater melt from Greenland lowers the salinity of the surface ocean at the critical drop points for the global circulation, we are getting very close to the "cliff" that shuts down the current entirely.

Climate Crisis "Keep/Stop/Start" Actions

Below for your consideration, are several author-suggested thought, behavior, and action items. Action nourishes inspiration! For a further explanation of the Keep/Stop/Start organizational tool, please see How to Read This Book.

KEEP:

Keep educating yourself on the ocean's challenges. This World Economic Forum article is a good start: https://www.weforum.org/agenda/2018/10/we-can-save-our-ocean-in-three-steps-if-we-act-now/ and Rick Nolthenius' PowerPoint: "The Ocean/Atmosphere Connection and Climate" is a valuable reference: http://www.cabrillo.edu/~rnolthenius/Apowers/A7-K36-OceanAtm.pdf

STOP:

1) *Stop supporting or voting for any elected representative* who does not take the climate crisis seriously. There are too many of them in state and federal offices now as it is.

2) *Stop using as much as you can plastic products,* including plastic bags. Recycle as much of it you can, to then be made again into new products. Fossil fuels are used in the production of plastic and reducing plastic use will also reduce fossil fuel demand. Plastics are another dramatic poisoning of the oceans not related to climate.

START:

1) *Start financially supporting organizations* that study and advocate technologies that fight the effects of the climate crisis within the ocean such as Oceana, Save Our Seas Foundation, and Greenpeace.

2) *Start being motivated by those who are fighting* for the ocean's complex challenges and begin by watching Sam Waterston's speech at an Oceana event: https://oceana.org/blog/now-time-stand-and-be-counted-sam-waterstons-inspiring-speech-will-fire-you-oceans

Start making lifestyle changes such as are found in this National Geographic article: https://www.nationalgeographic.com/environment/oceans/take-action/10-things-you-can-do-to-save-the-ocean/

~ ~ ~

Jill Cody, M.P.A is an authorized presenter (2007 4th class) of Al Gore's "An Inconvenient Truth" slideshow. She is also the author of the award-wining book *America Abandoned: The Secret Velvet Coup That Cost Us Our Democracy* and is producer and host of KSQD 90.7FM "Be Bold America!" radio program. Jill conceived of the *Climate Abandoned: We're on the Endangered Species List* anthology when realizing friends and acquaintances did not fully comprehend the breath, dynamics, and complexity of the climate crisis. Jill is a distinguished alumni of San Jose State University's College of Applied Sciences and Arts and currently serves on the university's Emeritus and Retired Faculty Association Executive Board.

~ ~ ~

Richard Nolthenius, Ph.D. is the head of the Astronomy Department at Cabrillo College in Aptos, California. He earned his degree in astronomy and astrophysics after doctoral work at Stanford University and UCLA. He was a member of the Thermodynamics Group at General Dynamics in their space program in San Diego and performed thermal analysis and design for the Atlas/Centaur rocket missions and space satellites, and was thermal systems designer on their proposal for what became the International Space Station. His post-doctoral work involved galaxy clustering algorithms and comparisons between numerical cosmological simulations and real observations. He was a visiting researcher and lecturer at UC Santa Cruz, in the city in which he now resides. Since 2009, Rick's focus has been on the science of climate change and its relation to political/economic systems. Read more at: www.climateabandoned.com.

Chapter 6

Climate Crisis & Putting a Price on Carbon

by

Robert Mullins

"They frack. They mine. They earn astronomical profits" – **Leonardo DiCaprio**

The politics of perpetual gridlock in Washington has become tiresome to many as a proposal from one side of the aisle is immediately denounced by people from the other side. But among the 435 members of the House of Representatives and 100 members of the Senate, surely there must be a few people there who are well-intentioned and aim to do the right thing. Consider the possibility that, say, a Republican can usually vote the party line but can actually risk being labeled "a moderate" for breaking ranks and supporting legislation to address climate change. Yes, there have been such people in Congress; one of them is Carlos Curbelo.

In spite of his Republican title, the Florida congressman stuck his political neck out to tackle the threat of climate change. Curbelo is among the founders of the Climate Solutions Caucus, a bipartisan group of members of the House of Representatives who are willing to work together to draft legislation to address the undeniable threat.

In lobbying fellow GOP representatives, Curbelo quipped, *"When my district is under water, I'll move to your state and run against you."* [1]

In spite of the deadlock-by-default position of the U.S. Congress during the last few years, the existence of Republicans like Curbelo is a hopeful sign of climate change cooperation. The Climate Solutions Caucus is bipartisan by design; if a member wants to join, he or she must "bring a date" from the opposite political party. Curbelo started things off by inviting Democratic Representative Ted Deutch, also of Florida. At the time of this writing, the caucus had ninety members, forty-five from each of the two major parties.

Unfortunately, Curbelo was one of several Republicans voted out of Congress in November of 2018 when a Democratic surge in mid-term elections gave them control of the House. Other members of the Caucus were also voted out of office and, according to a CCL spokesman, there's going to be some reshuffling of Caucus membership to include newly-elected representatives.

However, there was some exceedingly good news for CCL advocates in the New Year when the Energy Innovation and Carbon Dividend Act (EICDA) was reintroduced in the House on Jan. 24, 2019. It differs in some details from CCL's carbon fee and dividend proposal but is still basically advocating the same solution to dangerous climate change. The EICDA would impose a carbon fee on fossil fuels, which would reduce demand, but return the fee money to taxpayers as a dividend.

The Climate Solutions Caucus was created, in part, with political support from a grass-roots organization called Citizens' Climate Lobby (CCL), which was established in 2007. As its name indicates, the mission of CCL is to help concerned citizens band together to lobby Congress to take action to protect our climate. CCL currently has more than 100,000 members from all 50 states, with chapters in nearly all congressional districts, as well as many other supporters from abroad.

CCL deploys a multi-pronged approach to reach Congress members and enlist their support to counter climate change. Via email, snail mail, and even tweets, we in CCL urge Congress to support our crucial cause.

---

[1] "Miami's Fight against Rising Seas" BBC.com. April 4, 2017.

After all, if President Donald Trump can garner so much attention with only 280 Twitter characters, we can certainly do the same! We visit representatives in person in their district offices and even travel to their offices in Washington, D.C. We pen heartfelt letters to editors and write op-ed columns in our local newspapers, offering our rebuttals and reactions to the climate change stories we read. We actively lobby business leaders in our communities to support climate change legislation, and we take advantage of every speaking opportunity, sharing our story with various community groups. We also set up tables and kiosks at local events and distribute CCL literature to invite others to join our cause.

A Carbon Fee and Dividend Explained

Let's start by laying out the CCL legislative policy, and then we'll explain it in more familiar terms.

Essentially, the CF&D rule would impose a carbon fee on all fossil fuels collected in the U.S or imported from abroad. The fee would begin at $15 per ton of CO_2 emissions implicit in the carbon entering the U.S., as well as the CO_2 equivalent (CO2e) warming potential of non-CO_2 greenhouse gases entering, such as methane, nitrogen oxides, and the refrigerant HFC's. The fee would rise by $10 per ton each year. The abbreviated way this is expressed is "$15 per ton of CO2e". This fee would be imposed at the source, at the wellhead where oil is pumped from the ground, at the mine from which coal is extracted or at the port where fossil fuel-based products enter the U.S. from abroad. The reason for this *border adjustment*, this import tax, is that fuel from abroad cannot undercut the price of U.S.-sourced energy unless the exporting country also imposes an equivalent carbon fee.

The idea behind the carbon fee is to increase the retail price of fossil fuel-based energy to more fully reflect the cost of carbon dioxide (CO_2) emissions because of the damage these cause to the environment, public health, and the economy. Today, the price of carbon-based products does not reflect the damage they do. If the carbon fee is implemented, yes, it will cost more for us to fill up our cars, or heat and light our homes. In fact, the price of the vast majority of products we buy will rise, since fossil fuels comprise roughly 80% of our power including the energy to make those products. However, the rising cost of fossil fuels is expected to encourage the purchase of a hybrid or all-electric vehicles, the utilization of more economical and smart public transit, the installation of rooftop solar panels on our homes. In general, CF&D will raise the price of high carbon intensity products more than it will increase the price of low carbon intensity products. This will motivate stronger efforts to de-carbonize our economy, as stabilizing the climate requires.

Including taxpayer dividends in the plan is politically critical, as it is meant to offset the financial impact on Americans of the added expense for energy derived from fossil fuels. The money raised by the carbon fee would not remain in the hands of the federal government; instead, it would be returned to taxpayers as monthly dividends.

The amount of the dividend would rise along with the rise in the carbon fee at the energy source. Simultaneously, the monthly dividend would rise each year from $125.00 per month in 2018 – for a typical family of two adults and two minor children — to close to $400 per month in 2035, when the carbon fee would then be $175 per ton. On an annual basis, that dividend would add up to $1,500 per household in 2018 and $4,800 in 2035, in inflation-adjusted dollars.[2] The statistics are included in a report on climate change from the research firm Regional Economic Models, Inc. (REMI).

The Founding Fathers of CCL

The CCL was founded by Marshall Saunders, a San Diego real estate broker and a philanthropist who has provided microloans to those in the developing world. By 2006, Saunders was becoming increasingly concerned

---

[2] REMI Report. CitizensClimateLobby.org

about climate change and its impact on the environment and the world, and he gave public talks on the subject to any group who would listen.

While Saunders's message was well received, he soon realized that knowing is only half the battle; results require action. Although his audience acknowledged the reality of climate change, that alone could never have a significant impact on the issue. He realized that public policy changes, advocated by citizens, were necessary to enact more effective solutions. For instance, when Saunders first began speaking out about climate change, the U.S. Congress extended a law that gave $18 billion in subsidies to oil and coal companies.[3]

"It seemed to me that Congress was doing things exactly backward. Why? Because it's dominated by special interests—in this case, the fossil fuel industry," Saunders wrote on the CCL website, www.citizensclimatelobby.org.

To counter the energy industry's hold on Congress and the government in general, Saunders encouraged citizens to develop their skills in creating legislative proposals to combat global warming and learn how to lobby Congress to get them enacted. After all, the government is not just "*of the people*" but is also supposed to be "*by the people, for the people*"; if we demand it, they should provide it.

"Citizens' Climate Lobby's purpose is to ... empower individuals to have breakthroughs in exercising their personal and political power and by gaining the tools to be effective with our government," Saunders wrote. In other words, American citizens need to take our power back!

CCL is structured with two essential parts. As its name implies, it consists of a group of citizens who are concerned with climate change. CCL members lobby their representatives to act on this issue, specifically to enact the carbon fee and dividend program to reduce CO_2 emissions.

The second part of CCL, Citizens' Climate Education (CCE), is a not-for-profit organization that educates CCL volunteers on science and the potential solutions to climate change. Volunteers use that knowledge to teach politicians, the news media, and the general public about the climate change problem and the CF&D solution. Obviously, this is easier said than done, but CCE works hard to reach out to foundations, non-governmental organizations, and other groups to share the latest knowledge on climate change solutions. CCL/CCE is decidedly bipartisan, and they speak respectfully to all, even to climate change skeptics and deniers. Eric Tucker, a CCL member, puts it this way: "CCL taught me how to see the best in others, even in those who oppose me."

CCE declares on its website that its worthwhile mission is to "empower individuals to understand their own political and personal power, thus discovering and celebrating their own capacity to make a significant difference in this world."

Another founding father of CCL is James Hansen, a climatologist who has spoken out several times concerning climate change, including his testimony to Congress.[4] In 2013, he retired from NASA after forty-six years with the space agency, notably within the NASA Goddard Institute for Space Studies.

Hansen became more politically active by challenging government and fossil fuel company actions that he believed aggravated global warming. He was arrested in 2009, along with thirty other protesters including well-known actress and star of *Final Days of Planet Earth,* Darryl Hannah, for his role in a protest against mountaintop removal at a coal mine in West Virginia. Hansen found himself in police cuffs again a year later, in 2010, for another mountaintop removal mining protest that involved 100 people outside the White House.[5] It was around that time, however, that Hansen became a major supporter of Citizens' Climate Lobby and its proposal for a carbon fee and dividend policy for fossil fuels.

---

[3] Citizens' Climate Lobby's Founder. CitizensClimateLobby.org.
[4] James Hansen profile on Wikipedia.com.
[5] Ibid.

Besides James Hansen and Marshall Saunders, the CCL Advisory Board also includes leading conservative voices in government (more on that later) and well-known celebrities: Don Cheadle (*Hotel Rwanda*) and Bradley Whitford (*The West Wing*). Whitford appeared in a brilliant TV documentary series on the National Geographic Channel, *Years of Living Dangerously*. One episode, aired in December of 2016, profiled the work of Citizens Climate Lobby.[6]

A Drill-Down on the Economics of CF&D

It is very beneficial to dig into the details of some of the positive economic outcomes the CF&D could produce.

In 2014, the CCL commissioned a study of the economic impact of a carbon fee and dividend in the U.S. from Regional Economic Models, Inc. (REMI), also cited earlier. The REMI report states that the CF&D would achieve the following results between 2016 and 2035.

- a 50 percent reduction of carbon emissions below 1990 levels

- the addition of 2.8 million jobs above the baseline, driven by the steady economic stimulus of the energy dividend

- the avoidance of 230,000 premature deaths due to a reduction in air pollution that often accompanies carbon emissions

REMI, founded in 1980, has created economic impact studies for governmental and private-sector clients in a wide array of industries, including the energy industry.[7] Its structure accounts for both the costs and benefits of any policy options, including the implications for job creation, gross domestic product (GDP), personal income, and population.

The REMI report on CF&D incorporated data from the Energy Information Administration (EIA) and the National Renewable Energy Laboratory (NREL). REMI has also conducted carbon tax plan analyses for the states of California, Massachusetts, and Washington.

Opponents of legislation to address climate change or pollution in general often condemn such initiatives as jobs killers, as they increase the cost of doing business. In the case of CF&D, however, the risk will be more than offset by the stimulative effect of the dividend on the economy.

Remember that the carbon fee would increase by $10 per ton of CO_2 annually, which will automatically increase the size of dividend checks to households. REMI calculates that 53 percent of U.S. households and 58 percent of individuals would receive a net financial benefit because the dividend would exceed their estimated increase in energy costs. In other words, while the carbon fee would increase the cost of gasoline, home energy, and other fossil fuel-based products, it would be more than offset by the size of the dividend checks. Notably, the amount of the dividend check would be the same for everyone, regardless of income. This means the dividends, when viewed as a percentage of annual income, would be greater for low- to middle-income households. Higher-income families would likely pay out more in carbon fees than they would receive in dividends because they probably consume more energy; it is presumable that these families own more than one car, have larger and even multiple homes, and fly more often than people of more modest means. Of course, the carbon fee would also apply to jet fuel, putting upward pressure on airline ticket prices.

---

[6] "Years of Living Dangerously." National Geographic Channel. Dec. 7, 2016
[7] REMI Report. CitizensClimateLobby.org

The CF&D proposal offers more advantages than other considered solutions by comparison:

- Previous proposals to reduce CO_2 emissions with carbon fees have been labeled as job killers because they would not recycle fee revenue back into the economy as the CF&D proposal would. The REMI study calculates a $70 to $85 billion increase in gross domestic product (GDP) from 2020 on, with a cumulative increase in national GDP due to CF&D of $1.375 trillion.[8] To clarify, this means the economy would grow by $1.375 trillion *more* over 20 years than if there were no CF&D. In other words, we would make more money implementing CF&D than if the entire country of Luxembourg decided to give us their national earnings over the next two decades!

- The greatest employment gains would be in health care, retail, and other services, albeit excluding public administration. This is because people would have more pocket money to spend, and these industries are labor-intensive, responding to increased consumer spending by hiring more people.

CF/D: A tax that some conservatives are embracing

Why should CF&D appeal to conservatives? It's nothing personal, but I think it's important to acknowledge and address the political rift climate change has caused. We must discuss why CF&D is a beneficial solution for all sides, donkeys and elephants alike.

Part of the appeal of a carbon fee and dividend policy to conservatives is that it will replace a regulatory solution (rules that restrict fossil fuel emissions from power plants generating electricity) with a market-based solution.

Replacing a regulatory approach with a market-based solution is at the core of the conservative approach to the role of the government. For this reason, some notable, well-respected conservative voices in the U.S. are leading the CF&D bandwagon.

Gary Becker, a 1992 Nobel laureate in economics, was a professor of economics at the University of Chicago and served alongside George Shultz (secretary of the treasury for President Richard Nixon, and secretary of state under President Ronald Reagan, and on the CCL advisory board today) at the Hoover Institution at Stanford. Becker died in 2014 at age 83. Together, Becker and Shultz wrote an article published in April of 2013, *"Why We Support a Revenue-Neutral Carbon Tax."* The two wrote, *"We argue for revenue neutrality on the grounds that this [carbon] tax should be exclusively for the purpose of leveling the playing field, not for financing some other government programs or for expanding the government sector."*

More conservative voices joined the chorus in February of 2017, with the release of a manifesto, of sorts, from the Climate Leadership Council. The report, "The Conservative Case for Carbon Dividends," mirrors the carbon fee and dividend approach advocated by Citizens' Climate Lobby.

Besides Secretary Shultz, the Climate Leadership Council also includes James A. Baker III, who served as secretary of state under President George H.W. Bush (aka, Bush 41), as well as treasury secretary under President Reagan. Henry M. Paulson, who served as treasury secretary under President George W. Bush (Bush 43) also serves, among many other leaders in industry, government, venture capital, and the like.

The Climate Leadership Council makes the case that climate change is real and that there really does exist a conservative approach to the threat, one that should be embraced by the GOP. *"For too long, many Republicans have looked the other way, forfeiting the policy initiative to those who favor growth-inhibiting command-and-control regulations, and fostering a needless climate divide between the GOP and the scientific, business, military, religious, civic and international mainstream,"* the Council states. This last statement is so important

---

[8] Another chart possibility. "Additional GDP from CF&D 2020-2015." Source: REMI Study..

it should be yelled from the mountaintops – that is, the ones that we aren't blowing to smithereens with our mining!

The economic stimulus forecast of the carbon fee and dividend under the Climate Leadership Council proposal is similar to that in the CF&D backed by Citizens' Climate Lobby. The dividend would start at $2,000 a year for a family of four and rise each year as the carbon fee rises.

The Leadership Council believes that its CF&D plan would negate the need for government regulations, such as those that limit CO2 emissions from power plants. If these emissions are reduced by the CF&D, Environmental Protection Agency (EPA) regulations would no longer be necessary. *"Much of the EPA's regulatory authority over carbon dioxide emissions would be phased out, including an outright repeal of the Clean Power Plan,"* the Leadership Council report states, referring to an EPA policy enacted under President Barack Obama in August of 2015.[9] While this claim is highly controversial, to say the least, it is gratifying that they support a CF&D plan.

Unfortunately, the Clean Power Plan was an early target of President Trump. His executive order, issued on March 28, 2017, mandated an EPA review of the plan. Simultaneously, the Clean Power Plan is being challenged in a lawsuit that is making its way through the federal courts. Also, Trump's choice for EPA administrator at the start of his administration, Scott Pruitt, insisted on dismantling several other EPA pollution regulations. In his previous job as secretary of state for Oklahoma, Pruitt filed several lawsuits against the EPA challenging its air pollution regulations. The current EPA administrator is Andrew Wheeler.

The Case Against a Carbon Fee and Dividend

Can a case be made against CF/D? "97 percent of the world's climate scientists find the evidence convincing that humans are causing global warming," according to the United Nations Intergovernmental Panel on Climate Change (IPCC), in its 5th assessment report published in 2013.[10]

Since the release of the scientists' IPCC report, all sorts of groups with access to the Internet, cable TV news, or public relations firms have called the figure "fake". These include the American Enterprise Institute, the American Legislative Exchange Council, Americans for Prosperity, the Competitive Enterprise Institute, Heartland Institute, and the Institute for Energy Research, just to name a few.[11]

"The case for a U.S. carbon tax is weaker than the carbon tax proponents claim," says a study published in October of 2016 by the CATO Institute, another climate change skeptic organization, heavily supported financially by Koch Industries, Inc[12].: *"Policymakers and the general public must not confuse the confidence of carbon tax proponents with the actual strength of their case."*

The CATO report claims that the "social cost of carbon" (SCC) that climate change believers embrace is a subjective and "malleable" concept that can be influenced by the researchers' "largely arbitrary initial assumptions."

For instance, CATO takes issue with a climate change report from the Obama administration's Interagency Working Group (IWG), released in 2010 and updated in 2013. CATO argues that the IWG report presented a gloomier forecast for the SCC based on an analysis of *global* temperature changes in its research model. In doing so, IWG allegedly ignored a directive from the federal Office of Management and Budget (OMB) that federal regulatory analyses should focus on *domestic* temperature trends, not global ones.[13] *"Parameters are needed to calculate the social cost of carbon that, by their very essence, are subjective,"* the CATO report states.

[9] Clean Power Plan. Wikipedia.com.
[10] "Quantifying the consensus on anthropogenic global warming in the scientific literature." IOP Publishing. May 15, 2013
[11] "Global Warming Skeptic Organizations," Union of Concerned Scientists
[12] "The Case Against a U.S. Carbon Tax" Cato Institute, Oct. 16, 2016.
[13] Ibid.

Defenders of the climate science identify the CATO Institute as an organization dedicated to a concerted "disinformation campaign" designed to confuse the public and stall any progress on addressing climate change. *"Powerful coal, oil, and gas interests are trying to confuse us all about renewable energy and global warming. [They do it] not with facts or reasoned argument but with disinformation,"* says the Union of Concerned Scientists (UCS).

In its report, the UCS compares the energy industry disinformation campaign against climate change to the tobacco industry campaign in the 1950s and 1960s, which was aimed at discrediting scientific research that showed how smoking tobacco could increase the risk of lung cancer and heart disease. In fact, many of the same bad actors in the tobacco scandal are also involved in using dis-information to try to discredit climate scientists and their research.

The connection was also made in a documentary film released in 2014, *Merchants of Doubt*[14], based on a 2010 book co-authored by American science historians Naomi Oreskes and Erik M. Conway[15]. The book and film not only compare disinformation campaigns focused on climate change and tobacco use but also on acid rain, DDT, and the hole in the ozone. The basic strategy of these campaigns was to spread doubt and confusion about an issue even after scientific consensus had been reached. "Contrarian scientists joined forces with conservative think tanks and private corporations to challenge the scientific consensus on many contemporary issues," Oreskes and Conway wrote.

It is also noteworthy that the CATO Institute is opposed to smoking bans and mandatory use of seatbelts, which should bring them even more into question.

President Trump and the Art of Repeal

The political climate for climate change action received another jolt with the 2016 election of Donald Trump as U.S. president. Trump famously called climate change "a hoax" perpetrated by the Chinese.

Trump's election and his withdrawal from the Paris Accord have prompted more environmentalists to get more involved in opposition to him. "CCL membership has grown exponentially in recent months," said Steve Valk, Director of Communications. Before the presidential election, CCL averaged 395 new supporters per week. In the 19 weeks since the election, it averaged a weekly intake of 960 new supporters. The numbers tell us that CCL have recently experienced 2.5 times the growth it recorded before Trump took office.

"Prior to the election, there were many people who were concerned about climate change, but they assumed it was being handled by those who run our government," Valk wrote in an email. "When Trump was elected, these folks realized they needed to get in the game, and that's why we saw such a big spike in enrollment."

Citizens' Climate Lobby held its annual conference in Washington DC just two weeks after Trump made his Rose Garden announcement of withdrawal from the Paris Accords June 1, 2017. This conference is a pilgrimage for CCL members make to Capitol Hill every June, to lobby senators and representatives in person about the urgency of the carbon fee and dividend solution.

"Nothing substantive can happen unless it has support from both parties," said Jay Butera a CCL activist from Pennsylvania who helped organize the Climate Solutions Caucus in the House of Representatives.[16] "It feels as if the momentum is finally shifting in our favor."

---

[14] Merchants of Doubt Internet Movie Database. www.imdb.com
[15] Merchants of Doubt profile, Wikipedia.com.
[16] "Climate lobbyists make progress with Republicans on climate change." StateImpact.com. June 14, 2017.

Climate Caucus Puts Some Points on the Board

Building the Climate Solutions Caucus to 52 members is encouraging, but there are a total 435 members in the House of Representatives, so there is still some way to go before any real leverage will occur. Nonetheless, climate change advocates do have something to celebrate.

On July 13, 2017, the House voted down an amendment to a Defense Department authorization bill that would have blocked a study on the climate change impact on national security. The total vote to defeat the amendment was 185 to 234; 22 of the 24 Republicans on the Climate Solutions Caucus at that time voted against the amendment.

Congressman Ted Deutch, the Democrat from Florida and a co-founder of the Caucus, believes that such a vote was an indication that both sides really can work together to save the country from climate change: "This vote is proof that there is now a bipartisan majority in Congress of members who understand that climate change is a real threat to our communities, our economy, and our military readiness." The term "bipartisan majority" refers to the 234 votes against the amendment that would have blocked a climate change study by the Department of Defense – 234 is a majority of the 435 members of the House. The 22 votes against the amendment by Republicans who were on the Climate Solutions Caucus makes that a "bipartisan majority" against the amendment.

My experience, and those of many of my friends in CCL supports Marshall Saunders' wise counsel that CCL is about "empowering individuals to have breakthroughs in exercising their personal and political power." We feel that empowerment when our letters to the editor are published, when our city councils vote to endorse the carbon fee and dividend policy, or when another House member joins the Climate Solutions Caucus and brings a date from the other party with them.

Most importantly, *you* must understand that you *can* do something on your own to address climate change. You *can* have a positive effect on your community, your planet, and your fellow human beings. You *can* make this world a better place!

Climate Crisis "Keep/Stop/Start" Actions

Below for your consideration, are several author-suggested thought, behavior, and action items. For a further explanation of the Keep/Stop/Start organizational tool, please see How to Read This Book.

KEEP:

Keep reading about climate change, studying the threat, and learning what scientific steps must be taken to control climate change. However, while scientific facts are compelling, they are not always persuasive as some people have trouble absorbing too many facts as their eyes glaze over. What can be more convincing are the personal stories from advocates who talk about climate change in the context of securing a future for their children or grandchildren.

STOP:

Stop driving gas-guzzlers. Trade those SUVs in for a hybrid or electric vehicles or use mass transit more often. Stop driving alone in your car when ride-sharing services like Uber or Lyft are available.

START:

1) *Start addressing climate change* by joining organizations such as Citizens' Climate Lobby (https:// citizensclimatelobby.org). Chances are that there is a CCL chapter already up and running in your community or congressional district; if not, start your own.

2) *Start contacting your elected officials* and demanding that they support carbon fee and dividend legislation in the House of Representatives. Make sure they bring a date from the other party! Lobby your Congress members via Twitter and other social media, and encourage your friends to do the same.

3) *Start writing postcards* and letters to the editor of your local newspaper, op-ed responses or rebuttals, and columns.

4) *Start perusing websites* such as www.carbonlist.com to gain information and opinions on how to capture, use, reduce, and sequester carbon, or learn how to not put it there in the first place.

~ ~ ~

Robert Mullins is a longtime journalist and marketing writer based in Silicon Valley. In 2013, he joined Citizens Climate Lobby, a bipartisan grassroots organization whose mission is to lobby the U.S. Congress to pass legislation to impose a carbon fee on fossil fuels to reduce demand, thereby reducing CO2 emissions, a major contributor to climate change. Robert worked as a newscaster and news director at the National Public Radio (NPR) station in Milwaukee, WUWM-FM, and occasionally appeared as a reporter on NPR news programs such as *All Things Considered* and *Morning Edition*. Robert lives in Oakley, California. Read more at: www. climateabandoned.com.

Chapter 7

Climate Crisis & Extreme Weather

by

Jill Cody, M.P.A. and Peter M.J. Hess, Ph.D.

"Men argue. Nature acts." **– Voltaire**

Introduction

One thing scientists know for sure is that climate change creates *extremes,* something Americans seem to latch on to quickly. There are extreme Frisbee games, extreme hot yoga studios, and extreme burritos, but the climate extremes experienced in the world are now *extreme warnings.*

Hurricanes Harvey and Irma were most definitely extreme. In fact, Irma was the first hurricane to maintain 185 MPH for 42 hours, making it the longest-lasting Category 5 hurricane since the first weather satellite was shot into the sky on April 1, 1960. This fact hit close to home when the *Climate Abandoned* web designer, who lives in Naples, Florida, lost his house, and nearly all his possessions save his computer equipment and his cat, Buppy, to Irma's veracity. To protect himself (and Buppy!), he hid out in an old bomb shelter for three days.

While Americans were glued into the arrival of Hurricane Harvey and its aftermath in Houston, another hurricane in India made Harvey look like child's play when 21,000 people were pulled out of the Brahmaputra River floodwaters. The Hatimura dike was breached for the first time in forty-six years, and chapter author Raj Phukan had his hometown on red alert for a week. Raj saw this horrifying storm floodwater come within 200 meters of his home causing great alarm. Stay? Run?

Hurricane Maria, in 2017, did make the news in the United States since the Puerto Rican people are American citizens. Maria was such an extreme hurricane event that it changed a beautiful, tropical isle, and tourist hot spot into something of a warzone, so defoliated that it became hardly recognizable. Hurricane Maria is regarded as the worst natural disaster on record to affect those islands. One *Climate Abandoned* chapter author, Maria Santiago-Valentin who is from Puerto Rico, for days had no idea whether or not her family was alive.

The California wildfires have surpassed extreme and have become fierce. The 2018 wildfire season was the most deadly and destructive in the state's history with 8,527 wildfires destroying 1,893,913 acres, surpassing the previous most deadly and destructive wildfire season of 2017. The 2018 season killed 104 people, including 6 firefighters and, in addition to the tragic loss of life, the fires cost $3.5 billion in damages and $1.8 billion in fire suppression expenses. If you wonder where some of your tax money is spent, consider the damage climate change is costing[1]. The entire City of Paradise was burnt to the ground. Think about that for a minute … a bustling city of stores, restaurants, and services vanished overnight.

Just as we cannot entirely blame the climate crisis for hurricanes, we can neither blame it for wildfires But, in both these cases, the climate crisis is responsible for adding to the intensity of these natural disasters. While this book was being drafted in 2017, a wildfire in the uniquely beautiful Napa, Sonoma, and Mendocino Counties wine country required over 10,000 firefighters to face off wildfires as large as all five boroughs of New York City. This three-county fire directly affected two additional chapter authors. One author, Betsy Rosenberg, provided shelter in her home for five evacuees and another, Dr. Peter Hess, experienced the total loss of his family home. When

---

[1] https://calmatters.org/articles/cost-of-california-climate-change/

his father's house in Santa Rosa burnt to the ground, he was so devastated that he suffered a post-traumatic stress disorder (PTSD) reaction. Why? Because he had previously lost three other mountain area homes to climate change exacerbated California wildfires.

The photo above is an aerial view of the Santa Rosa, CA neighborhood where Peter Hess' family home had been and is a visual example of what climate change looks like and will look more like in the future.

Heat

Bats are boiling alive. That's right. The planet has become too hot in many places for wildlife to survive. In Australia, the temperature reached 116 degrees, and 400 flying foxes were found dropping to the ground: dead from dehydration. Kate Ryan, Campbelltown flying fox colony manager, said, "It would be like standing in the middle of a sandpit with no shade." Flying Fox Conservation Fund director, Scott Heinrich said, "In a way, they're kind of boiling in their bodies.[2]" Ringtail possums in Australia are also dropping dead out of trees. Many suffering from dehydration, left their trees for the beach to drink salt water. Melanie Attard, a wildlife rescuer with Aware Wildlife in Frankston, said, "We assume they've come out due to the heat stress heading for the water in desperation. It's not nice seeing a possum throwing itself into the beach and drinking seawater. It's really desperate.[3]"

The University Bridge in Seattle, WA is about 100 years old and can't handle the heat due to its age. When the city began experiencing 90-degree weather, the Suquamish Police Department closed the bridge in the middle of the afternoon to give it a cooling bath. The street crews sprayed cold dechlorinated water on the bridge to prevent the metal from expanding and damaging the span.

---

[2] https://news.nationalgeographic.com/2018/01/australian-heat-wave-flying-fox-deaths-koala-spd/
[3] https://www.theguardian.com/environment/2019/mar/07/falling-out-of-trees-dozens-of-dead-possums-blamed-on-extreme-heat-stress

In July 2018, all-time heat records were set around the world due to vast areas of heat domes scattered around the Northern Hemisphere. In Siberia, a frigid region of Russia reported temperatures of 90 degrees, 40 degrees over normal and meteorologist Nick Humphrey explained that "It is absolutely incredible and really one of the most intense heat events I've ever seen for so far north." In Algeria, the town of Ouargla reported temperatures of 124.3 degrees. It may have been the highest temperature measurement in all of Africa, and the ordinarily mild regions of Ireland (Shannon 89.6 degrees), Scotland (Glasgow 89.4 degrees) and Canada (Montreal 97.9 degrees) looks like they saw the hottest summer ever recorded. The City of Los Angeles set an all-time record of 111 degrees on July 6.[4]

How on earth do we know about all that's happening? Well, it isn't all on earth. As the World Meteorological Organization (WMO) describes the vast technology:

> Currently, well over 10,000 manned and automatic surface weather stations, 1,000 upper-air stations, 7,000 ships, 100 moored and 1,000 drifting buoys, hundreds of weather radars and 3,000 specially equipped commercial aircraft measure key parameters of the atmosphere, land and ocean surface every day. Add to these some 16 meteorological and 50 research satellites to get an idea of the size of the global network for meteorological, hydrological and other geophysical observations. Once collected, observations are quality-controlled, based on technical standards defined by the WMO Instruments and Methods of Observation Programme (IMOP), then made freely available to every country in the world through the WMO Information System (WIS).[5]

Things are getting too hot to handle. Why? Earth has become a greenhouse full of heat and humidity (look up "Greenhouse Effect"). There is no escape and research from the Intergovernmental Panel on Climate Change is showing us that we haven't seen anything yet. More than 1,300 U.S. and other scientists have studied statistical models that foretell, in 80 years, the temperature on the planet will rise 2.5 – 10 degrees Fahrenheit. Don't think that the climate today will be the same and then all of a sudden, 80 years from now, that Earth will be unlivable. No. The world will become unlivable for many along the way. The ride towards the end of the century will be miserable.[6]

Why will life on Earth become more and more miserable? Will Steffen, an earth systems scientist with the Australian National University, explains:

> We normally think that the temperature rise and the level of climate change we will experience is proportional to our emissions. And that's probably true at low levels of emissions and temperature rise, but we argue that the Earth as a complex system has inbuilt feedback processes. And if we start to trigger these feedback processes, they act like a cascade, like knocking down a row of dominoes. And once they start tipping and falling, we can't stop them. That would take us to a much different climate, a much warmer climate, a much more difficult climate for humans to live in[7].

Wildfires

Fire is a chemical process of the rapid oxidation of carbon-based material, and it is integral to the recycling of plant material in earth's ecology. At some point in the last 200,000 years, evolving *Homo sapiens* domesticated it, but a fire is not easily controlled in the wild. As humans migrated around the globe, they discovered that some

---

[4] https://www.washingtonpost.com/news/capital-weather-gang/wp/2018/07/03/hot-planet-all-time-heat-records-have-been-set-all-over-the-world-in-last-week/?noredirect=on&utm_term=.5f49213939b9

[5] https://public.wmo.int/en/our-mandate/what-we-do/observations

[6] https://climate.nasa.gov/effects/

[7] https://therealnews.com/stories/new-climate-study-warns-of-dangerous-hothouse-earth-scenario

regions are natural "fire ecologies" carrying a much higher risk of wildfire than others.[8] These regions include brush and pine woodlands around the Mediterranean and Eastern Europe, forests and chaparral in western North America, eucalyptus groves in Australia and dry, hilly regions in Chile and South Africa. In the American West wildfire has been a constant natural hazard that has influenced public policy.[9]

While wildfire in all these regions began to change in nature and size during the past two decades paralleling the gradual rise in global mean temperature measured by climatologists, they also became larger and more intense, more difficult to contain, and more deadly, as exemplified by recent high-fatality conflagrations in locations as diverse as Australia[10], California[11], Greece[12], and Portugal[13]. The Arctic Circle battled wildfires in the summer of 2018 with at least eleven wildfires burning in Greenland, Sweden, Canada, Alaska, and Siberia! A new term "megafire" has been coined for a wildfire that has reached a size of 100,000 acres or more. In 2018 a bush fire in Australia burned over 2 million acres, gaining the dubious distinction of being the first "gigafire.[14]" In 2019, Australia suffered record-breaking heat wave delivered the hottest start to March on record for the southern third of the country.

To be sure, not all of this increase is climate-related. But anthropogenic climate disruption is, in fact, exacerbating the conditions that predispose some regions to wildfire[15]. Indeed, many of the factors contributing to fire are made worse by an increasingly erratic climate. As atmospheric greenhouse gas levels rise, landmasses retain more thermal radiation and become hotter. Hotter continents can build up high-pressure ridges that push moisture-laden storms offshore for months at a time during the rainy season:

> In 2013, a high-pressure zone formed over the Pacific Ocean, diverting precipitation towards Alaska. While these zones are common, they normally change position quickly. The recurring high-pressure "ridge" is unprecedented in modern weather records in that it remained in place for many months at a time, held by a large, static bend in the jet stream[16].

Using California as an example, in the 2013 case, climate disruption of the jet stream was at least partly responsible for the unprecedented weather pattern named the "ridiculously resilient ridge" that blocked storms from the state, leading to an increase in hot and dry weather. This pattern exacerbated a drought that set in a few years before. High-pressure ridges stall winter rains from coming onshore and contribute to the desiccation of regional ecologies. As water tables drop during long-term drought, trees lose moisture, which weakens their resistance to pathogens and insects. Drought-stressed Ponderosa and Lodgepole pine readily fall victim to massive epidemics of bark beetle infestation that can kill 50-80% of pines in a forest, leaving standing and fallen dead trees in a disastrous buildup of fuel ripe for a wildfire[17].

The average fire season in California is now 78 days longer than it was in 1970, and wildfires in the United States burn more than twice the area they did half a century ago[18]. Climate change disrupts historically stable seasonal weather patterns, creating summer heat waves and autumn wind events that generate explosive fire conditions. October and November can be perilous especially when strong and erratic winds arise, in southern California called "Santa Ana Winds" and in northern California "Diablo Winds." Such winds make the critically

---

[8] Ecological Society of America, "Fire Ecology," https://www.esa.org/esa/wp-content/uploads/2012/12/fireecology.pdf

[9] Stephen J. Pine, *Fire in America: A Cultural History of Wildland and Rural Fire*

(University) of Washington Press; rpt. Ed. March 1, 1997); Timothy Egan, *The Big Burn: Teddy Roosevelt and the Fire that Saved America* (Mariner Books, rpt. ed. 2010).

[10] Victorian Bushfires Royal Commission 2009 Report, http://royalcommission.vic.gov.au/finaldocuments/summary/PF/VBRC_Summary_PF.pdf

[11] https://www.sfchronicle.com/california-wildfires/article/Camp-Fire-Death-toll-rises-to-86-after-13458956.php

[12] https://www.nytimes.com/2018/07/24/world/europe/greece-fire-deaths.html

[13] https://www.theguardian.com/world/2017/jun/22/portugal-forest-fires-under-control

[14] *Wildfire Today: News and Opinions* (October, 2018) https://wildfiretoday.com/tag/megafire/

[15] Center for Climate and Energy Solutions, https://www.c2es.org/content/wildfires-and-climate-change/

[16] Climate Signals.org, "Atmospheric blocking increase," http://www.climatesignals.org/climate-signals/atmospheric-blocking-increase

[17] Dahr Jamail, *The End of Ice*, 140-148.

[18] The western fire season is growing months longer https://www.climatecentral.org/library/climopedia/the_western_fire_season_is_growing_months_longer

important job of wildland fire fighting increasingly difficult and dangerous, and conditions will only get worse as the planet warms further. In recent years these winds have gusted to sixty miles per hour or more, and have been responsible for exacerbating wildfires such as Oakland's Tunnel Fire (1991), San Diego's Cedar Fire (2003)[19], and the Tubbs (2017) and Camp (2018) Fires.

Recent wildfires in California have now included a once-rare phenomenon: fire tornados! In the 2017 Tubbs Fire that slammed into many neighborhoods in Santa Rosa, tornadic winds flipped and even flung automobiles. Even more dramatic was a fire tornado generated by the Carr Fire that gutted parts of the city of Redding.

> The Redding tornado was the worst of any kind ever seen in California, with a base the size of three football fields, winds up to 165 miles an hour and temperatures of at least 2,700 degrees — nearly double the temperature of a typical wildfire[20].

Quite apart from the human emotional cost and the senseless and tragic loss of life, is it sustainable to build vast tracts of wooden houses in flammable ecosystems and count on replacing them every twenty-five years? Accelerating wildfire risk is a function of at least two intersecting trends: 1) climate-change-induced droughts, bark beetle timber kills, intensification of fires, and lengthening of fire seasons, 2) Relentless human population growth pushing people farther out into the Wildland Urban Interface (WUI). Houses may be more affordable but they are more exposed to the risk of fire, and their inhabitants also may serve as more sources of ignition.

Hurricanes

There have always been hurricanes ... but nothing like these in human memory. A hurricane is a rapidly rotating storm system with the wind blowing counterclockwise in the Northern Hemisphere and clockwise in the Southern Hemisphere (called Tropical Cyclones). Hurricanes and Tropical Cyclones form over large bodies of warm ocean water, and they draw their energy through the evaporation of water from the ocean surface. It's straightforward. The warmer the ocean, the stronger the rotating storm system becomes. It is essential to remember that a hurricane is not only a wind event but also a rainfall event with accompanying storm surges. In any given year there may be several or just a few, but regardless of the number of occurrences a year their intensity is derived from the temperature of the ocean. Obviously, coastal and island regions are of the highest risk.

Islands are the most threatened at this early stage of climate change. Hawaii recently lost East Island, located in the northernmost part of the Hawaiian island chain, due to a direct hit by Hurricane Walaka. East Island was a treasured spot because it was a nursery for Monk Seals (already an endangered seal) and hosted 96% of Hawaii's green sea turtle nesting grounds. The National Oceanic and Atmospheric Administration (NOAA) estimated that nearly 20% of the green sea turtle eggs had not hatched yet and were totally destroyed. Chip Fletcher, a University of Hawaii climate scientist, had visited the area three months before the hurricane and thought that it would take decades of sea level rise to claim it. When he revisited the area again, post-hurricane (please forgive the language because it is a direct quote) he exclaimed, "I had a holy shit moment, thinking 'Oh my God, it's gone. It's one more chink in the wall of the network of ecosystem diversity on this planet that is being dismantled.[21]"

The Caribbean states and territories are a favorite vacation destination for people from all over the world and entice 25 million a year to visit bringing $49 billion to the economy. The Caribbean islands are often referred to as the most tourism-dependent region in the world. Now, these beautiful islands are facing an existential threat.

---

[19] See the gripping account by Sandra Millers Younger, *The Fire Outside My Window: a Survivor Tells the True Story of California's Epic Cedar Fire* (Rowman and Little-field, 2013 http://www.sandramillersyounger.com/books.html

[20] https://www.sfchronicle.com/california-wildfires/article/Catastrophic-fire-tornadoes-a-terrifying-new-13162374.php.

[21] https://motherboard.vice.com/en_us/article/evw337/hurricane-walaka-caused-remote-east-island-hawaii-to-disappear-sea-level

In 2017, the most destructive hurricanes they had ever seen rumbled through. Hurricanes Irma and Maria, both a Category 5 (highest rating), slammed into the islands causing thousands of deaths and obliterating the island of Dominica. Puerto Rico, nearly two years later is still suffering the tragic aftermath. The prime minister of Dominica plainly stated, "Heat is the fuel that takes ordinary storms ... storms we could normally master in our sleep ... and supercharges them into a devastating force. Now, thousands of storms form on a breeze in the mid-Atlantic and line up to pound us with maximum force and fury. We as a country and as a region did not start this war against nature. We did not provoke it. The war has come to us."

These island states feel abandoned by the United States. Puerto Rico is an American territory, yet received no help of any consequence. Also, there has been no moral leadership from the U.S. government. In fact, the decision to withdraw from the Paris Agreement and dismantle of the Clean Power Plan (passed during the Obama administration) has caused worldwide disappointment and dissatisfaction. The Bahamas' foreign minister, Darren Henfield, expressed, "The US is a major player in the world, and it needs to lead, we depend on it to be a moral voice on issues where people are vulnerable. We really hope the U.S. readjusts its position. It seems there will be doubters until we start completely losing islands."

For those who think in terms of money and not always in terms of human life and beauty, it would be essential to realize that hurricane property damage costs are exorbitant. Hurricane Katrina was the #1 costliest at $161 billion. Together Hurricane Irma and Maria cost $140 billion. The price tags are only going to go up from here.

Drought

There have always been droughts, but not like the one Syria suffered.

There are four types of droughts: Meteorological, Agricultural, Hydrological, and Socioecological[22]. Meteorological droughts are specific to a region. Those living in a particular region understand their normal rain patterns, and when there isn't the expected precipitation aligned with their historical records, then it meets the conditions for a meteorological drought. An Agricultural drought considers the water needs of crops during their growing stages, and if there isn't the water available for healthy growth and germination, then it is regarded as an agricultural drought. Hydrological droughts are when there are continuously low amounts of water in rivers, streams, and reservoirs. Lastly, Hydrological droughts are when water demands exceed the supply. Droughts occur in all types of climate and, next to hurricanes, can cause the most physical and economic damage. It can be caused not only by a lack of precipitation but also by overpopulation and overuse.

The most famous and disastrous U.S. drought, until the Syrian drought, was in the 1930s. At its worst, it covered 60 percent of the United States and caused millions of people to leave their homes, and many went to the West Coast. Scientists believe that the "Dust Bowl" years was caused by an unrelenting high-pressure ridge over the West Coast forcing precipitation off to other regions. Since the increase in CO2 in the atmosphere, between the 1980 and the mid-2000s, there were 16 droughts, with thousands of people dying and with the economic cost $210 billion.

The Dust Bowl drought was relatively small compared to the one that ignited Syria's civil war. That drought was likely the worst seen in 900 years and was a significant factor in causing the conflict[23]. Farmers, no longer being able to survive, moved into the urban areas. Seventy-five percent of their farms failed, and 85 percent of their livestock died forcing their migration. When political grievances were not addressed by the government, disruption and violence broke out. After 15 years of war anywhere from 368,000 - 560,000 of men, women, and children have died. The University of Arizona and NASA scientists studied tree rings in the area and discovered that the drought Syria experienced was far outside the natural Meteorological precipitation cycle.

---

[22] https://www.livescience.com/21469-drought-definition.html
[23] https://news.vice.com/en_us/article/3kw77v/the-drought-that-preceded-syrias-civil-war-was-likely-the-worst-in-900-years

The Pentagon has stated that climate change is a "threat magnifier" and links climate-exacerbated droughts with civil unrest. Syria became a prime example.

Climate Crisis "Keep/Stop/Start" Actions

Below for your consideration, are several author-suggested thought, behavior, and action items. Action nourishes inspiration! For a further explanation of the Keep/Stop/Start organizational tool, please see How to Read This Book.

KEEP:

Keep informing yourself about the relationship between the broad concept of climate change and more specific instances of extreme weather. Many people confuse weather with climate change. Weather is what is being experienced on any particular day. Climate is the pattern of weather over a period of time, often a 30-year span.

STOP:

Stop allowing others ignorance to distort your view of climate change. While it is true that governments, companies, and private landowners need to become much more educated and proactive about implementing mitigations of threat from heat, wildfire, hurricanes, and drought, never allow the deniers claim to dominate, namely that climate has *nothing* to do with the extreme weather the planet is now experiencing.

START:

1) *Start following* trusted websites and writers on the subject of climate change such as:

    - NASA Global Climate Change - *"Vital Signs of the Planet"*: https://climate.nasa.gov

    - Real Climate – *"Climate science from climate scientists"*: http://www.realclimate.org

    - Union of Concerned Scientists – *"Confronting the Realities of Climate Change"*: https://www.ucsusa.org/global-warming

    - Dr. Michael Mann, Distinguished Professor of Atmospheric Science, Penn State: http://www.michaelmann.net/blog

    - Bill McKibben, Founder 350.org and Distinguished Scholar at Middlebury College http://billmckibben.com/articles.html

    - Yale Climate Connections: https://www.yaleclimateconnections.org

    - Vice News - *"Tipping Point Blog"*: https://news.vice.com/en_us/topic/tipping-point

2) *Start reading up on the Earth's temperature* and its causes and consequences, and talk with others about it:

    - National Geographic – "What is global warming, explained: The planet is heating up – and fast": https://www.nationalgeographic.com/environment/global-warming/global-warming-overview/

- Michael Mann, *The Hockey Stick and the Climate Wars*, Columbia University Press, 2012.

3) *Start reading up on wildfire* and its causes and consequences, and talk with other people about it:

- Gary Ferguson, *Land on Fire: The New Reality of Wildfire in the West*, Timber Press, 2017.

- Michael Kodas, *Megafire: The Race to Extinguish a Deadly Epidemic of Flame,* (Houghton Mifflin Harcourt, 2017).

4) *Start reading up on hurricanes* and they're causes and consequences, and talk with others about them:

- Natural Resources Defense Council- *"Hurricanes and Climate Change: Everything You Need to Know"*: https://www.nrdc.org/stories/hurricanes-and-climate-change-everything-you-need-know

- Union of Concerned Scientists – *"Hurricanes and Climate Change"*: https://www.ucsusa.org/global-warming/science-and-impacts/impacts/hurricanes-and-climate-change.html

5) *Start reading up on droughts* and they're causes and consequences, and talk with others about them:

- Union of Concerned Scientists – *"Causes of Drought: What's the Climate Connection?"*: https://www.ucsusa.org/global-warming/science-and-impacts/impacts/causes-of-drought-climate-change-connection.html

- Climate Communication - *"Droughts"*: https://www.climatecommunication.org/new/features/extreme-weather/drought/

6) *Join the Climate Reality Project* and fight like your world depends on it: www.climaterealityproject.org

~ ~ ~

Jill Cody, M.P.A is an authorized presenter (2007 4th class) of Al Gore's "An Inconvenient Truth" slideshow. She is also the author of the award-wining book *America Abandoned: The Secret Velvet Coup That Cost Us Our Democracy* and is producer and host of KSQD 90.7FM "Be Bold America!" radio program. Jill conceived of the *Climate Abandoned: We're on the Endangered Species List* anthology when realizing friends and acquaintances did not fully comprehend the breath, dynamics, and complexity of the climate crisis. Jill is a distinguished alumna of San Jose State University's College of Applied Sciences and Arts and currently serves on the university's Emeritus and Retired Faculty Association Executive Board.

~ ~ ~

Peter M.J. Hess, Ph.D. is a Roman Catholic theologian who specializes in issues at the interface of science and religion. He has served as international director for the Center for Theology and the Natural Sciences (CTNS) in Berkeley, California and director of Outreach to Religious Communities with the National Center for Science Education (NCSE) in Oakland, promoting dialogue in the areas particularly of evolutionary biology and climate change. Peter Hess earned an MA in philosophy and theology from Oxford University and a PhD in historical theology from the Graduate Theological Union in Berkeley. He is co-author of *Catholicism and Science* (Greenwood Press, 2008) and writes on the religious and ethical implications of a growing human population in light of climate change. Peter lives in Berkeley, California. Read more at: www.climateabandoned.com.

Chapter 8

Climate Crisis & Archeology

by

Mike Newland, M.A.

"Historical sites across the globe are at risk due to rising sea levels, but without reliable data we can't even assess the full scope of the problem, let alone solve it." **– Patty Hamrick**, Archaeologist at New York University

Even before we got there, I could see that we had a big problem. As we hiked through the sand, Mark turned eastward and pointed to a twenty-foot dune, one that looked like as if a giant fist had punched through it, splitting it in two. "That's new," he said. "The wind comes in from another direction now, and it knocked a hole in the dune. We're having all sorts of problems here. Wait till you see the site."

Mark Hylkema, an archaeologist for the California Department of Parks and Recreation, had invited me to travel down to **Año** Nuevo State Park. He claimed he had the so-called smoking gun, the evidence necessary to prove the impact of climate change on ancient archaeological sites along the California coastline. It was 2014, and at the time, I was among only a handful of archaeologists in the area who'd just started to wrap our heads around the problem. Mark was on the front lines, as much of the territory he covered was along the California central coast, south of San Francisco to Monterey.

Even after I saw the dune, I was unprepared for what I saw when we arrived at the dig site. Walking out onto the bedrock, I saw artifacts everywhere: stone scrapers, sea lion bone, shells, the hand stones used for grinding food, and burnt rock, the remains of long-ago cooking pits, all just strewn about.

"The waves are high enough now to sweep over the site," Mark said with a frown. "The winds and storm surge come in from a different direction, and all the dirt and sand is gone. This site, thousands of years old, is totally destroyed. All that's left of this village is what you see here, the few artifacts too heavy to be swept by the last storm, but even these won't be here for long."

The dilapidated site was home to ancestral Salinan people, and it was unsettling because it is rare for a site to be utterly wiped off the face of the Earth. Usually, there are more substantial bits left behind, something jumbled or a little pocket missed by a bulldozer. Seldom is a 3,000-year-old locale stripped down to the bedrock, the irreplaceable evidence of millennia of human lives gone forever.

I picked up a piece of Monterey chert, a smooth, blackish-brown rock the Salinan people used to fashion tools. "Take a good look at it," Mark advised. "We'll be the last to see this place."

To be frank, I almost turned the job down. I was an archaeologist working at the Anthropological Studies Center at Sonoma State University in early 2012, and I had a slew of other projects on my plate, so I wasn't all that interested in taking on more work. However, I had to jump at the chance when I saw that it involved Point Reyes.

The National Park Service (NPS) aimed to complete a pilot study on the impact of climate change on the archaeological sites at the Point Reyes National Seashore. I had done a lot of research there already, and I had a good working relationship with the native Coast Miwok community. Located near the border of Sonoma and Marin Counties, some twenty or so miles north of Golden Gate in California, Point Reyes is a stunning place. Conducting research there was a privilege, and I couldn't possibly pass up the opportunity. The fact that I knew nothing about climate change at the time didn't faze me; I knew the archaeology and figured I could learn the rest as I went along. I accepted the project, then did what every archaeologist does at the start of a big project: I sat down to read all I could about the subject.

For the next three months, I filled my head with information from field reports and journal articles penned by scientists who collected data from Morocco, Greenland, the Great Barrier Reef, the Channel Islands, Greece, Netherlands, Scotland, and British Columbia. I spent time perusing some of the first decent computer models on climate-change impact and projected what San Francisco and Los Angeles might look like in the wake of a six-foot sea level rise. I was awed by the projections for New Orleans, Miami, and New York. I read about pending temperature shifts, ocean acidification, mass forest die-offs, and their possible causes. No matter how many articles and studies I read, the brilliant climatologists, ocean scientists, and geographers all seemed to come to the same conclusion: The oceanic acid levels were best explained by dissolved CO_2, the dramatic increase in global temperature melting polar and glacial ice, and subsequent sea level rise. There was a significant, undeniable increase in greenhouse gases, and ice cores clearly showed that the damaging spike began at the beginning of the Industrial Revolution. In other words, the only arguable source of those gases was human activity. Over and over again, that grim and telling conclusion was reached, from what seemed like a thousand different viewpoints and angles, one independent data set after another all pointing in the same direction.

One of the most remarkable aspects of archaeology as a field of study is that archaeologists have the opportunity to examine data sets no one else can. After all, one cannot feasibly be issued a National Institute of Health grant to conduct a 500 year-long study on malnutrition amidst an island population. Archaeological sites are more than just great repositories of artifacts that reveal much about human behavior. In fact, they contain data about what the *environment* used to be like, including past climate-change events. Coastal sites are the most immediately vulnerable to climate change. There, the high shell content reduces soil acidity, increasing natural preservation of thousands of years of untapped environmental and climate data. Archaeological sites are our only access to this critical data, and those sites are quickly disappearing as a result of climate change.

Several years ago, I was going through a collection of 4,000-year-old artifacts. They were excavated from California State Park lands in Lake County some twenty years prior, and we had been asked to organize the collection and discover any additional analysis of value. Greg White, an expert on the Native American archaeological sites in the region, generously shared his knowledge with me, so we discussed me sending some samples for radiocarbon dating from the deeper parts of the excavation. Radiocarbon (C14) dating is a process by which a lab can assess how much radioactive carbon, which all living things inhale naturally from the atmosphere every day, is left in charcoal, bone, organic-rich soil, and wood samples. Radioactive isotopes have a very steady half-life, and the material starts to degrade the instant something dies because only living things take in fresh C14. If one measures the remaining C14, one can gain an idea of how long ago that material was part of a living thing. Because it seemed like a relatively deep deposit, I was excited about the possibility of gleaning some early dates.

Greg looked at the notes and just shook his head. "Don't bother," he said. When I asked him why, he went on to tell me about the Middle Holocene.

Between roughly 4,000 and 7,500 years ago, Earth saw a much different climate than that of today. In California, where I do nearly all of my work, air temperatures were around 1-2° C hotter than they are now. A long, brutal drought settled over California for much of that time, and it changed the landscape dramatically. In Mojave Desert archaeology, we refer to Middle Holocene Hiatus, because the Mojave and Sonoran Deserts appear almost wholly abandoned, save for campsites around the few rivers and lakes that managed to survive.

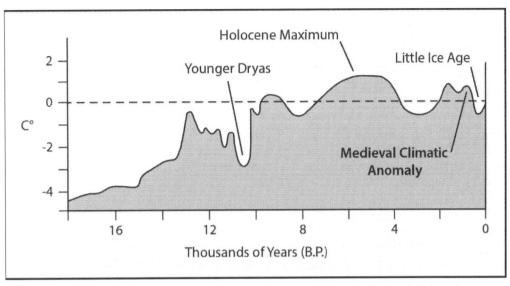

Late Pleistocene-Holocene temperature fluctuations. (courtesy of ASC 2016; after Tausch et al. 1993).

Greg explained the impact in Lake County: Over time, the temperature increase was enough to kill off soil development. Plant life died, and organic material ceased to be introduced at a rate that would allow the soil to form. This, in turn, proved inhospitable to the introduction of additional plants. Over time, the soil degraded into inorganic sediment (sand, gravel, silt, etc.), and the archaeology of thousands of years deflated to one exposed surface. Material from 5,000 years ago was jumbled with material from 10,000 years ago. The only rain came in the form of dramatic storms, drenching areas that were prone to poor soil, subject to erosion, so sites incurred even more damage. Most sites older than 4,000 years were a jumbled mess, and the environment looked like a desert. "Probably like the Mojave areas you've been working," Greg cautioned about the desertification and erosion that likely occurred throughout much of central California.

This is a dire warning for us. The Intergovernmental Panel on Climate Change prediction for temperature increases for most of North America (the most recent being 2013 at the time of this writing) is that the *best*-case scenario by the year 2100 is a 1.5 to 2°C temperature increase. The worst is 4 to 5°C hotter. These temperature shifts would be devastating to global agriculture, even in the so-called best-case scenario, a fact made evident by what happened to the native populations and their landscape of 5,000 years ago. The Middle Holocene is one of the most enigmatic, relevant time periods for archaeologists to study, and part of our roadmap forward lies in what our ancestors did or failed to do in terms of their survival through such events.

That, however, was assuredly not the only time period in prehistory during which climate shifts caused major problems. While not as dramatic or as long a temperature shift as the Middle Holocene saw, the Medieval Climatic Anomaly (MCA) appears to have had a devastating effect on the native population of California. The evidence for this is much better represented in archaeological sites throughout northern California. The MCA consisted of several centuries of drought and warmer weather that spanned roughly 800 to 1100 AD. Recent studies into this time period suggest that, throughout much of California, Native American village sites were abandoned, with populations appearing to concentrate in central, advantageously located villages near reliable water and food resources. What can we glean from these discoveries? Simple: Climate change not only limited available resources, but it also changed human behavior as a trickle-down effect, populations shifted across a broader landscape to survive these stressed times. Maintaining balance not only with nature but also with one's neighbors, creating a process through which resources are shared during times of stress, is one of the most critical aspects of culture, profoundly shaping societies of the past and present.

Don Steinruck makes the best-smoked salmon I've ever tasted, and I was lucky enough to receive five small jars of it. I gave one to my parents, with great reverence. The salmon is fished from the Klamath and Smith Rivers by Tolowa and Yurok Native Americans, in the northwestern corner of California, and Don smokes it in the traditional Tolowa way, which he learned from his mother-in-law. The Tolowa and Yurok have fished along those rivers that have been owned and handed down to them through many generations, so the extensive knowledge their elders have about the environmental health and productivity of the waterways, as well as the changes to the health and productivity over time, is absolutely encyclopedic.

I had the honor of sitting in on some interviews with Tolowa and Yurok elders as they discussed the changes they'd witnessed in their ancestral fishing and collection areas. A wide range of topics was covered, from the salmon population to seaweed availability to changes in sand particle size at certain beaches where smelt was caught. Shifts in the amount of snowpack were recounted, as were the changing dynamics between elk and deer herds, the productivity of different basketry plants and acorns, and the near absence of the pileated woodpecker, the source of the red feathers for Tolowa regalia. Elders sadly recalled the devastating impact on local flora and fauna from mid-century logging and agriculture and mentioned how slow the healing process was going in their scarred landscape.

I was there to study the impact of climate change on traditional Tolowa resources, aided by a collaborative grant from the National Park Service (NPS), Tolowa Dee ni' Tribal Nation, and Elk Valley Rancheria. Climate change is a profoundly complex topic for many tribal people. At the heart of most discussions like this is one question: Is climate change the will of the Creator or an act of mankind? If it is Creator's will, then the response from the tribal community is to let it unfold, whatever the outcome. If it is an act of mankind, on the other hand, the only acceptable and responsible response is to protect tribal natural resources and their cultural heritage to the best of our ability. How tribal governments and elder councils answer this spiritual question has a direct effect on how tribes choose to respond with concrete action, policy, and advocacy.

I've met some tribal people who don't think climate change is caused by human action, but they still agree that major changes are underway. Most have noticed environmental shifts during their lifespans, changes of which environmental scientists are just now taking note. Suntayea Steinruck, Tribal Historic Preservation Officer, an incredible traditional Tolowa singer, and a good friend, had this to say about climate change, from her tribal perspective:

> The concept and reality of climate change is very frustrating for my community, as well as for other tribal communities around the world. In our view of cultural heritage, there is balance. You never take more than you need, and you never take too little, or that balance is broken. Every choice has a reaction, and it is the duty of the Tribal Heritage Preservation Office to look at choices and weigh the action that, in good faith, will make sure balance is not disturbed. Climate change has done this. Our world is unhinged, and the balance is broken. The poor choices of man and outcome to Mother Earth have resulted in immeasurable costs and long-term disastrous effects. At the forefront of climate change's wrath are our non-renewable tribal cultural resources. Our heritage has lasted since the beginning of time. Now it is at the mercy of climate change.

During that trip, Suntayea and I visited two significant archaeological sites along the coast, *Shin-yvslh-sri*, meaning "in summer (where) they dry (surf fish)," and *Lht'vsr-me'*, or "sand in," a launching place for boats. Both were ancestral village and camp locations to Tolowa people, used till the early twentieth century. Amelia Brown, a Tolowa elder, was recorded in 1978 as having fished and camped at Lht'vsr-me' as a child. Both were symbolic locations to the Tolowa, but both were also rapidly disappearing into the ocean.

On the way to the sites, we stopped by a currant bush. It was noted that the plant was blooming early; that was important because currant blossoms signal the start of smelt season. It was still early spring, and smelt do not typically run until summer. Throughout the trip, we discussed other aspects of smelt fishing: the quality of beach sand that dictates suitability for spawning, the type of fishing gear required, and how shorebirds were tracked to discover the smelt runs. When we arrived at Shin-yvslh-sri, we reached the edge of the ocean terrace and looked down at the narrow beach below. The cliff edge, some twenty feet above the water, was a raw wound, bleeding rock and soil into the sea. Suntayea squinted and pointed at a protrusion, and I carefully reached down and pulled up a polished elk-antler gauge, a tool the Tolowa used to meter the spaces in their nets for fishing. As we gazed over the edge, we wondered how much of the site and, therefore, the Tolowa ancestral history of the place had already fallen away.

Lht'vsr-me' was in even worse condition. Near the beach, the waves at high tide and storm surge had already gouged out much of the site, stripping it to bedrock. As we investigated the damage, I was asked what I thought would be an appropriate action. All I could do was shake my head. Truthfully, the only activity I could think of was for Tolowa spiritual leaders to create a ceremony to properly let go of the crumbling site and others like it. I knew from looking at the coastal erosion models that it was not only that site that was in danger; the whole hillside, stretching hundreds of feet, is likely to be gone by the year 2100.

Some valuable lessons must be learned by the loss of these sites. NPS has focused their salmon habitat restoration efforts near the area on a few of the major waterways that empty out of the steep North Coast Range into the Pacific. As they see it, the smaller streams aren't large enough to support salmon; however, Amelia Brown recalled fishing salmon out of the stream at Lht'vsr-me', a stream mistakenly thought to be too small to support salmon habitat. Rescue excavations at Lht'vsr-me' recovered salmon vertebrae. The traditional ecological knowledge of native people (TEK) is often at odds with Western science-based understanding of the environment. Archaeology can, under the right conditions, help bridge the two worldviews by supporting TEK through field evidence that Western science will accept.

On the drive home, I mentally reviewed my conversations with the elders. For the Tolowa, the smelt was a critically important fish, with a complex cultural practice related to harvesting. When I got back, I dug through the literature, all those anthropological studies of the Tolowa people over the past century, and I began to map out everything needed to harvest smelt. I asked Suntayea for clarification on various ceremonial plants and tools. Then, one Sunday morning, I sat at the kitchen table and stared rather helplessly at a graphic showing all the connections, trying to find a way to fit them all together. Fortunately, my 8-year-old daughter who was in my lap easily took the pencil out of my hand and straightened several of the connections. The finished product, a collaborative work between myself, Suntayea, and Caitie, looked something like this:

Ceremony is at the center of the web because it connects everything and dictates when and how events occur. Of course, we could extend this web into other aspects of plant and animal harvesting, like salmon fishing and acorn gathering, but ceremony would still be in the central position, for it is the cultural glue that holds everything together.

I wondered what such a web might look like today if we were to choose a single resource, something simple, and map it out. For a simple apple, the web might look something like this:

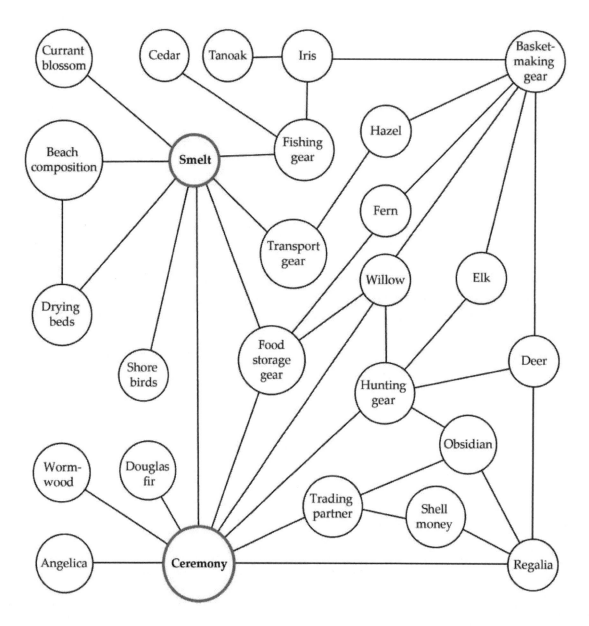

Web showing the relationships between smelt, ceremony, and the necessary gear and materials needed to fish for smelt (Newland and Steinruck 2016)

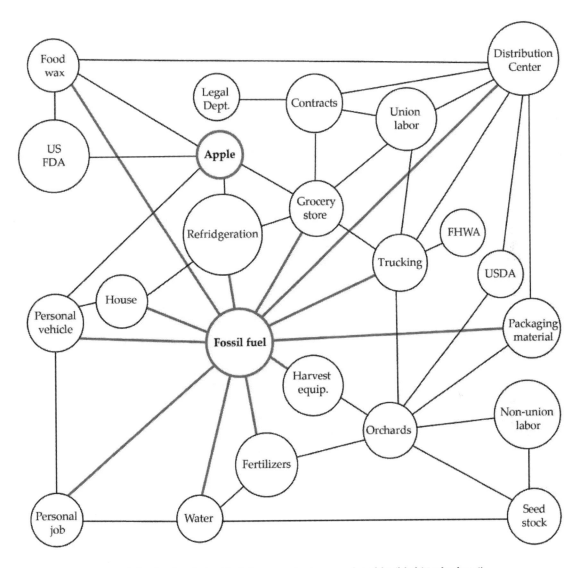

Web showing the relationships between purchasing an apple and fossil fuel (Newland 2016)

It is worthy to note that in this web, what sits in the middle is fossil fuel, not ceremony, or, alternatively (see next page):

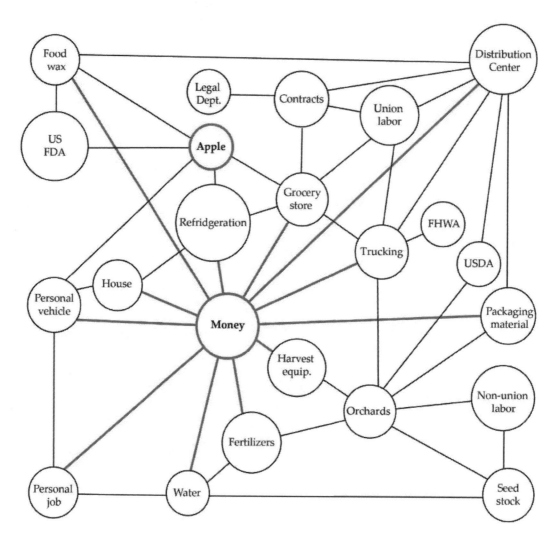

Web showing connection between purchasing an apple and money (Newland 2016)

In truth, fossil fuel and money coexist in that spot, and they are undeniably interconnected. At this point in time in prevailing American culture, we have traded ceremony for fossil fuel and money, but this is not the case for the Tolowa people of today. For them, the ceremony still holds the central position, even if money plays an important role. The two can coexist, and money doesn't have to be in the driver's seat. Nevertheless, the prevailing American way of life has been to remove ceremony from nearly all aspects of daily activity. In doing so, we've removed an important regulating component of our culture. We have disregarded the responsibility and connection we have to our environment and our responsibility to others within our broader interaction sphere.

Is this what we want? The archaeological evidence is pretty damning: Break the connection from your environment, either through over-exploitation or manmade climate change, and culture collapses. Our survival has often relied not on adaptation but on small pockets of the population harboring pre-adaptive traits, such as a more extensive network of freshwater sources, a deeper food resource base, a hunting-and-gathering strategy better than that of our neighbors. These allow small populations to survive climate and natural resource transitions.

As a species, we are about to lose the entirety of our maritime heritage. Every beach where a boat was launched, every village on a coastal edge, and every ceremonial fishing location is about to be destroyed, inundated, or irrevocably changed during our lifetimes and those of our grandchildren. Thousands more critical archaeological and historic sites are threatened by wildfire. All evidence of past climate-change events and human survival of those events (or lack thereof) is found in archaeology. There is no other way to retrieve that crucial data, yet we are about to bid it all farewell. It is essential that we prevent this, and there are many good reasons why.

One of the arguments that those who oppose climate-change regulation make is that we simply can't afford it, that our culture can't survive the transition and the change is too great. In the short term, this does bear some truth. Yes, the necessary changes will be difficult, and they will devastate some industries and cause the economic collapse of some communities. Change on such a broad scale will be enormously painful, so it comes as no surprise that some of the most vocal opponents are those who reside in the portions of the country that will bear the brunt of abandonment of fossil fuel economy and those who own businesses that depend on fossil fuels.

Still, change is possible, and it may need to be rapid and disruptive. Most of what we might consider the important underlying aspects of the culture will survive. We know this because Western civilization has forced such a change on nearly every indigenous culture it has come in contact with over the past millennia. Their resource webs were torn apart, their access to vital plants and food denied, either through land grabbing, relocation, forced enculturation, or environmental destruction; yet such cultures survived and continue to thrive by rebuilding their webs to accommodate new situations and reclaim as much of their ancestral heritage as possible. The world needs a better grasp of how the indigenous people of the world lived, how they functioned within their natural environment, the condition of that natural environment, how they adapted and changed when enculturation or genocide was forced upon them, and what worked and what didn't as they carved their own way forward. Ironically, understanding how indigenous people have prevailed under the rapid cultural and economic change forced upon them by Western civilization may also be Western civilization's only way forward. This time, though, we have to turn the tables and force it upon ourselves.

These issues, our general failure as a species to respond well to climate change, the success of small, well-positioned populations with pre-adaptive traits, the ability of TEK to inform new ways for us to build resiliency into our modern allegedly natural landscape, our complicated placement of money and fossil fuel in the central position of our functioning resource web, and the resiliency of culture in the face of disruption, present us with some difficult challenges. It may seem we are working against insurmountable odds too great to overcome, but we have many critical resources at our disposal. Not only that, but we, as citizens and members of the general public, of human culture as a whole, can take concrete actions moving forward.

For the first time in the history of our species:

1) We *can* understand the causes and effects of climate change. We can make predictions about likely future scenarios. We can steer away from the worst scenarios while preparing for the changes we can't resolve.

2) We *can* consult with the rest of our species about the best approach going forward. Communication is instantaneous in today's modern world, and we can almost immediately discuss this problem with every densely populated portion of the planet.

3) We *can* organize in such a way that action can be taken on a global scale.

These resources are game-changers for us. Climate change is a global illness, and rather than wait for a surgeon or a silver bullet, we, the body, must begin the healing process. This will be accomplished cell by cell, person by person, community by community, and it is our ability to communicate with each other that will facilitate that healing. Yes, technological advances can dramatically reduce our carbon footprint, but only if we, like the body, support its development, take in that technology as it matures, absorb it, and promote it throughout our living system. Yes, prominent leaders can affect change, but only if we, like the body, support their mission and direct them to the problem areas that really need work. In the past, our species has survived because of individual efforts and the actions of small groups. This continues to be the case, and our efforts at the national and international level should be complementary to, not in lieu of, our own individual actions.

Like it or not, we, as a species, each and every one of us, have embarked on the most massive, most ambitious, most contentious human experiment of all time. The stakes have never been higher. We are all guinea pigs, and every single person on the globe will have to live with the results. There is no getting out of the experiment, nor is there any going back to the way things were before. We must get this right or risk collapsing the ecosystem of the entire planet. We cannot abandon our responsibility in stemming climate change. If we do get it right, we will be forever changed. We will improve where we get our energy from, how we distribute our resources, and how we work together collectively as a global population to achieve our goals. To succeed, we must have a better understanding of past climate change events and how they affected the environment, what strategies humans used to survive those events, what worked and what didn't, and how we can use that data to ensure our own survival.

Climate Crisis "Keep/Stop/Start" Actions

Below for your consideration, are several author-suggested thought, behavior, and action items. For a further explanation of the Keep/Stop/Start organizational tool, please see How to Read This Book.

KEEP:

1) *Keep voting!* Local elections may, in the long run, be as or more important than state or national elections in terms of building resiliency in your own community.

2) *Keep volunteering.* Do what you can to help your community prepare by investing your time and money into projects that benefit the ecosystems and watersheds of our local land base.

3) *Keep contacting your local public officials.* Let them know your concerns and remind them that they work for the public and that their actions are visible to all. Tell them when they are doing a good job, because positive reinforcement can be as beneficial as complaints.

STOP:

1) *Stop relying on the federal government* and the international community to solve the climate crisis if you are not committed to strong local engagement in the process.

2) *Stop waiting for technological advances* that will supposedly absolve us of having to make harsh cutbacks in energy consumption. The time to make those cutbacks is now.

3) *Stop building local infrastructures* like housing developments, roadways, and business parks that don't take climate change into account. Don't build on a coastline that will only be inundated by sea level rise or collapse as the cliffs give way. Don't build new buildings that aren't energy efficient.

START:

1) *Start making your own web.* What holds your web together? How can you cut ties with or lessen the importance of fossil fuels in your web based on your own home and life?

2) *Start building resiliency* into your local environment.

   a) Make sure local timber harvesting and land management include planning for forest fires. If you are a property owner with acreage, how prepared are you for fire? How do you staff your vegetation and fuel load? Do your neighbors have good land management practices or are they putting you at risk?

   b) Ensure the quality of your local aquifer. What contaminants invade your water system? Do the creeks and rivers maintain sufficient water quality, temperature, and sediment load to support native flora and fauna?

   c) Stabilize your slopes. Areas that are susceptible to landslide may need to be reassessed as to their vegetation populations, so plant communities are re-created to help anchor the hill and reduce fire hazard.

3) *Start upgrading* your community's infrastructure.

   a) Make wind, solar, and other renewable energy projects priorities. This not only reduces your community carbon footprint, but it also lessens community reliance on the larger electrical grid, a grid that is aging in much of the nation and will undoubtedly be overextended as temperatures rise. Localizing your power helps ensure that your business and home are less likely to be disrupted by regional power outages, and it reduces overall regional and state fossil fuels needs for electricity.

   b) Make water conservation and recycling priorities in any new construction projects. Look carefully and critically at landscaping plans for new developments to see if plants are appropriate for your location and are drought tolerant.

   c) Install electric vehicle charging stations at government centers and invest in electric vehicles for government use.

   d) Create a bicycle- and pedestrian-friendly community and encourage everyone to use alternate means of transportation that are healthier for body and environment.

   e) Reuse and rehabilitate old buildings; the carbon footprint of renovation to aid an older structure to meet modern energy efficiency is less than creating a new building that meets those same standards.

   4. *Start signing up* for 100 percent renewable energy from your local energy provider. Vote with your pocketbook!

   5. *Start purchasing hybrid and electric* vehicles if it these make sense where you live.

~ ~ ~

Mike Newland, M.A. is the director of the Northern California Cultural Resources Group for Environmental Science Associates, one of the nation's oldest, most successful professional environmental firms. He has been a professional archaeologist for over twenty years and has worked throughout California and the Great Basin. For the past five years, his research efforts have focused on the effects of modern climate change on cultural resources. Mike is a regular commentator on the topic of archaeology and its intersection with everyday life for the KQED *Perspectives* series. Mike Lives in Santa Rosa, California. Read more at: www.climateabandoned.com.

Chapter 9

Climate Crisis & Eco-Restoration

by

Lois Robin, M.A.

The mountains... I become part of it.
The herbs, the fruit tree... I become part of it.
The morning mists, the clouds, the gathering waters...
I become part of it.
The wilderness, the dew drops, the pollen...
I become part of it.

(Navajo Chant)

The Dirt on Climate Change

The Dirt on Climate Change is the name of a film our team is creating, and by "dirt," we also mean soil. By "soil," we mean excellent soil, soil that is porous and spongy, soil that crumbles when you handle it, topsoil that extends two feet deep or more into the earth. This soil supports a vast number of microorganisms: mycorrhizal fungi, earthworms, and a diverse spectrum of microscopic creatures that spend their lives creating this valuable material for the plants and, thus, for us.

When this soil receives water, a cycle begins which allows plants to flourish. Woody and perennial plants put down roots that would enable carbon to be sequestered deep in the soil, and more water is retained therein. Enough sequestered carbon can turn the tide of global warming.

Renewing cycles of water, soil, and earth can revitalize and restore the planet.In the life of this planet, humans have drastically altered these cycles; for centuries prior, they were the cycles that kept the planet humming and surviving, but they have now been severely impacted and no longer allow Earth to recover as well as it once did.

We used forests for construction and burning. We dried up the planet by sending all our storm water into drains that carried it to rivers that dumped into oceans that are now rising rapidly. Did we consider that the concrete we contentedly smoothed over everything, creating our roads and parking lots and sidewalks and building foundations, would prevent rainwater from entering the earth and radiating heat when touched by the sun? Sadly, no, and just as sadly, we still don't fully understand the impact of human practices on the integrity of the planet.

The list of damages continues. Today in cool Santa Cruz, I walk along my asphalt street. It is hot. When I walk under a tree that shadows the road, I feel cool again. Try it. Notice how the sun's heat is reflected off the sidewalks and streets you travel. Then imagine heat rising from concrete roads, parking lots, and construction all over the globe. No wonder Earth is warming! Today, San Francisco set a record of 106 degrees.

Tourists from inland cities flock to Santa Cruz this time of year because the added heat from global warming has already made summers unbearable inland. Most of the people living on the globe hug the coasts when they can, because large bodies of water keep us cool. Can we imagine civilization without impervious, heat-creating infrastructure, without all those streets, roads, parking lots, and roofs? Can we start de-paving the globe and use other coverings such as decomposed granite, stone, inlaid bricks, cobbles, and other cooler materials that allow the rain to penetrate? In the view of ecologists, it is not a matter of whether we can; rather, we must, unless we want to fry ourselves. Can we reduce rather than increase the number of nonporous roads and parking lots and plant more trees to shade and cool the areas we've already developed? Urban forests will cool us and the planet, and we need more of them.

Did we know that killing predators like wolves and lions would leave herds of animals to graze without pattern or restraint? Were we aware that casually removing all of the allegedly unneeded plant matter, all that valuable grass and native plant life underfoot, would leave the soil dry and bare and desertified? Fortunately, today, bold, ingenious leaders are finding ways to rectify these problems.

Allen Savory, an African-turned-American, was probably was the first to identify the problem and the solution for successfully grazing herds in Africa. A lover of big African predators, he began to notice that as the local people dispatched the animals, herds of prey sporadically moved across Africa, chewing the plants to nubs, leaving the former grasslands desertified. He decided to reverse the process, allowing vegetation to regrow by preventing such severe grazing, and land that was previously desertified slowly began to recover.

Soon, the soil improved, water was retained, and land that was formerly dried and desiccated became verdant and meadow-like. Rivers and streams returned! Allen brought those same practices to the United States, where he continues to demonstrate the benefits of holistically managed grazing.

In California, the great Central Valley was once quite wet. When it was occupied by Native Americans, in fact, it was so moist that the people had to construct their homes on stilts. Now, when we drive through the Valley on Highway 99, we see agricultural fields that must be served by irrigation pipes, pumping from deep aquifers to satisfy the thirst of almond, pistachio, peach, and plum orchards and a myriad of other crops. The desertification of the Central Valley didn't happen so long ago, but it can be healed. Through eco-restoration, even desiccated land can be restored to productive functioning, without depleting underground aquifers.

In an attempt to capture the potential for grazing in order to restore land and capture carbon, our film crew visited Joe Morris, a local rancher who has been grazing his cattle in the foothills that frame the town of Watsonville.

A cowboy and several lively, talented shepherd dogs help move the cattle along. Joe does not allow his grasses to be nibbled lower than four inches, and he knows what he's doing; he has been doing this work for 26 years and grazes 11,000 acres of land. We rode along in Joe's truck with his border collies to a spot high in the hills. The springs and brooks sparkled and glimmered, and the doggies splashed and luxuriated in the warm sun and cool water. Joe encouraged us to look at the variety of native and non-native plants and grasses that proliferated there, and we recognized most, but some were new species to Joe. Although it was August, so most of the grasses were brown, in that arena of managed grazing, there was a green glow of new growth that could only be attributed to the ample water provided from his grazing practices. One lovely plant drenched with blue flowers, evidently a native of the mint family, has seldom been seen before in these hills.

How does the land become wetter with managed grazing? Simple! Animal hooves open the soil up to accept rainwater and wet cow urine and dung. In turn, the water in the soil encourages deep-rooted grasses and plants that sequester carbon. The plants release cool moisture that evaporates, and the humid air condenses, further moistening the land. With continual repetition, the natural cycle of water/soil/plants is restored.

As a final gesture, Joe took us to a clump of cow dung and used his pocketknife to dig into it. I admit to some squeamishness, but I was rewarded for my fortitude with the sight of squirming beetles, which I captured on camera. Dung beetles are another sign that grazing is productive. They eat the fly eggs that distress the cattle, and, most importantly, they join the microbes and other insects to open the soil, rendering it more permeable. Those beetles in that pile of dung are just another sign of diversity returning to the soil. Joe regularly measures the carbon that is sequestered through his grazing practices, and the increases in carbon assure him that he is on the right track.

Deeply philosophical about his business of providing good protein and great eating for his customers, Joe carefully considers the relationship between cattle, people, and soil. He knows this balance must include good lives for his cattle, with cows, bulls, and calves grazing side by side, nudged into a herd by a cowboy and dogs. He, his wife Julie, and their helpers deliver his excellent products to designated places in Monterey, Santa Cruz, and San Benito Counties, and his are the best meat products I've ever eaten. If you would like to see a six-minute interview with Joe, visit climatechangehitshome.com and view our previous film, *Climate Change Hits Home: Santa Cruz*. A more extended interview with him on the website reveals the thoughts of a dedicated, articulate man, a true spokesperson for the values of managed grazing.

People who genuinely think about the causes of climate change often point to cows as part of the problem, as the animals are said to release methane in unsupportable ways. The usual message is that because of the methane problem, we should all be vegetarians or vegans. This is indeed the situation with cows raised in feedlots. In her unsurpassed *Cows Save the Planet*, Judith Schwartz says of food lots:

According to Andrea Malberg, director of research at the Savory Institute, when cattle are in crowded feedlots (concentrated animal feeding operations or CAFO's) the manure is handled in liquid form in lagoons or holding tanks—an ideal scenario for the production of methane. If, however, cattle graze on fields in lower concentration, the manure becomes fertilizer. She adds, "These are human decisions, not cows'."

The arguments over the chemistry have convinced me that the methane issue is bogus when animals are grass fed. Although cattle are most commonly grazed, sheep and even pigs can be raised in the same way, as they are at the Markegard Ranch near Pescadero and with Sally Calhoun at her Paicines Ranch.

Managed or rotational grazing is often a tool of regenerative agriculture, which requires that:

1) No tilling be done to the land. Instead, a layer of cover crop or compost is allowed to remain on the land for some time. The land is never left bare.

2) No fertilizer is added though manure may be used.

3) Pesticides, herbicides are not used. These products kill valuable microbia and insects that are perfecting the soil.

4) There is some way for rain to penetrate easily, either through grazing in which the cattle hooves open up the soil or with a tool that slices the ground but does not turn it over.

Soil

Plainly you can see that regenerative agriculture differs from simply growing food organically, which is increasingly common in Santa Cruz and adjoining counties. It is similar conceptually to permaculture but is an actual practice, not merely a concept. The food that is produced is, of course, organic.

The soil's ability to capture carbon and store water has led to an upsurge of interest in this often-overlooked natural resource.

In California, a new program, the Healthy Soils Initiative, plans to put unorthodox farming practices to the test. With modest grants of up to $50,000, administered by the California Department of Food and Agriculture (CDFA), a network of farmers and ranchers throughout the state will embark on a series of experiments in carbon farming. Carbon farming refers to improving soil health by biological processes that limit the amount of synthetic chemicals applied to crops and adopting techniques aimed to reduce nutrient loss. Happily, many other states are implementing similar programs to promote carbon farming.

So that we may learn more about regenerative agriculture, our film crew visited Sally Calhoun at Paicines Ranch in San Benito County on a hot August day. Her sprawling ranch encompasses 7,600 acres of land in the rolling, tree-dotted hills near Hollister. Sally, a vivacious woman in her sixties, greeted the three of us warmly, and she was eager to show us what she believes and hopes will be the agricultural wave of the future. She began an enthusiastic account of regenerative agriculture as we drove to her new vineyard. The small grape vines were planted last April, in untilled ground, and Sally explained that tilling has been the cause of desertification and loss of soil vitality. Prior to the planting, a cover crop was grown on the land to allow the rainwater to penetrate the soil deeply, and the soil to begin producing an army of fungi and insects.

Wending their way to the site was a convoy of sheep, ready to nibble away at the grasses and plants that covered the vineyard, but they would not be allowed to chew it to the ground. Sally believes sheep, not cows, will save the planet, as they are expert grazers, and since she also grazes cows on her land, she is in a position to compare. Sheep also nibble the low-lying grapevine tendrils, ensuring stronger and better leaves and fruit at the top. Electrified fences guide the animals so they do not feast too much in any one place. Sally anticipates that the vineyard will be productive in three years.

She then showed us another field that is being prepared for agriculture with a cover crop. Kale, broccoli, and other plants provided a green and gray tapestry across Sally's acreage. In a season or so, Paicines Ranch will support various crops, mostly forage. The San Benito River runs through the property and raises the water table, keeping her land from broiling under the summer sun. On a hot summer day, with little concrete and plenty of greenery, we were not uncomfortable standing in the sun for several hours as we listened to Sally's hopes and dreams for success with this new style of agriculture. She said that those of us interested in the paradigm for climate change described in this chapter should call ourselves "regenerators," and I am happy to report that henceforth, I will refer to myself as just that!

Water

A vital part of these cycles is water, yet in almost all communities, rainwater is quickly ushered to drains, only to be whisked away to rivers and into the oceans. Such is the case with our local rivers: the San Lorenzo, Soquel Creek, Aptos Creek, and especially the Pajaro. The Pajaro River is actually encased for twelve miles with levees to stop any flooding. Agriculture is intense on both sides of the river and is conducted by irrigation through pumping the groundwater. So much pumping has occurred that saltwater intrusion from the bay into the underground aquifers is causing the loss of those aquifers for usable water. The State of California has formed Groundwater Sustainability Management Agencies in localities where groundwater loss is significant.

I asked Dr. Andrew Fisher, Professor of Earth and Planetary Sciences at UC Santa Cruz, who has done influential research on groundwater and aquifers, what can be done to provide more direct access to rainwater. He said, "I agree with you completely that this should be done, but under current policies, restrictions, private rights, agency regulations, laws, etc., it is virtually impossible to do anything of that kind." He sighed and added, "I know, because I've tried."

The Pajaro Valley Water Management District *is* sponsoring a Recharge Net Metering program which has the goal of infiltrating over 1,000 acre-feet per year within 5 years. The program provides a rebate for landowners as an incentive to infiltrate rainwater into the ground.

Raspberries and other bush berries currently being grown require even more water than strawberries and lettuce. Apples were once the crop of choice in this area because they could thrive with more or less water. Apples are still grown, albeit much less often, because they are not as remunerative as berries. Yet, we all pay a big price for berry farmers to continue their profits and for us to gorge on berries: Our water supply is jeopardized.

In making matters worse for eco-restoration, berries are now grown under tent-like hoops. The plastic coverings largely prevent rainwater from reaching the ground. When I was up in the hills with Joe Morris and looked down to the valley below, I saw the sea of white tents, erected to protect the berries, tents that remain in place all year to protect plants from spoilage due to rainwater. Farmers instead use drip irrigation with water from aquifers, and much water is necessary. Since the crops can be very remunerative, there is little incentive to change these practices. Perhaps the new California Groundwater Sustainability Agencies will find a way to stop the overreliance on aquifer water.

If all this is folly, though, what are the answers? For the sake of our film, we sought small efforts to allow rainwater to fall directly on the soil: a swale, a rain garden, and a check dam. We have filmed a beautiful front yard filled with booming vegetables, watered by greywater and rooftop water held in barrels, so all rainwater finds its way into this garden. We will film properties where the overflow from creeks is allowed to flood the yards of these homeowners temporarily. We filmed a commercial parking lot made entirely of neatly laid bricks, so water can trickle through the cracks to the aquifer. Several elegantly planted swales keep rainwater from rushing to the curb at the same establishment, a beer garden.

A small home in Santa Cruz has a sump that collects rainwater under a cobbled riverbed that turns through the shrubs in the front yard; this allows the percolation of rainwater into the soil instead of wasting it at the curb and gutter. It would be helpful if curbside vegetated strips were created to absorb rainwater before it runs from sidewalks to the gutter. As these ideas gain understanding and acceptance, more and more small projects will occur around the globe, throughout our, and in Europe as well. Not only will these impact the course of climate warming, but they will drastically improve the quality of landscapes in the cities and towns.

One of my pet peeves is the proliferation of artificial lawns, encouraged by our local water company as a way to save water during drought years. Those who install them are rewarded with rebates on their water bills. While artificial lawns are looking more and more natural and claim to allow passage of rainwater, the ground under them becomes so hard that little water can penetrate the product. Missing is the soil ecosystem engendered by perennial plant roots that create macrospores through which water passes and loosens soil density. Also missing is the evaporative cooling of real green cover. As the product is improved to look more natural, the artificial turf industry is ever expanding, constantly covering more significant amounts of soil. I have never personally been a fan of it, and soccer players often refuse to play on it, because it is so painful and damaging to the knees. It also retains heat, like concrete and asphalt, and what can we anticipate doing with it when it needs replacing twenty or so years from now? Most turf is non-recyclable. Who needs it? Let's find solutions that save the planet, not pave it.

As I write, Hurricane Harvey is totally overwhelming the coastal cities with floodwaters. Several TV and radio broadcasts have provided revealing reports about the jeopardy folks in these cities have placed on themselves through poor regulatory practices.

For example, PBS ran the following highly instructive interview:

KIAH COLLIER, REPORTER, TEXAS TRIBUNE: So, the projects were to focus on Houston's vulnerability to hurricanes and then flooding and how flooding has been exacerbated by unchecked development in Houston. Building rules are not too strict, and developers are just paving, you know, over prairie land and wetlands, and not thinking too much about it. And when rains like this hit, it just makes it so much worse because those flood waters don't have anywhere to go.

And, of course, you know, with hurricanes comes rain. And, you know, there's a big question whether Harvey would hit Houston as a hurricane. It was never projected to do that. But we wrote about a specific hurricane that if it hits at a particular point on the coast, it would send an enormous storm surge into a highly populated area and the U.S.'s largest refining and petrochemical complex, which is here in the bayou city.

HOST SREENIVASAN: One part of your story mentioned that Houston had, what, the most urban flooding in the last forty years?

COLLIER: Yes, that's right. Flooding kills more people here than anywhere else in the U.S. and in general. I mean, flooding kills a lot of people, more than have been hurricanes actually. So, it's the number one killer, in terms of natural disasters.

SREENIVASAN: So, when you look at these beautiful sets of maps that you have and you look at sort of 100-year storms and 500 year storms, you can kind of layer this over into where these flood zones are and you see the dots just kind of line up and they go right over areas that are expected to flood.

COLLIER: Right, exactly. And these recent floods... So, this is the third major flood Houston has seen in the last three years I guess. So, there's been historic floods, you know, basically once a year for the past three years, and a lot of that flooding has occurred outside the zones that FEMA considers most likely to flood.

And today, we're seeing flooding that exceeds 500-year flood levels, which is really rare.

SREENIVASAN: You know, Houston has a flood control board. I mean, is there tension there between what scientists predict and what they say is necessary to prevent these things and how Houston develops?

COLLIER: Right. So, our project was based on, you know, a lot of interviews with scientists who said, you know, Houston is not doing enough to mitigate flooding. It's not leaving open green space in these developments. And it's only going to get worse because of climate change.

And so, we talked to the head of the flood control district for our story. And he basically didn't agree with any of that and told us that they can kind of fight concrete with concrete. You know, he's an engineer and said, you know, we can engineer our way out of this problem with this big flood retention, you know, kind of basically massive concrete, you know, public works projects that funnel the floodwaters out to the Gulf of Mexico.

And, you know, we don't have... Houston doesn't have enough of those and he admitted that this was behind on creating those projects or building those projects. But scientists that we talked to say you're not going to be able to engineer out of the problem. What Houston really needs is smarter development rules.

SREENIVASAN: All right. Kiah Collier of *The Texas Tribune*, joining us via Skype form Houston today, she's one of the authors of *Boom Town, Flood Town* and *Hell or High Water*... You can find a link to that from our website. Thanks so much for joining us.

COLLIER: Thank you.

Another reporter on *Tom on the Beach* website commented: "Texans are paying for their short-sighted independence. Nobody can tell a Texan what to build and where."

The situation is not so different here in the Pajaro Valley, where the towns and farm fields are prevented from flooding by levees. The Army Corps of Engineers has been asked to build the levees higher to protect the farm fields; instead, it might be wiser and cheaper to buy neighboring farm fields to allow some flooding. However, that is unlikely to happen, considering the high investments farmers have made in their land. Wetlands in the delta area of the Pajaro River, however, have been preserved by the persistence of the non-profit Watsonville Wetlands Watch.

You may wonder how these ideas of eco-restoration are being disseminated. Voices of Water for Climate has been bringing the ideas of scientists in eastern Europe to the United States. A leading hydrologist/engineer, Michal Kravčík, received a Goldman Environmental Prize for his contributions to the water management of the Torysa River in Slovakia.

Another group in the United States, Biodiversity for a Livable Climate, has just published a significant compendium of many ideas and projects being promulgated in this country based on science and practicality.

The Savory and Rodale Institutes are experimenting and sharing the results of various projects and experiments. Also, the Marin Carbon Project (MCP) achieves carbon sequestration in rangeland and agricultural soils through research and development of scalable, repeatable carbon farming techniques.

The environmental community has been primarily focused on efforts to get carbon out of the air, to support alternative clean energy, and to oppose pipeline and fracking. Although those efforts are of the utmost importance, many environmental leaders still report a stasis, a lack of improvement beyond a certain point using the means we now have to pull CO_2 from the air.

Paul Hawken's new and influential *Carbon Drawdown* assigns a point value to 100 different options for carbon drawdown. The highest points are granted to managed grazing and regenerative agriculture. Although the cycles of carbon/water and plants I have described are connected and interrelated, and Hawken isolates them, his inclusion of these in his excellent book has piqued the interest of many in the climate change movement.

Trees and Forests

As we well know, trees and forests are mighty sequesterers of carbon, yet the cutting of trees continues in the Sierra. Even in our coastal communities, large numbers of redwoods are surrendered each year to the needs of builders and the construction industry. These forests have an enormous mass of underground natural activity, and the web of myccorhizal fungi in the soil sequesters the carbon.

Our video team knows trees and forests are part of the equation of eco-restoration, so we interviewed Dr. Betsy Herbert, now retired. Dr. Herbert has engaged with forestry issues her entire career, and she is awed by the amount of carbon held in trees. Meeting with her in the awe-inspiring Henry Cowell Redwood State Park, its giants all around us, she said she especially values the old, big trees, the redwoods along with many others. She said that protecting these old trees and not allowing clear cutting of any forests is the most criticalaction citizens can take. It's always great to plant trees, as they provide beauty and cooling shade, but the biggest bang for our ecological buck is guarding the lives of big old trees. "Lend support to all the nonprofits and agencies defending existing forests," she advises.

Most hopeful is a program of forest carbon offsets run by the State of California. I was impressed to read in the *Nature Conservancy* publication of November 2016 that dozens of projects nationwide are selling offsets on the California market. The Yurok Indians of Northern California were among the first to get involved. The tribe acquired some 22,000 acres of forest from Green Diamond Resource Company in 2011. Working with an investment group, Forest Carbon Partners, the tribe applied for offset credit on 7,760 acres of these lands, committing to stable forest management for the next century. They were issued more than 800,000 in offsets, estimated at worth six to eight million dollars, by the Air Resources Board. Nevertheless, Dr. Herbert warned that not all aspects of this program of carbon offsets are as successful.

Eco-restoration is happening in various places around the globe, and there are glimmers of it on the Central Coast. More is being done in Marin and Sonoma Counties, where regenerative agriculture is widely practiced, and in Los Angeles, where water is being better utilized and trees are being planted. This approach to climate change is ongoing. When it is united with efforts to remove carbon from the air through the use of alternative energy, the people who live on this planet may be inspired to recover what has been lost and to notice the life lessons from observing where they have strayed. Perhaps this conclusion will provide the ending we seek for our film. May we all belong to a cooler, wetter, more productive and enjoyable planet!

Climate Crisis "Keep/Stop/Start" Actions

Below for your consideration, are several author-suggested action items. For a further explanation of the Keep/Stop/Start organizational tool, please see How to Read This Book.

KEEP:

1) *Keep the big old trees, especially redwoods.* Every year Redwood trees put on girth, and since they are already so huge, great quantities of carbon are sequestered within them.

2) *Keep the soil covered at all times*, either with compost or crops. We must stop leaving soil bare to dry out and blow away, desert-tifying vast areas of land.

3) *Keep planting trees*—you can never have too many, and eventually, the little ones will grow big and sequester carbon while providing shade and cooling. The towns and cities need our discerning help as well as the rural places. We can create more urban forests to cool the cities and towns and sequester carbon.

4) *More Keeps*:

   - Wetlands, swamps, sumps, and swales
   - Trees, especially old and big ones
   - Forests
   - A cover of plant material on your soil at all times
   - Eating grass-fed sustainably grazed meat and organic foods
   - Worms and insects in the soil
   - Growing your gardens
   - Predators such as wolves, bears, and lions

STOP:

1) *Stop feeding animals in feedlots* but instead encourage cattle, sheep and pigs to be "grass" or "plant" fed. Grazing needs to be managed so that plants are never bitten to the ground but are allowed to regrow by leaving several inches intact. In this way, we will reduce the methane in the atmosphere and begin to recover land whose healthy plants will sequester carbon for us. If we are meat eaters, and most of us are, we should buy only grass-fed meat from animals that have been appropriately managed to allow regeneration of plant life.

2) *Stop sending precious water down drains* to the rivers and then to the seas. We must let rainwater seep into the ground where it falls, and when there is concrete, we must halt the water from draining away by capturing it in swales, barrels or catchments for use during times when there is no rain. We must put out barrels to collect water off the roof if we have not already done so.

3) *More Stops*:

- Clearcutting of our forests

- Leaf blowers from blowing away cover on dirt

- Sending rainwater through drains and rivers to the ocean

- Animal feedlots

- Covering soil with plastic

- Using petrochemical plastic

- Tilling the soil

- Paving with non-permeable materials

- Using pesticides and herbicides

- Using artificial lawn

START:

1) *Start learning what our state agencies* are doing to regulate and hopefully protect the big, old trees. And when they favor industry too much, we must sign petitions or write letters that keep them on a protective course. By all means, Start or Keep visiting the magnificent old groves throughout our state and country and provide them with the respect and wonder that is their due.

2) *Start encouraging insects* to build soil instead of routinely killing them. The insects, the microscopic creatures, the bacteria, the fungi and all the other denizens of the biome need to grow and thrive on building deep, loamy, porous, nutrient-rich soil. To encourage this kind of soil, we need to stop tilling. Tilling disrupts valuable biome and causes carbon loss. Plant by digging holes or slits into which we place plants or seeds. New tools are available now to assist with this task. Although this method may be difficult

and challenging in many places, "no till" is the key to future successful soil building agriculture. We must cease the use of pesticides, fungicides, and fertilizer and make compost available and ample.

3) *Start bringing back* the beavers to their previous haunts for they can produce waterways for us that cool the planet and store our water. In fact, most wildlife has a crucial role to play in our ecosystems, and we need to retain the ones we have and stop killing them while we continue to learn about the crucial roles they play in our ecosystems. We must keep and treasure the lowliest, like dung beetles so they can keep revitalizing the soil.

4) *More Starts*:

   • Letting rainwater be absorbed by the soil where it falls
   • Capturing and using rainwater
   • Eating sustainably-raised beef, pork, and lamb and primarily organic foods
   • Planting trees and urban forests
   • Use swales, sumps, curbside gardens, and rain gardens
   • Plant new gardens
   • Supporting regenerative agriculture
   • De-paving and using only permeable pavement
   • Encouraging soil life (insects, fungi)
   • Becoming conscious of the problems created by too much concrete

Finally, we must sing daily the praises of the natural world, of the fantastic life-giving cycles of water/soil and plants. We must hold the naysayers at bay while we build a lasting ecosystem that relates to climate change but does not capitulate to it and start to deal with our ecosystems consciously and respectfully.

We have completed our film. We hope our viewers will be inspired by it and take away some hopeful concepts. For further information, please visit: http://www.climatechangehitshome.com/2018/04/06/the-dirt-on-climate-change-2018-film/

~ ~ ~

Lois Robin, M.A. has been a multi-media artist for the last thirty years. She has produced three films, published articles and photographs in several publications including TimeLife Books, and publicly exhibited her photographs. Her last film was *Climate Change Hits Home: Santa Cruz.* Most of her work has been environmental or about California Indians. She wrote a book *Mamita's House: A True Tale of Tortilla Flat* about an Indian family living in Carmel at the turn of the last century. At the same time, she has been an environmental activist and Chair for twenty years of the Sierra Club's Pajaro River Committee. Today she is working on a film on an alternative paradigm for climate change. She has been a Commissioner on the Santa Cruz County Fish and Wildlife Commission for five years. Read more at: www.climateabandoned.com.

Chapter 10

Climate Crisis & Selling Doubt
Part I: Exxon & The Benefit of (Selling) the Doubt

by

Cathy Cowan Becker, M.A.

"To see what is in front of one's nose needs a constant struggle" – **George Orwell**

On June 1, 2017, President Donald Trump withdrew the United States from the Paris climate agreement. The move was not altogether a surprise, considering Trump had previously tweeted that climate change is a concept "created by and for the Chinese in order to make U.S. manufacturing non-competitive[1]." All told, Trump tweeted climate change denial at least 115 times since from 2011 to 2017[2].

That a sitting president could repeat such falsehoods — when 97 percent of climate scientists say climate change is real, caused by humans, and poses an existential threat to all life on earth[3] — is a testament to the power of climate denial. For three decades a concerted campaign to raise doubt about climate change and question the integrity of climate scientists has been waged by fossil fuel barons, cold war ideologues, free-market fundamentalists, and a slate of conservative foundations and think tanks that together have spent billions of dollars to confuse the public about climate science. Their purpose was simple: to delay or derail regulations on carbon emissions that threatened the profits of the fossil fuel industry. And the unfortunate outcome is that it worked.

In this chapter, we will examine the multibillion-dollar industry of climate denial, modeled on the campaign by tobacco corporations to raise doubt that cigarettes cause lung cancer. We will trace the funding of the decades-long climate denial campaign by fossil fuel corporations, which knew climate change was real from their own scientific research, and the spread of denial through a network of foundations, front groups and conservative media outlets.

The product of doubt

In 1969, the tobacco industry had a problem. Scientific evidence linking cigarettes to lung cancer had been building for decades, and in 1964 the surgeon general issued a report outlining the ill effects of smoking on public health[4]. The next year Congress passed a law requiring warning labels and cigarette packages, and in 1969 cigarette advertising on television and radio was banned[5]. As a result, public opinion went from not thinking that smoking causes lung cancer to believing it does[6].

Tobacco companies were looking at a massive drop in their profits and needed to figure out a way to fight back. A public relations executive at Brown & Williamson outlined a proposal to do just that in a famous memo now

[1] Trump, Donald J. Twitter, November 6, 2012. https://twitter.com/realdonaldtrump/status/265895292191248385?lang=en

[2] Matthews, Dylan. "Donald Trump has tweeted climate change skepticism 115 times. Here's all of it." Vox, June 1, 2017. https://www.vox.com/policy-and-politics/2017/6/1/15726472/trump-tweets-global-warming-paris-climate-agreement

[3] Cook, John, et al. "Quantifying the consensus on anthropogenic global warming in the scientific literature." Environmental Research Letters, May 15, 2013. http://iopscience.iop.org/article/10.1088/1748-9326/8/2/024024

[4] "Smoking and Health: Report of the Advisory Committee to the Surgeon General of the Public Health Service," Reports of the Surgeon General, 1964. https://profiles.nlm.nih.gov/ps/retrieve/ResourceMetadata/NNBBMQ

[5] "The 1964 Report of Smoking and Health," The Reports of the Surgeon General, Profiles in Science, U.S. Laboratory of Medicine. https://profiles.nlm.nih.gov/ps/retrieve/ResourceMetadata/NNBBMQ

[6] Saad, Lydia. "Tobacco and Smoking." Gallup News, August 15, 2002. http://news.gallup.com/poll/9910/tobacco-smoking.aspx

available through Truth Tobacco Industry Documents, a collection of internal corporate documents produced during litigation between U.S. states and seven major tobacco industry organizations[7].

"Doubt is our product, since it is the best means of competing with the 'body of fact' [linking smoking with cancer] that exists in the mind of the general public," the 1969 memo advised. "It is also the means of establishing a controversy. ... If we are successful in establishing a controversy at the public level, there is an opportunity to put across the real facts about smoking and health. ... If in our pro-cigarette efforts we stick to well documented fact, we can dominate a controversy and operate with the confidence of justifiable self-interest.[8]"

In other words, the Brown & Williamson memo suggested the tobacco industry should fight the scientific consensus that cigarettes cause lung cancer not by disproving the science, but by raising doubt among the public. For the next 35 years, tobacco corporations engaged in a series of deceptive tactics designed to delay further regulations and keep selling cigarettes. Among their claims were that cigarettes are not addictive, low tar and light cigarettes are less hazardous, secondhand smoke is not harmful, nicotine levels can be controlled, and science linking smoking and lung cancer is not conclusive[9]. In fact, smoking-related health conditions are the leading preventable cause of death in the United States, accounting for nearly one in five deaths annually, more than 1,300 deaths per day[10].

What finally stopped the tobacco industry's misinformation campaign were two lawsuits. In 1997, 46 state attorneys general led by Mississippi sued seven tobacco organizations to recover Medicaid costs for treating health problems due to smoking. Under the resulting Tobacco Master Settlement Agreement[11], the tobacco industry was ordered to pay states $206 billion over 25 years, disband three of its lobbying and trade groups, and stop marketing to children[12]. Then in 1999, the Department of Justice sued several major tobacco companies under the Racketeer Influenced and Corrupt Organizations Act, charging the industry had engaged in a decades-long campaign to misrepresent and cover up the ill effects of smoking. Judge Gladys Kessler's 1,683-page opinion[13] found the industry liable, stating "Defendants have engaged in and executed – and continue to engage in and execute – a massive 50-year scheme to defraud the public.[14]"

Although the tobacco industry was eventually brought to heel, it wrote the playbook of deception for other industries that wanted to continue making a profit by selling harmful products without changing their business model. Industries fighting over acid rain, the ozone hole, even flame retardants in furniture used this model to spread confusion about science[15]. But no industry took the tobacco playbook more to heart than fossil fuels – oil, gas, and coal – whose product creates the carbon emissions that are disrupting the climate and threatening all life on earth. And not only did the oil industry lift many of the same tactics from tobacco, it also hired many of the same people to carry them out[16].

---

[7] See a discussion of this memo at Center for Media and Democracy, "Smoking and Health Proposal,"SourceWatch, Tobacco Portal. https://www.sourcewatch.org/index.php/Smoking_and_Health_Proposal
[8] Brown & Williamson, "Smoking and Health Proposal," Truth Tobacco Industry Documents, University of California San Francisco. http://news.gallup.com/poll/9910/tobacco-smoking.aspx
[9] Bates, Clive, and Andy Rowell. "Tobacco Explained: The truth about the tobacco industry in its own words, "Action on Smoking and Health. http://www.who.int/tobacco/media/en/TobaccoExplained.pdf
[10] Centers for Disease Control and Prevention. "Tobacco-Related Mortality." https://www.cdc.gov/tobacco/data_statistics/fact_sheets/health_effects/tobacco_related_mortality/index.htm
[11] National Association of Attorneys General. "Master Settlement Agreement." Archived at https://web.archive.org/web/20080625084126/http://www.naag.org/backpages/naag/tobacco/msa/msa-pdf/1109185724_1032468605_cigmsa.pdf
[12] Office of Attorney General Josh Shapiro, "Summary of Key Points in the Master Settlement Agreement," Commonwealth of Pennsylvania. https://www.attorney-general.gov/Consumers/Tobacco_Enforcement_Section/Summary_of_Key_Points_in_the_Master_Settlement_Agreement/
[13] Judge Kessler's opinion is available at http://www.publichealthlawcenter.org/sites/default/files/resources/doj-final-opinion.pdf
[14] Public Health Law Center, "United States v. Philip Morris (DOJ Lawsuit)," Mitchell Hamline School of Law. http://www.publichealthlawcenter.org/topics/tobacco-control/tobacco-control-litigation/united-states-v-philip-morris-doj-lawsuit
[15] Oreskes, Naomi, and Erik M. Conway, Merchants of Doubt: How a Handful of Scientists Obscured the Truth on Issues from Tobacco Smoke to Global Warming (Bloomsbury Press, 2011).
[16] Oreskes and Conway.

Exxon knew

But it didn't start out this way. Back in the 1970s, not only did Exxon, America's largest fossil fuel corporation, know about climate change, it sponsored much of the original scientific research that showed burning of fossil fuels would lead to climate change. A groundbreaking 2015 investigation by "Inside Climate News", including research into internal company documents and interviews with former employees, found that Exxon conducted cutting-edge climate research from the mid-1970s to the mid-1980s – but then, without calling public attention what it had found, the company began funding climate denial to manufacture doubt about what its own scientists had confirmed[17].

In 1977, Exxon scientist James Black told company executives that emerging evidence showed the world faced grave risks from climate change caused by burning fossil fuels, and Exxon scientist Henry Shaw discussed the environmental effects of carbon dioxide with the Carter administration[18]. To better understand how much carbon was being absorbed by the oceans, Exxon spent over $1 million to outfit one of its supertankers with carbon dioxide sampling equipment, a project that ran from 1979 to 1982[19]. When federal grant money for the tanker project ran out, Exxon turned to climate modeling. An internal 1981 memo predicted that that doubling CO_2 from pre-industrial levels would increase global temperatures by 3° Celsius and over 10°C at the poles – which would be catastrophic for the climate[20].

At the same time, Exxon was planning to develop the Natuna natural gas field in the South China Sea, a project its scientists determined would produce twice the greenhouse gas emissions as the equivalent amount of coal. Exxon scientists studied several options for getting rid of the excess carbon – venting it into the atmosphere, which would cause climate change, dumping it in the ocean, which would release it within a decade, or injecting it into the ground, which was prohibitively expensive. In the end, Exxon's own scientists distributed a 43-page internal company memo that concluded the only way to reduce the greenhouse gas effect was through major reductions in the use of fossil fuels[21].

Climate change goes global

Exxon's scientists were not the only ones to conclude that burning fossil fuels would lead to drastic climate change. In 1988, NASA scientist James Hansen, then director of the Goddard Institute for Space Studies, testified before Congress during a heat wave in Washington that climate change had arrived. To get his point across, Hansen held up a pair of large cardboard dice to represent the climate during the previous 30 years, with two panels in white for an average summer, two panels in blue for a cooler than normal summer, and two panels in red for a hotter than average summer. Then he held up a pair of dice with four panels in red to represent climate for the next 30 years. "It is obvious that the man in the street will notice the dice are loaded," Hansen said[22].

That same year, the Intergovernmental Panel on Climate Change, or IPCC, was established at the request of world governments with the purpose of assessing "the scientific, technical, and socioeconomic information relevant for the understanding of the risk of human-induced climate change[23]." The IPCC grew out of the Advisory Group on Greenhouse Gases, set up in 1985 by the International Council of Scientific Unions, United

[17] Banerjee, Neela, Lisa Song, and David Hasemyer, "Exxon: The Road Not Taken," InsideClimate News, September 16, 2015. https://insideclimatenews.org/content/exxon-the-road-not-taken

[18] Banerjee, Neela, Lisa Song, and David Hasemyer, "Exxon's Own Research Confirmed Fossil Fuels' Role in Global Warming Decades Ago," InsideClimate News, September 16, 2015. https://insideclimatenews.org/news/15092015/Exxons-own-research-confirmed-fossil-fuels-role-in-global-warming

[19] Banerjee, Neela, Lisa Song, and David Hasemyer, "Exxon Believed Deep Dive Into Climate Research Would Protect Its Business," InsideClimate News, September 17, 2015. https://insideclimatenews.org/news/16092015/exxon-believed-deep-dive-into-climate-research-would-protect-its-business

[20] Banerjee, Song and Hasemyer, "Exxon's Own Research Confirmed Fossil Fuels' Role in Global Warming Decades Ago."

[21] Banerjee, Neela, and Lisa Song, "Exxon's Business Ambition Collided with Climate Change Under a Distant Sea," October 8, 2015. https://insideclimatenews.org/news/08102015/Exxons-Business-Ambition-Collided-with-Climate-Change-Under-a-Distant-Sea

[22] Shabecoff, Philip, "Global Warming Has Begun, Expert Tells Senate," The New York Times, June 24, 1988. http://www.nytimes.com/1988/06/24/us/global-warming-has-begun-expert-tells-senate.html?pagewanted=all

[23] IPCC Facts, "IPCC History," United Nations Foundation. http://www.ipccfacts.org/history.html

Nations Environment Programme, and World Meteorological Organization. Due to pressure from the Reagan administration, the IPCC was designed to be independent from the United Nations, with leading climate scientists from all over the world, and their reports had to be approved by every participating government before being released[24].

Despite such strict approval requirements, the IPCC has produced a series of reports on climate change, starting with the First Assessment Report in 1990, Second Assessment Report in 1995, Third Assessment Report in 2001, Fourth Assessment Report in 2007, and Fifth Assessment Report in 2013[25]. Thousands of scientists and other experts contribute to writing and reviewing reports on various aspects of climate change. The IPCC does not do any original research; rather, it pulls together all existing climate research. The most scrutinized part of each report is the Summary for Policymakers, which must get line-by-line approval by all delegates and participating governments[26].

IPCC reports support the United Nations Framework Convention on Climate Change, or the UNFCCC, which is the main international treaty on climate change. Introduced at the Earth Summit in Rio in 1992, the chief objective of the UNFCCC is "stabilization of greenhouse gas concentrations in the atmosphere at a level that would prevent dangerous anthropogenic interference with the climate system." In practical terms, this means peaking and then reducing carbon emissions in a way that allows ecosystems to adapt naturally to climate change, doesn't threaten food production, and enables economic systems to develop in a sustainable manner[27]. Parties to the convention have met annually to assess progress since 1995.

Even without today's sophisticated climate modelling equipment, projections from the IPCC's First Assessment Report in 1990 were remarkably accurate[28]. The report stated that "emissions resulting from human activities are substantially increasing the atmospheric concentrations of the greenhouse gases carbon dioxide, methane, chlorofluorocarbons (CFCs) and nitrous oxide. These increases will enhance the greenhouse effect, resulting on average in an additional warming of the Earth's surface." Specifically, the authors predicted that under business as usual, in which countries do nothing to lower carbon emissions, average global temperatures would rise 0.3°C per decade above pre-industrial times, resulting in a likely increase of 1°C by 2025 and 3°C by 2100. Other scenarios that assumed increasing control over carbon emissions projected temperature rises of 0.2°C or 0.1°C per decade. Concentration of CO_2 in the atmosphere was thought to hit about 425 parts per million by 2025 and 825 ppm by 2100, with 400 ppm to 500 ppm for lower emission scenarios. In reality, as of 2018 temperatures have risen almost 1.2°C with 411 ppm of CO_2[29].

Launching denial

The world's meetings and discussions on what to do about global climate change did not escape the notice of Exxon and other fossil fuel corporations. In a cycle that has repeated itself over the past three decades, whenever the UNFCCC began to reach an agreement or the U.S. government came close to passing a law that would require lowering carbon emissions, the denial campaign would swing into action. This began in 1989 when Exxon, under new president Lee Raymond, joined with the American Petroleum Institute, Chevron, Shell, Texaco, Amoco, National Coal Association, Cyprus AMAX Minerals, Ford, General Motors, Edison Electric, American Forest and

[24] Weart, Spencer, "The Discovery of Global Warming: International Cooperation" Center for History of Physics, American Institute of Physics. https://history.aip.org/history/climate/internat.htm
[25] Find all the IPCC Assessment Reports at https://www.ipcc.ch/publications_and_data/publications_and_data_reports.shtml
[26] Intergovernmental Panel on Climate Change, "Understanding Climate Change: 22 years of IPCC assessment," November 2010. https://www.ipcc.ch/pdf/press/ipcc_leaflets_2010/ipcc-brochure_understanding.pdf
[27] United Nations, "United Nations Framework Convention on Climate Change," 1992. https://unfccc.int/resource/docs/convkp/conveng.pdf
[28] Nuccitelli, Dana, "Contrary to Contrarian Claims, IPCC Temperature Projections Have Been Exceptionally Accurate," Skeptical Science, December 27, 2012. https://skepticalscience.com/contary-to-contrarians-ipcc-temp-projections-accurate.html
[29] IPCC Working Group I, "Policymakers Summary," First Assessment Report. https://www.ipcc.ch/ipccreports/far/wg_I/ipcc_far_wg_I_spm.pdf

Paper Association, U.S. Chamber of Commerce, National Association of Manufacturers, and more than 40 other corporations to form the Global Climate Coalition[30].

As the largest industry group active in climate policy, the GCC's official mission was "to coordinate business participation in the international policy debate on the issue of global climate change and global warming[31]." What that meant in practical terms was running campaigns designed to create public opposition to policies that would require reductions in greenhouse gas emissions. From 1989 to 2002, the GCC led aggressive lobbying and advertising efforts aimed at raising doubt about the integrity of the IPCC and the scientific evidence that carbon emissions from burning fossil fuels cause global warming. The GCC lobbied government officials, sent out press releases, ran advertising, participated in international climate conferences, critiqued climate models, and engaged in personal attacks on climate scientists[32]. Its policy positions included outright denial of climate change, emphasizing uncertainty in climate science, advocating for additional research, highlighting the benefits and downplaying the risks of climate change, and stressing the economic costs of proposed regulations[33].

Among the GCC's actions was an attack on Benjamin Santer of the Program for Climate Model Diagnosis and Intercomparison at Lawrence Livermore National Laboratory in California, a story told by Naomi Oreskes and Erik Conway in their landmark book *Merchants of Doubt*[34]. Santer was lead author of the chapter on detection and attribution for the IPCC's Second Assessment Report – the chapter that discussed how scientists know that climate change is happening and what is causing it[35]. The chapter, drafted with 31 participants[36], used a "fingerprint" method to determine the cause of climate change. If the cause was the sun, then warming would start at the top of the atmosphere and work its way down. But if the cause was greenhouse gas emissions trapping heat on the surface, the lower atmosphere would warm up while the upper atmosphere would remain cool. The latter is the pattern Santer's lab found.

During the process of writing and revising the chapter, an early version was leaked. The chapter went through three more rounds of review and commentary by scientists and governments, including a debate over whether the human influence on global climate should be described as "appreciable" or "discernible." Santer also removed a summary at the end of the chapter so it would have the same format as other chapters in the Second Assessment Report.

That was enough for the deniers to attack. The GCC issued a report accusing the IPCC of "scientific cleansing" and claiming Santer had altered the text without approval to downplay uncertainty about the science. Frederick Seitz, former president of the National Academy of Sciences and a central figure among climate change deniers, repeated these charges in an op-ed in *The Wall Street Journal* calling the revisions a "disturbing corruption of the peer review process[37]." Santer countered in *The New York Times* that the chapter contained several pages discussing uncertainty, and the revisions were not politically motivated but the result of peer review[38]. Still, the GCC ran newspaper ads claiming that unless the IPCC "promptly undertakes to republish the printed versions ... the IPCC's credibility will have been lost.[39]"

[30] Membership lists for the Global Climate Coalition can be found at DeSmogBlog, https://www.desmogblog.com/global-climate-coalition; SourceWatch, https://www.sourcewatch.org/index.php/Global_Climate_Coalition; and ExxonSecrets, https://exxonsecrets.org/html/orgfactsheet.php?id=38

[31] "About Us," Global Climate Coalition website, archived at https://web.archive.org/web/20010210211514/http://www.globalclimate.org:80/aboutus.htm

[32] Lieberman, Amy, and Susanne Rust, "Big Oil braced for global warming while it fought regulations," Los Angeles Times, December 31, 2015. http://graphics.latimes.com/oil-operations/

[33] Mooney, Chris, "Some Like it Hot," MotherJones, May/June 2005. http://www.motherjones.com/environment/2005/05/some-it-hot/

[34] Oreskes and Conway, pp. 197-213.

[35] Santer, Benjamin, et al., "Detection of Climate Change and Attribution of Causes," Climate Change 1995: The Science of Climate Change, Contribution of Working Group I to the Second Assessment Report of the Intergovernmental Panel on Climate Change, pp. 407-444. Accessed at https://www.ipcc.ch/ipccreports/sar/wg_I/ipcc_sar_wg_I_full_report.pdf

[36] The authors of Chapter 8 are listed at Climate Change 1995: The Science of Climate Change, Contribution of Working Group I to the Second Assessment Report of the Intergovernmental Panel on Climate Change, p. 547.

[37] Seitz, Frederick, "A Major Deception on Global Warming," The Wall Street Journal, June 12, 1996. https://www.wsj.com/articles/SB834512411338954000

[38] Stevens, William K., "At Hot Center of Debate on Global Warming," *The New York Times,* August 6, 1996.

[39] Levy, David L., and Sandra Rothenberg, "Corporate Strategy and Climate Change: Heterogeneity and Change in the Global Automobile Industry," Global Environmental Assessment Project, Belfer Center for Science and International Affairs, October 1999.

The GCC was also instrumental in keeping the United States from joining the Kyoto Protocol after it was negotiated in 1997. The Kyoto Protocol extended the framework of the 1992 Earth Summit in Rio by establishing legally binding commitments to reduce greenhouse gas emissions for 37 industrialized nations and the European Community[40]. The U.S. commitment, had it been ratified, would have been to reduce emissions 7 percent below 1990 levels by 2012[41]. Although the Vice President Al Gore signed the Kyoto Protocol 1998[42], the Clinton administration did not submit it for ratification in the Senate, which had already passed the Byrd-Hagel resolution disapproving any international agreement that "would result in serious harm to the economy of the United States" by not also requiring developing countries to make emission reductions[43]. In 2000, George W. Bush was elected president; one of his earliest acts was to reject to Kyoto Protocol for good[44], prompting international condemnation[45].

Although the GCC had previously budgeted about $1 million a year on campaigning against carbon regulations, starting in 1997 it spent $13 million in opposition to the Kyoto Protocol[46]. The coalition funded the Global Climate Information Project, which hired the same public relations firm that created the notorious "Harry and Louise" ads credited with defeating the Clinton health care proposal. Using the theme "It's not global and it won't work," the anti-Kyoto ads claimed that Americans would pay "50 cents more for every gallon of gasoline" even though there was no proposal for such a tax[47]. The GCC was also behind passage of the Byrd-Hagel resolution, lobbying not only Sen. Robert Byrd, a Democrat, and Chuck Hagel, a Republican, but meeting with members of Congress to get a vote of 95-0[48]. Later the GCC held a secret meeting with Undersecretary of State Paula Dobriansky. "POTUS rejected Kyoto, in part, based on input from you," Dobriansky told the GCC[49].

By this time, the GCC had increasingly been recognized as an industry front group, and campaigns by media and environmentalist groups convinced many of its member corporations to leave the coalition. In 1997 British Petroleum and DuPont withdrew; the next year Shell followed. Ford withdrew in 1999, followed by Chrysler, Texaco, Southern Company, and General Motors in 2000[50]. By 2001, the GCC disbanded, claiming it had achieved its mission after the United States withdrew from Kyoto[51].

The Exxon echo chamber

The downfall of the Global Climate Coalition did not end Exxon's climate denial campaign; on the contrary, Exxon was just getting started. In 1998, Exxon (now merged with Mobil Corporation) created a Global Climate Science Communications Team that included Randy Randol, senior lobbyist; Joe Walker, public relations representative of the American Petroleum Institute; and Steve Milloy from Advance of Sound Science Coalition, which had been created by Philip Morris to manufacture doubt about risks from secondhand smoke[52]. A memo

[40] UNFCCC, "Fact Sheet: The Kyoto Protocol." https://unfccc.int/files/press/backgrounders/application/pdf/fact_sheet_the_kyoto_protocol.pdf

[41] All Politics, "Clinton Hails Global Warming Pact," CNN/Time, December 11, 1997. http://www.cnn.com/ALLPOLITICS/1997/12/11/kyoto/

[42] Cushman, John H., "U.S. Signs a Pact to Reduce Gases Tied to Warming," The New York Times, November 13, 1998. http://www.nytimes.com/1998/11/13/world/us-signs-a-pact-to-reduce-gases-tied-to-warming.html

[43] Byrd-Hagel Resolution, Senate Resolution 98, 105th Congress. Archived at https://web.archive.org/web/20100626110143/http://www.nationalcenter.org/KyotoSenate.html

[44] Borger, Julian, "Bush kills global warming treaty," The Guardian, March 29, 2001. https://www.theguardian.com/environment/2001/mar/29/globalwarming.usnews

[45] Sanger, David E., "Bush will Continue to Oppose Kyoto Pact on Global Warming," The New York Times, June 12, 2001. http://www.nytimes.com/2001/06/12/world/bush-will-continue-to-oppose-kyoto-pact-on-global-warming.html

[46] Farley, Maggie, "Showdown at Global Warming Summit," Los Angeles Times, December 7, 1997. Archived at https://web.archive.org/web/20160118234039/http://articles.latimes.com/1997/dec/07/news/mn-61743/2

[47] SourceWatch, "Global Climate Information Project." https://www.sourcewatch.org/index.php/Global_Climate_Information_Project

[48] van den Hove, Sybille, Marc Le Menestrel, and Henri-Claude De Bettignies, "The oil industry and climate change: Strategies and ethical dilemmas," Climate Policy, 2: 2002, pp. 3-18.

[49] Greenpeace Research, Briefing Memorandum, U.S. Department of State, Global Climate Coalition Meeting (06-21-01). http://www.greenpeace.org/usa/wp-content/uploads/legacy/Global/usa/report/2009/10/global-climate-coalition-meeti.pdf

[50] Brown, Lester, "The Rise and Fall of the Global Climate Coalition," Worldwatch Issue Alert, Earth Policy Institute, July 25, 2000. Archived at https://web.archive.org/web/20020202133543/http://www.worldwatch.org/chairman/issue/000725.html

[51] Home page, Global Climate Coalition website, archived March 30, 2002, at http://web.archive.org/web/20020330015411/http://globalclimate.org/

[52] Shulman, Seth, et al., "Smoke, Mirrors and Hot Air: How ExxonMobil Uses Big Tobacco's Tactics to Manufacture Uncertainty on Climate Science," Union of Concerned Scientists, January 2007, pp. 9-10. https://www.ucsusa.org/sites/default/files/legacy/assets/documents/global_warming/exxon_report.pdf

from this team outlined its strategy: "Victory will be achieved when average citizens understand (recognize) uncertainties in climate science" and when "recognition of uncertainties becomes part of the 'conventional wisdom.[53]'"

The team's communications plan included recruiting "independent scientists," developing "papers that undercut the 'conventional wisdom' on climate science," and distributing a "steady stream of climate science information." It also called for spending $5 million over two years to "maximize the impact of scientific views consistent with ours on Congress, the media and other key audiences.[54]" To achieve these goals, ExxonMobil ratcheted up funding for climate denial, ultimately sending $16 million to dozens of front groups between 1998 and 2005, in many cases accounting for more than 10 percent of their budgets[55]. These front groups employed many of the same spokespeople and touted much of the same junk science in order to create an "echo chamber" that amplified and sustained misinformation about climate change, climate science, and climate policy.

Among the front groups supported was Frontiers of Freedom, for which ExxonMobil provided more than $1 million from 1998 to 2005[56]. Its 2002 grant of $232,000, nearly a third of the annual budget, went to fund the organization's Center for Science and Public Policy, which promoted a non-peer-reviewed report by executive director Robert Ferguson called "Issues in the Current State of Climate Science: A Guide for Policy Makers and Opinion Leaders.[57]" This publication pushed many typical fallacies of climate deniers, such as:

- Current temperature changes are not abnormal.

- Current storms are not more extreme than usual.

- Elevated carbon dioxide helps plants.

- Animal species are expanding rather than going extinct.

- Coral bleaching is normal and reefs are in excellent health.

- Ice sheets in Greenland and Antarctic are growing.

- Glacier retreat and sea level rise cannot be established.

- Mosquito-borne diseases are not on the rise[58].

Among the references cited by Ferguson was an article by Willie Soon, a scientist at the Harvard-Smithsonian Center for Astrophysics who claimed that the sun, rather than carbon dioxide, is responsible for changes to the climate. Soon has taken more than $1.2 million from the fossil fuel industry for his research[59]. In this paper, Soon argued that temperature anomalies in the Arctic are due to solar irradiance[60], a claim that practically no

[53] Sawin, Janiet, Kert Davies, et al., "Denial and Deception: A Chronicle of ExxonMobil's Efforts to Corrupt the Debate on Global Warming," Greenpeace, May 2002, p. 4. http://www.greenpeace.org/usa/wp-content/uploads/2015/11/exxon-denial-and-deception.pdf

[54] Cushman, John H. Jr., "Industrial Group Plans to Battle Climate Treaty," The New York Times, April 26, 1998. http://www.nytimes.com/1998/04/26/us/industrial-group-plans-to-battle-climate-treaty.html

[55] "Smoke, Mirrors and Hot Air," pp. 10-11.

[56] "Smoke, Mirrors and Hot Air," p. 31.

[57] The report is no longer available on the Frontiers of Freedom website, but can be found at http://ilovemycarbondioxide.com/archives/Climate_Science.pdf

[58] Ferguson, Robert, "Issues in the Current State of Climate Science: A Guide for Policy Makers and Opinion Leaders," Center for Science and Public Policy, March 2006, pp. 4-24.

[59] Gillis, Justin, and John Schwartz, "Deeper Ties to Corporate Cash for Doubtful Climate Researcher," The New York Times, February 21, 2015. https://www.nytimes.com/2015/02/22/us/ties-to-corporate-cash-for-climate-change-researcher-Wei-Hock-Soon.html

[60] Soon, Willie W.-H., "Variable solar irradiance as a plausible agent for multidecadal variations in the Arctic-wide surface air temperature record of the past 130 years," Geophysical Research Letters, August 27, 2005. http://onlinelibrary.wiley.com/doi/10.1029/2005GL023429/full

other climate scientist agrees with[61]. The research was funded by the ExxonMobil Corporation, along with the Charles G. Koch Foundation and American Petroleum Institute. In a previous paper co-authored with colleague Sallie Baliunas, Soon alleged that the 20th century was not the warmest in the past 1,000 years nor unique in the climate record[62]. Sen. James Inhofe (R-Okla.), cited the paper to claim climate change is a due to natural causes, not human activity[63]. Three editors of *Climate Research*, which published the paper, resigned over it[64], and 13 scientists cited in the paper said Soon and Baliunas had misinterpreted their research[65]. The research was funded by American Petroleum Institute, with consultation from Craig and Keith Idso of Center for the Study of Carbon Dioxide and Global Change, another Exxon-funded front group[66].

Another organization funded by ExxonMobil was the George C. Marshall Institute. Founded in 1984 to advocate for President Reagan's Strategic Defense Initiative, or "Star Wars," GMI became a home base for climate denial, supported with $630,000 in grants from ExxonMobil. In 1997, its chairman Frederick Seitz along with the Oregon Institute for Science and Medicine launched the "Oregon Petition," a document claiming to have signatures of 30,000 scientists who do not believe in manmade climate change. In fact, less than 1/10th of 1 percent of signers had a background in climatology[67], and the petition was riddled with fake names such as fictional characters from the TV show MASH, the fictional attorney Perry Mason, and one of the Spice Girls[68]. The petition was formatted to look like an article from *Proceedings of the National Academy of Sciences*, of which Sietz had once been president. In response, the National Academy of Sciences issued a statement denouncing the petition and disassociating themselves from it[69]. Seitz was another prominent climate denier with roots in the tobacco industry. While president of Rockefeller University in the 1970s, he became a permanent consultant for R.J. Reynolds Tobacco Company, overseeing their medical research program until 1988[70].

Also funded by ExxonMobil was the Heartland Institute, a Chicago-based free market think tank that between 1998 and 2006 received $676,500[71], much of it designated for climate denial projects[72]. Among Heartland's actions during this period were to create a website, www.globalwarmingheartland.org, that asserts there is no scientific consensus on climate change, and to distribute "Energy Policy for America: A Guidebook for State Legislators," which advised legislators to oppose carbon taxes and cap and trade programs[73]. Like so many, Heartland also got its start with the tobacco industry, working with Philip Morris in the 1990s to question

---

[61] See a list of studies showing no relationship between solar irradiance and temperatures on earth at "Sun and climate: moving in opposite directions," Skeptical Science. https://www.skepticalscience.com/solar-activity-sunspots-global-warming-intermediate.htm. Other discussion of the topic are at "A blanket around the Earth," Global Climate Change: Vital Signs of the Planet, NASA, https://climate.nasa.gov/causes/, and "How Does the Sun Affect Our Climate?", Union of Concerned Scientists, https://www.ucsusa.org/global-warming/science-and-impacts/science/effect-of-sun-on-climate-faq.html

[62] Soon, Willie, and Sallie Baliunas, "Proxy climatic and environmental changes of the past 1000 years," *Climate Research,* 23: 2003, pp. 89-110. http://www.int-res.com/articles/cr2003/23/c023p089.pdf

[63] Mooney, Chris, "James Inhofe proves 'flat Earth' doesn't refer to Oklahoma," *The American Prospect*, May 7, 2004. Reposted at http://www.heatisonline.org/contentserver/objecthandlers/index.cfm?ID=4647&Method=Full

[64] More on the Climate Research controversy can be found in "Smoke, Mirrors, and Hot Air," p. 15.

[65] American Geophysical Union, "Leading climate scientists reaffirm view that late 20th century warming was unusual and resulted from human activity," July 7, 2003. http://www.agu.org/sci_soc/prrl/prrl0319.html

[66] Global Warming Disinformation Database, "Center for the Study of Carbon Dioxide and Global Change," DeSmog Blog. https://www.desmogblog.com/center-study-carbon-dioxide-and-global-change

[67] Grandia, Kevin, "The 30,000 Global Warming Petition Is Easily-Debunked Propaganda," *Huffington Post*, May 25, 2011. https://web.archive.org/web/20150815222009/http://www.huffingtonpost.com/kevin-grandia/the-30000-global-warming_b_243092.html

[68] Hebert, Josef H., "Odd Names Added to Greenhouse Plea," Associated Press, May 1, 1998. http://www.apnewsarchive.com/1998/Odd-Names-Added-to-Greenhouse-Plea/id-aec8beea85d7fe76fc9cc77b8392d79e. Also see Curry, Rex, "Exxon's Climate Denial History: A Timeline," Greenpeace, http://www.greenpeace.org/usa/global-warming/exxon-and-the-oil-industry-knew-about-climate-change/exxons-climate-denial-history-a-timeline/

[69] Stevens, William K. "Science academy disputes attack on global warming," *The New York Times*, April 22, 1998. http://www.nytimes.com/1998/04/22/us/science-academy-disputes-attack-on-global-warming.html

[70] Oreskes and Conway, pp. 10-11. Also Hevesi, Dennis, "Frederick Seitz, 96, Dies; Physicist Who Led Skeptics of Global Warming," *The New York Times*, March 6, 2008. http://www.nytimes.com/2008/03/06/us/06seitz.html

[71] ExxonSecrets, "Factsheet: Heartland Institute," Greenpeace. https://exxonsecrets.org/html/orgfactsheet.php

[72] "Smoke, Mirrors, and Hot Air," p. 32.

[73] Bast, Joseph L. "Ten Principles of Energy Policy," Heartland Institute, p. 11. https://www.desmogblog.com/sites/beta.desmogblog.com/files/bast-Ten-Principles-of-Energy-Policy.pdf

the science linking smoking to lung cancer. Heartland had a board member from Philip Morris until 2008 and maintained a "Smoker's Lounge[74]" on its website until 2016[75].

The American Legislative Exchange Council (ALEC), an organization designed to link state legislators with corporations and create templates for state legislation, received more than $1.6 million from ExxonMobil[76], much of it earmarked for climate denial projects. In 2002 ALEC disseminated a report by the CATO Institute's Patrick Michaels that argued the Kyoto Protocol would have no detectable impact on the Earth's temperature but would impose grave economic damage on the United States[77]. ALEC also published *Energy, Environment, and Economics*, a guidebook to explain climate change to busy legislators and provide sample legislation for their states. The publication claimed climate science is unsettled, computer models are inaccurate, increased carbon is not causing glaciers to retreat, higher temperatures are not causing sea levels to rise, global warming is not leading to more extreme storms, polar bears can adapt to melting ice, and temperature drives carbon dioxide levels, not the other way around[78]. To deal with these supposed uncertainties, ALEC provided model legislation that would prohibit control of carbon emissions by citizens or corporations, promote offshore oil and gas drilling, and give states sovereignty over regulation of air pollution[79]. ALEC's most troubling model legislation was the Animal and Ecological Terrorism Act, which defines "any politically motivated activity intended to obstruct or deter any person from participating in an activity involving animals or ... natural resources" as terrorism[80]. The AETA was signed into law in 2006[81].

The most prominent organization funded from ExxonMobil was the Competitive Enterprise Institute, which received more than $2 million from 1998 to 2005[82]. Founded in 1984 to fight government regulation, CEI began attracting significant grants from ExxonMobil after economist Myron Ebell moved there from Frontiers of Freedom in 1999. Ebell proudly touted his climate denial on his CEI staff page, noting that he has been named "enemy #1 to the current climate change community" by Business Insider[83] and "Villain of the Month" by the Clean Air Trust for successfully lobbying President Bush to reverse a campaign pledge to control carbon emissions from electric utilities[84]. At CEI, Ebell was director of energy and environment and chair of the Cooler Heads Coalition, a group of organizations that "question global warming alarmism and oppose energy rationing policies.[85]" Members included all the ExxonMobil-funded organizations discussed so far, along with a host of others such as Committee for a Constructive Tomorrow, Fraser Institute, JunkScience.com, and more[86]. Among other things, the coalition lobbied successfully to defeat the McCain-Lieberman Climate Stewardship Act of 2003 and 2005[87], and Waxman-Markey's American Clean Energy and Securities Act of 2009[88].

---

[74] See a capture from May 11, 2016, of the Smoker's Lounge page on the Heartland Institute website at https://web.archive.org/web/20160511152708/https://www.heartland.org/policy-documents/welcome-heartlands-smokers-lounge

[75] Global Warming Disinformation Database, "Heartland Institute," DeSmog Blog. https://www.desmogblog.com/heartland-institute

[76] ExxonSecrets, "Factsheet: ALEC – American Legislative Exchange Council," Greenpeace. https://exxonsecrets.org/html/orgfactsheet.php?id=10

[77] Michaels, Patrick J., "Global Warming and the Kyoto Protocol: Paper Tiger, Economic Dragon," The State Factor, American Legislative Exchange Council, 2002. https://web.archive.org/web/20030325103312/http://www.alec.org/meSWFiles/pdf/0208.pdf

[78] Simmons, David R., "Climate Change Overview for State Legislators," *Energy, Environment, and Economics*, American Legislative Exchange Council, 2007, pp. 12-15. https://www.sourcewatch.org/images/7/7c/Energy5th.pdf

[79] Energy, Environment, and Economics, pp. 66-88.

[80] Holland, Joshua, "Creating a Right-Wing Nation, State by state," Alternet, November 15, 2005. https://www.alternet.org/story/28259/creating_a_right-wing_nation%2C_state_by_state

[81] Civil Liberties Defense Center, "Animal Enterprise Terrorism Act (AETA)," https://cldc.org/resources/animal-enterprise-terrorism-act-aeta/

[82] ExxonSecrets, "Factsheet: Competitive Enterprise Institute CEI," Greenpeace. https://exxonsecrets.org/html/orgfactsheet.php?id=2

[83] Wiesenthal, Joe, "The 10 Most-Respected Global Warming skeptics," Business Insider, July 31, 2009. https://www.businessinsider.com.au/the-ten-most-important-climate-change-skeptics-2009-7#myron-ebell-3

[84] Clean Air Trust, "Trust Names 'Cooler Heads' Chairman Clean Air 'Villain of The Month," March 2001. https://archive.is/P2yXa

[85] Competitive Enterprise Institute, "Myron Ebell," December 29, 2011. https://archive.is/mxjzo

[86] Climate Investigations Center, "Cooler Heads Coaltion." http://climateinvestigations.org/climate-deniers/cooler-heads-coalition/

[87] O'Harrow, Robert Jr., "A two-decade crusade by conservative charities fueled Trump's exit from Paris climate accord," Washington Post, September 5, 2017. https://www.washingtonpost.com/investigations/a-two-decade-crusade-by-conservative-charities-fueled-trumps-exit-from-paris-climate-accord/2017/09/05/fcb8d9fe-6726-11e7-9928-22d00a47778f_story.html

[88] Samuelsohn, Darren, "Lobbying Frenzy Begins as House Climate Bill Heads for Floor," The New York Times, June 23, 2009. https://archive.is/A0gmp#selection-403.1-418.0

Influencing policy

Funding front groups was not the only way ExxonMobil influenced climate policy. They also donated money to political candidates and lobbied government officials directly. From 2000 to 2006, ExxonMobil's PAC and employees gave more than $4 million to federal candidates and parties[89]. After the George W. Bush took office in 2001, ExxonMobil participated in Vice President Dick Cheney's Energy Task Force[90], and through the Global Climate Coalition successfully lobbied the Bush administration to withdraw from the Kyoto Protocol. But ExxonMobil's influence in the Bush administration did not stop with public policy. It also extended to personnel matters and even the administration's scientific documents.

In 2001, just after publication of the IPCC Third Assessment Report[91], ExxonMobil successfully lobbied the Bush administration to replace IPCC chair Robert Watson with Indian engineer-economist Rajendra Pachauri. Watson, chief scientist at the World Bank, strongly advocated for the idea that burning fossil fuels caused climate change[92]. "Can Watson be replaced now at the request of the U.S.?" wrote ExxonMobil lobbyist Randy Randol to the White House. By 2002, Robert Watson was out[93]. ExxonMobil also recommended that House Speaker Dennis Hastert send Harlan Watson (no relation), a staff member with the House Science Committee, to attend international meetings on climate change[94]. Bush then appointed Harlan Watson, who opposed U.S. engagement in the Kyoto Protocol, as lead climate negotiator[95].

Even more startling was the influence Exxon had over reports about climate change from the U.S. Climate Change Science Program, established by Congress in 1989 to evaluate and synthesize information from a dozen federal agencies about the impacts of climate change in the United States. ExxonMobil objected to the conclusion of the program's 2000 report that found climate change is caused by human activity and poses a threat[96]. Phil Cooney, chief of staff at the White House Council on Environmental Quality, reached out to Myron Ebell of Competitive Enterprise Institute for advice, and was told the administration should distance itself from the report and set up Christine Todd Whitman, head of the EPA, as the fall guy[97]. The Bush administration publicly denigrated the report[98] and ousted Whitman within a year[99]. In 2005 Rick Piltz, senior associate at program, filed whistleblower testimony[100] stating that Cooney had made changes to the program's scientific reports to play down the connection between greenhouse gases and climate change, and play up any scientific uncertainty. Cooney, previously climate team leader for the American Petroleum Institute, had no training in science[101]. Two months after his interference became public, he resigned to go work for ExxonMobil[102].

---

[89] "Smoke, Mirrors, and Hot Air," p. 19.

[90] Milbank, Dana, and Justin Blum, "Document Says Oil Chiefs Met With Cheney Task Force," *Washington Post*, November 16, 2005. http://www.washingtonpost.com/wp-dyn/content/article/2005/11/15/AR2005111501842.html

[91] Working Group I, Climate Change 2001: The Scientific Basis, Intergovernmental Panel on Climate Change, 2001. https://www.ipcc.ch/ipccreports/tar/wg1/index.php?idp=0

[92] Revkin, Andrew C., "Dispute Arises Over a Push to Change Climate Panel," *The New York Times*, April 2, 2002. http://www.nytimes.com/2002/04/02/world/dispute-arises-over-a-push-to-change-climate-panel.html

[93] Lawler, Andrew, "Battle Over IPCC Chair Renews Debate on U.S. Climate Policy," Science, April 12, 2002. http://science.sciencemag.org/content/296/5566/232.1.full

[94] Eilperin, Juliet, "Climate Official's Work is Questioned," The Washington Post, December 5, 2005. http://www.washingtonpost.com/wp-dyn/content/article/2005/12/04/AR2005120400891.html

[95] "Smoke, Mirrors, and Hot Air," p. 20.

[96] U.S. Global Change Research Program, "Climate Change Impacts on the United States: The Potential Consequences of Climate Variability and Change." 2000. https://www.globalchange.gov/browse/reports/climate-change-impacts-united-states-potential-consequences-climate-variability-and

[97] "Smoke, Mirrors, and Hot Air," pp. 21-22.

[98] Vries, Lloyd, "Bush Disses Global Warming Report," CBS News, June 3, 2002.

[99] Seelye, Katharine Q., "Often isolated, Whitman Quits as EPA Chief," *The New York Times*, May 22, 2003. http://www.nytimes.com/2003/05/22/us/often-isolated-whitman-quits-as-epa-chief.html

[100] Piltz, Rick, "On Issues of Concern About the Governance and Direction of the Climate Change Science Program," Climate Science and Policy Watch, June 2, 2005. http://www.climatesciencewatch.org/2005/06/02/on-issues-of-concern-about-the-governance-and-direction-of-the-climate-change-science-program/

[101] Revkin, Andrew, "Bush Aide Softened Greenhouse Gas Links to Global Warming," *The New York Times*, June 8, 2005. http://www.nytimes.com/2005/06/08/politics/bush-aide-softened-greenhouse-gas-links-to-global-warming.html

[102] Revkin, Andrew, "Former Bush Aide Who Edited Reports Is Hired by Exxon," *The New York Times*, June 15, 2005. http://www.nytimes.com/2005/06/15/politics/former-bush-aide-who-edited-reports-is-hired-by-exxon.html

Exxon also cultivated relationships with members of Congress. Joe Barton (R-Texas), received more than $1 million from the oil and gas industry from 2000 to 2006. He sponsored the Energy Policy Act of 2005, which provided subsidies for drilling in the Gulf of Mexico and exempted fracking from the Safe Drinking Water Act[103]. As chair of the House Science Committee, Barton launched an investigation into the work of Michael Mann, Raymond S. Bradley and Malcolm K. Hughes, the scientists behind the famous "hockey stick" study that used evidence from tree rings, glaciers, and coral growth to show that temperatures have skyrocketed in the last century compared to the five centuries previous[104]. Barton sought all data related to their research and held congressional hearings stacked with climate denier witnesses such as John Christy and Steven McIntyre, a mining executive with the George C. Marshall Institute, to attack their findings[105]. Twenty leading climate scientists sent a letter of concern, and Ralph Ciccerone, president of the National Academy of Sciences, offered to appoint an independent panel to assess the research, an idea Barton dismissed[106].

Exxon gets out

Then suddenly ExxonMobil disavowed its own climate denial. "In 2008 we will discontinue contributions to several public policy groups whose position on climate change could divert attention from the important discussion on how the world will secure energy required for economic growth in an environmentally responsible manner," Exxon said in its 2007 Corporate Citizenship report[107]. What caused ExxonMobil to back off the climate denial campaigns it had so ardently supported for almost a decade? Rex Tillerson, who took over as chairman and CEO in 2006, often gets credit for this shift, but other factors were more important.

First was public pressure from scientific and environmental groups. In September 2006, the Royal Society, Britain's most prestigious scientific society, sent an unprecedented letter to ExxonMobil, accusing the corporation of funding 39 climate denial front groups and asking it to stop making "inaccurate and misleading" statements about global warming[108]. Greenpeace also launched campaigns against ExxonMobil's climate denial, issuing reports[109], feature stories[110], and its ExxonSecrets database in 2004[111]. And in 2007 the Union of Concerned Scientists published a comprehensive report, "Smoke, Mirrors, and Hot Air: How ExxonMobil Uses Big Tobacco's Tactics to Manufacture Uncertainty on Climate Science," which found that from 1998 to 2005, ExxonMobil spent at least $15.8 million funding 43 climate denial organizations[112].

Second was pressure from shareholders, particularly the Rockefeller family that inherited a fortune from Exxon's forerunner, Standard Oil. In 2004 the family began privately trying to persuade ExxonMobil to stop funding climate denial and take a more proactive approach to climate change. In 2008 they went public, holding a press conference to express their concern and sponsoring several shareholder resolutions regarding climate change, including one to strip Tillerson of the chairman title and install an independent chairman of the board[113].

[103] Grunwald, Michael, and Juliet Eilperin, "Energy Bill Raises Fears About Pollution, Fraud," *Washington Post*, July 30, 2005.

[104] Eilperin, Juliet, "GOP Chairmen Face off on Global Warming," Washington Post, July 18, 2005. http://www.washingtonpost.com/wp-dyn/content/article/2005/07/17/AR2005071701056.html

[105] "Smoke, Mirrors, and Hot Air," p. 23.

[106] Eilperin, Juliet, "GOP Chairmen Face off on Global Warming."

[107] ExxonMobil, "Taking on the world's toughest energy challenges," 2007 Corporate Citizenship Report, p. 39. http://www.socialfunds.com/shared/reports/1211896380_ExxonMobil_2007_Corporate_Citizenship_Report.pdf

[108] Adam, David, "Royal Society tells Exxon: stop funding climate change denial," *The Guardian*, September 6, 2006. https://www.theguardian.com/environment/2006/sep/20/oilandpetrol.business

[109] See for example, "Denial and Deception: A Chronicle of ExxonMobil's Efforts to Corrupt the debate on Global Warming," Greenpeace, May 2002. http://www.greenpeace.org/usa/wp-content/uploads/2015/11/exxon-denial-and-deception.pdf?a1481f

[110] See for example, "What Exxon doesn't want you to know," Greenpeace, June 22, 2004. http://www.greenpeace.org/international/en/news/features/exxon-secrets/

[111] ExxonSecrets, "FAQ," Greenpeace. https://exxonsecrets.org/html/faq.php

[112] "Smoke, Mirrors, and Hot Air," pp. 31-33.

[113] Douglass, Elizabeth, "Exxon's 25 Years of 'No': Timeline of Resolutions on Climate Change," InsideClimate News, February 27, 2017. https://insideclimatenews.org/content/exxons-25-years-no-timeline-resolutions-climate-change

After this tactic failed, the Rockefellers divested their family fund of all fossil fuel assets, starting with ExxonMobil due to its "morally reprehensible conduct[114]."

After receiving the letter from Britain's Royal Society, ExxonMobil acknowledged publicly for the first time that fossil fuels cause climate change. "We recognize that the accumulation of greenhouse gases in the Earth's atmosphere poses risks that may prove significant for society and ecosystems," the company wrote[115]. Its current statement reads:

> The risk of climate change is clear and the risk warrants action. Increasing carbon emissions in the atmosphere are having a warming effect. There is a broad scientific and policy consensus that action must be taken to further quantify and assess the risks[116].

ExxonMobil did keep its promise to stop funding Competitive Enterprise Institute, Heartland Institute, and a handful of other climate denial organizations[117]. However, the company is still funding ALEC, the Manhattan Institute, U.S. Chamber of Commerce, and National Black Chamber of Commerce, all of which engage in climate denial[118]. ExxonMobil declared support for a revenue-neutral carbon tax in 2009, but refused to sign a letter of support for a global price on carbon at the Paris climate talks and opposed carbon tax bills in Massachusetts in 2015. Nor does the company agree with the conclusion of the IPCC, World Bank and International Energy Agency that most known carbon reserves must remain in the ground[119]. Instead, Tillerson has said climate change is an "engineering problem" and humans "will adapt[120]." Meanwhile, Democratic candidates Bernie Sanders, Martin O'Malley, and Hillary Clinton called on the Justice Department to investigate whether ExxonMobil committed fraud by obfuscating the reality of climate change[121], and 17 state attorneys general opened their own investigations[122].

~ ~ ~

Cathy Cowan Becker can boast a twenty-year career in journalism, communications, and public relations. She is currently communications coordinator for the Mershon Center for International Security Studies at Ohio State University, where she is also pursuing a dual master's in public administration and environment and natural resources. Cathy journeyed on environmental study-abroad trips in Costa Rica in 2015 and Iceland in 2016. She also traveled to Paris for the 2015 Climate Conference and to Oceti Sakowin camp at Standing Rock in 2016. She marched with 400,000 people for climate in New York City in 2014, as well as with 150,000 for science and 250,000 for climate in Washington, DC in 2017. Cathy lives in Grover City, Ohio. Read more at: www.climateabandoned.com.

[114] Ratner, Paul, "Why the Rockefellers, Who Made Their Money in Oil, Are Fighting the Fossil Fuel Industry," Big Think, January 19, 2017. http://bigthink.com/paul-ratner/why-the-rockefellers-are-battling-exxonmobil-over-climate-change
[115] Banerjee, Neela, "Rex Tillerson's Record on Climate Change: Rhetoric vs. Reality," InsideClimate News, December 22, 2016. https://insideclimatenews.org/news/22122016/rex-tillerson-exxon-climate-change-secretary-state-donald-trump
[116] ExxonMobil, "Our position on climate change." http://corporate.exxonmobil.com/en/current-issues/climate-policy/climate-perspectives/our-position
[117] Heger, Monica, "ExxonMobil Cuts Back Its Funding for Climate Skeptics," IEEE Spectrum, July 1, 2008. https://spectrum.ieee.org/energy/environment/exxonmobil-cuts-back-its-funding-for-climate-skeptics
[118] Negin, Elliott, "ExxonMobil is Still Funding Climate Science Denial Groups," Huffington Post, July 13, 2016. https://www.huffingtonpost.com/elliott-negin/exxonmobil-is-still-fundi_b_10955254.html
[119] Banerjee, Neela, "Rex Tillerson's Record on Climate Change: Rhetoric vs. Reality."
[120] Mooney, Chris, "Rex Tillerson's view of climate change: It's just an 'engineering problem'," Washington Post, December 14, 2016. https://www.washingtonpost.com/news/energy-environment/wp/2016/12/13/rex-tillersons-view-of-climate-change-its-just-an-engineering-problem/?utm_term=.50fc15764510
[121] Cama, Timothy, "Hillary joins calls for federal probe of Exxon climate change research," The Hill, October 29, 2015. http://thehill.com/policy/energy-environment/258589-clinton-joins-calls-for-federal-probe-of-exxon
[122] Hasemyer, David, and Sabrina Shankman, "Climate Fraud Investigation of Exxon Draws Attention of 17 Attorneys General," InsideClimate News, March 30, 2016. https://insideclimatenews.org/news/30032016/climate-change-fraud-investigation-exxon-eric-shneiderman-18-attorneys-general

Chapter 11

Climate Crisis & Selling Doubt
Part II: America's Koch Problem and Climate Denial Machine

by

Cathy Cowan Becker, M.A.

"To be truly radical is to make hope possible rather than despair convincing" – **Raymond Williams**

As ExxonMobil stepped back from funding climate denial, another source stepped up: Charles and David Koch. But rather than simply funding an echo chamber as ExxonMobil had done, the Koch brothers expanded their climate denial spending to build an entire political empire that rivals the two major parties in its influence. Through Koch Industries and their family foundations – the Charles G. Koch Foundation, David H. Koch Foundation, Claude R. Lambe Foundation, and Knowledge and Progress Fund – the Kochs have not only funded a series of conservative think tanks and front groups to spread doubt about climate science, but also spent millions on political action groups to elect climate-denying politicians to federal and state office, as well as on direct lobbying to get their legislative and policy agenda passed once their politicians were in office.

As outlined by Greenpeace, from 2005 to 2008 the Koch brothers spent $24.9 million funding 35 climate denial think tanks and front groups, while ExxonMobil spent $8.9 million – meaning the Kochs outspent ExxonMobil by a factor of three to one[1]. During the same period, the Koch Political Action Committee spent $5.7 million, about double what ExxonMobil spent, to elect a series of mostly Republican climate denial candidates to Congress, including Sen. James Inhofe (R-Okla.), and Rep. Joe Barton (R-Texas), who have ridiculed climate science and renewable energy[2]. And the brothers spent $37.9 million on lobbying for themselves and the rest of the oil and gas industry, just behind ExxonMobil and Chevron, by hiring firms such as Hunton and Williams, the Rhoads Group, and Pyle and Associates[3].

All told, from 1997 to 2015, the Koch brothers have sent more than $100 million to 84 groups that deny climate change[4]. This is part of the $558 million distributed by 140 foundations to 91 organizations between 2003 and 2010 in what sociologist Robert Brulle of Drexel University calls the "climate change countermovement[5]." This is defined as a well-funded and organized effort to undermine public faith in climate science and block action by the U.S. government to regulate emissions. This countermovement involves a large number of organizations, including conservative think tanks, advocacy groups, trade associations and conservative foundations, with strong links to sympathetic media outlets and conservative politicians[6].

Besides the Koch foundations, other important funders of climate denial during from 2003 to 2010 included the Scaife, Bradley, Howard, Pope, Searle, Templeton, and DeVos foundations. But the Kochs were arguably

---

[1] Greenpeace, "Koch Industries: Secretly Funding the Climate Denial Machine," 2010, pp. 4-5. http://www.greenpeace.org/usa/wp-content/uploads/legacy/Global/usa/report/2010/3/koch-industries-secretly-fund.pdf?9e7084

[2] "Koch Industries: Secretly Funding the Climate Denial Machine," pp. 31-32.

[3] "Koch Industries: Secretly Funding the Climate Denial Machine," pp. 29-30.

[4] Greenpeace, "Koch Industries: Secretly Funding the Climate Denial Machine." http://www.greenpeace.org/usa/global-warming/climate-deniers/koch-industries/

[5] Brulle, Robert, "Institutionalizing delay: foundation funding and the creation of U.S. climate change counter-movement organizations," Climatic Change, 122 (4), pp. 681-694. http://drexel.edu/~/media/Files/now/pdfs/Institutionalizing%20Delay%20-%20Climatic%20Change.ashx?la=en

[6] McKechnie, Alex, "Not Just the Koch Brothers: New Drexel Study Reveals Funders Behind the Climate Change Denial Effort," Drexel University, December 20, 2013. http://drexel.edu/now/archive/2013/December/Climate-Change/

the most important players for two reasons: they funded a long list of organizations, and they funneled a lot of climate denial money through untraceable sources, meaning they spent more than researchers like Brulle and Greenpeace have been able to trace.

One of Brulle's key findings is that an increasing share of climate denial funding – about 24 percent by 2010[7]– went through a group called DonorsTrust and its affiliate Donors Capital Fund. Dubbed the "dark-money ATM of the conservative movement" by Mother Jones reporter Andy Kroll[8], DonorsTrust is a not a foundation or nonprofit, but a "donor advised fund." Donors to such funds do not get final say over where their money goes, but they can make recommendations. In exchange, they get a larger tax deduction than if they were giving to a normal charity, and – this is key — they can remain anonymous. With all the heat brewing around climate denial, anonymity became a favored option for those who wanted to fund it.

By 2010, DonorsTrust had distributed $118 million to 102 think tanks and advocacy groups on record as denying human-caused climate change and opposing environmental regulations[9]. One recipient was the Heartland Institute, which received more than $17 million through 2014[10]. In 2005 Heartland began holding the International Conference on Climate Change, or ICCC[11], an annual gathering of climate denial activists and scientists. These meetings were used to produce four volumes in a series called Climate Change Reconsidered, credited to the Nongovernmental International Panel of Climate Change, or NIPCC. Patterned after the IPCC reports on climate change, the NIPCC publications included multi-authored reports on Physical Science[12] and Biological Impacts[13] that came to opposite conclusions as the IPCC. For example, the Summary for Policymakers claims that doubling carbon dioxide in the atmosphere does not present a threat, a warming of 2 degrees C would not be harmful to the environment, and more carbon dioxide will green the planet, feed humans, and result in a reduction of human mortality[14]. Another NIPCC publication, "Why Scientists Disagree About Global Warming,[15]" was sent to 200,000 K-12 science teachers across the country[16]. In response, the National Center for Science Education told teachers to keep Heartland's climate denial out of the classroom[17].

Another recipient of funding from DonorsTrust and Donors Capital Fund is Committee for a Constructive Tomorrow, or CFACT, which received $7.8 million of its $11.8 million budget between 2005 to 2014[18]. Started in 1985 as a conservative response to the progressive U.S. Public Interest Research Groups, CFACT "can be heard relentlessly infusing the public-interest debate with a balanced perspective on environmental stewardship[19]." In 2009, CFACT hired Marc Morano, former producer of "Rush Limbaugh, the Television Show" and climate

[7] Brulle, "Institutionalizing Delay," p. 690.

[8] Kroll, Andy, "The Dark-Money ATM of the Conservative Movement," *Mother Jones,* February 5, 2013. http://www.motherjones.com/politics/2013/02/donors-trust-donor-capital-fund-dark-money-koch-bradley-devos/

[9] Goldenberg, Suzanne, "How DonorsTrust distributed millions to anti-climate groups," *The Guardian*, February 14, 2013. https://www.theguardian.com/environment/2013/feb/14/donors-trust-funding-climate-denial-networks

[10] Global Warming Disinformation Database, "Donors Capital Fund," DeSmog Blog, https://www.desmogblog.com/donors-capital-fund. Including donations from both Donors Trust and Donors Capital Fund brings the amount Heartland got to more than $18 million. See Conservative Transparency, "DonorsTrust & Donors Capital Fund" American Bridge 21st Century, http://conservativetransparency.org/org/donorstrust-donors-capital-fund/?og_tot=7672&order_by=&adv=heartland+institute

[11] See presentations from the International Conferences on Climate Change held so far at http://climateconferences.heartland.org/

[12] Idso, Craig D., Robert M. Carter, S. Fred Singer, et al., *Climate Change Reconsidered II: Physical Science*, Nongovernmental International Panel on Climate Change, 2013. http://climatechangereconsidered.org/climate-change-reconsidered-ii-physical-science/

[13] Idso, Craig D., Sherwood B. Idso, Robert M. Carter, S. Fred Singer, et al., *Climate Change Reconsidered II: Biological Impacts,* Nongovernmental International Panel on Climate Change, 2013. http://climatechangereconsidered.org/climate-change-reconsidered-ii-biological-impacts/

[14] Idso, Carter, Singer, et al., "Summary for Policymakers," *Climate Change Reconsidered II: Physical Science,* Heartland Institute, 2013, p. 4. https://www.heartland.org/_template-assets/documents/CCR/CCR-II/Summary-for-Policymakers.pdf. See also Idso, Idso, Carter, Singer, et al., "Summary for Policymakers," *Climate Change Reconsidered II: Biological Impacts,* Heartland Institute, 2013, p. 3. https://www.heartland.org/_template-assets/documents/CCR/CCR-IIb/Summary-for-Policymakers.pdf

[15] Idso, Craig D., Robert M. Carter, and S. Fred Singer, "Why Scientists Disagree About Global Warming," Nongovernmental International Panel on Climate Change, 2015. http://climatechangereconsidered.org/why-scientists-disagree-about-global-warming/

[16] Worth, Katie, "Climate Change Skeptic Group Seeks to Influence 200,000 Teachers," Frontline, March 28, 2017. https://www.pbs.org/wgbh/frontline/article/climate-change-skeptic-group-seeks-to-influence-200000-teachers/

[17] Branch, Glenn, "Don't Let Heartland Fool Teachers1" National Council on Science Education blog, April 1, 2017. https://ncse.com/blog/2017/04/don-t-let-heartland-fool-teachers-0018504. Heartland defended its mailings to teachers at https://www.heartland.org/publications-resources/publications/why-scientists-disagree-about-global-warming

[18] Global Warming Disinformation Database, "Center for a Constructive Tomorrow (CFACT)," DeSmog Blog. https://www.desmogblog.com/committee-constructive-tomorrow

[19] CFACT, "Statement of Purpose." http://www.cfact.org/about/

researcher for Sen. James Inhofe (R-Okla.)[20], to run its Climate Depot website, which Newsweek called the "most popular denial site" and the conservative publication Townhall dubbed "the No. 1 online destination for the truth behind today's global warming hysteria[21]." In 2009 CFACT was a co-organizer of the Copenhagen Climate Challenge, a conference for climate deniers in conjunction with the COP 15 climate conference in Denmark[22], and in 2015 it released the film "Climate Hustle[23]" to coincide with the COP 21 climate conference in Paris.

Although DonorsTrust and Donors Capital Fund are not required to disclose their donors, we can surmise for several reasons that the Koch brothers are high on the list. First, follow the people. A long list of DonorsTrust and Donors Capital Fund employees have extensive histories with the Koch network, starting with their founder, Whitney Ball, an ardent libertarian from West Virginia who got her start as director of development for the Koch-founded Cato Institute[24] and served on the board of the Koch-funded State Policy Network[25]. Other employees with Koch-related histories include board chair Kimberly Dennis, who worked for the Institute for Humane Studies and Pacific Research Institute and served on the boards of George Mason University and Property and Environment Research Center; and board member Thomas Beach, who served as chairman of the Reason Foundation and of Property And Environment Research Center[26]. All these organizations are among the 84 Koch-funded groups listed by Greenpeace as denying climate change and opposing environmental regulations[27].

Second, follow the money. DonorsTrust and Donors Capital received large donations from Koch foundations such as the Knowledge and Progress Fund, which gave more than $13.7 million from 2005 to 2014[28]. Meanwhile, DonorsTrust and Donors Capital Fund have distributed funds to a long list of Koch-affiliated organizations, starting with Americans for Prosperity Foundation, which received $23.4 million between 2005 and 2014[29]. Americans for Prosperity Foundation is the financial arm behind Americans for Prosperity, founded by Charles and David Koch to act as an astroturf operation for their political activities[30]. Americans for Prosperity was instrumental in starting the Tea Party movement and defeating the American Clean Energy and Security Act of 2009, which would have established a cap and trade program for carbon emissions in the United States[31].

Another Koch network organization funded by Donors Trust and Donors Capital Fund is State Policy Network, a web of think tanks in every state that received $19.7 million. An investigation by Center for Media and Democracy found that State Policy Network and its member think tanks are "major drivers of the right-wing, ALEC-backed agenda in statehouses nationwide, with deep ties to the Koch brothers and the national right-wing network of funders, all while reporting little or no lobbying activities[32]." Donors Trust and Donors Capital Fund also gave $11.5 million to the Mercatus Center[33], a think tank at George Mason University founded by Richard Fink, executive vice president of Koch Industries, with a grant and ongoing support from the Charles G. Koch Foundation[34].

[20] CFACT, "CFACT Staff." http://www.cfact.org/about/staff/

[21] CFACT, "Climate Depot." http://www.cfact.org/cfact-programs/climate-depot/

[22] Gray, Louise, "Copenhagen climate summit: Behind the scenes at the sceptics' conference," The Telegraph, December 9, 2009. http://www.telegraph.co.uk/news/earth/copenhagen-climate-change-confe/6765032/Copenhagen-climate-summit-Behind-the-scenes-at-the-sceptics-conference.html

[23] Morano, Marc, and Mick Curran, "Climate Hustle," Fathom Events, 2015. http://www.climatehustle.org/

[24] Mayer, Jane, Dark Money: The Hidden History of the Billionaires Behind the Rise of the Radical Right, Random House Large Print, 2016, p. 329.

[25] Global Warming Disinformation Database, "State Policy Network," DeSmog Blog. https://www.desmogblog.com/state-policy-network

[26] Conservative Transparency, "DonorsTrust & Donors Capital Fund," American Bridge 21st Century, http://conservativetransparency.org/org/donorstrust-donors-capital-fund/

[27] Greenpeace, "Koch Industries: Secretly Funding the Climate Denial Machine." http://www.greenpeace.org/usa/global-warming/climate-deniers/koch-industries/

[28] SourceWatch, "Knowledge and Progress Fund," Center for Media and Democracy, https://www.sourcewatch.org/index.php/Knowledge_and_Progress_Fund

[29] Conservative Transparency, "DonorsTrust & Donors Capital Fund," American Bridge 21st Century, http://conservativetransparency.org/org/donorstrust-donors-capital-fund/?og_tot=7672&order_by=&adv=americans+for+prosperity

[30] SourceWatch, "Americans for Prosperity," Center for Media and Democracy, https://www.sourcewatch.org/index.php/Americans_for_Prosperity

[31] The defeat of American Clean Energy and Security Act of 2009 is covered in the Frontline episode "Climate of Doubt," October 23, 2012. https://www.pbs.org/wgbh/frontline/film/climate-of-doubt/

[32] Center for Media and Democracy, "State Policy Network: The Powerful Right-Wing Network Helping to Hijack State Politics and Government," November 2013. https://www.alecexposed.org/w/images/2/25/SPN_National_Report_FINAL.pdf

[33] Conservative Transparency, "DonorsTrust & Donors Capital Fund," American Bridge 21st Century, http://conservativetransparency.org/org/donorstrust-donors-capital-fund/?og_tot=7672&order_by=&adv=mercatus+center

[34] Global Warming Disinformation Database, "Mercatus Center," DeSmog Blog. https://www.desmogblog.com/mercatus-center

Koch Industries

Why were the Koch brothers so determined to deny climate science and delay action on climate change? To answer that, we must look first at the nature of the Koch brothers' business. Charles G. Koch and David H. Koch are two of the four sons of Fred C. Koch, a chemical engineer who co-founded an oil refinery firm that later became Koch Industries, now the second-largest privately held company in the United States with $100 billion in annual revenue[35]. The company operates in 60 countries, employing 100,000 people, with 60,000 in the United States[36]. Koch Industries has many subsidiaries in the fossil fuel industry, including:

- **Flint Hills Resources**[37], formerly Koch Petroleum Group, sells gasoline, diesel, jet fuel, ethanol, polymers, oils and asphalt. It employs 5,000 people and at three chemical plants, six asphalt plants, four ethanol plants, and three refineries, one of which processes 392,000 barrels of Canadian tar sands oil per day[38].

- **Koch Ag & Energy Solutions**, a holding company for **Koch Fertilizer**, one of the world's largest makers of ammonia, urea, phosphate and sulfur-based products; **Koch Energy Services**, which markets natural gas throughout North America; **Koch Methanol**, and **Koch Agronomic Services**.

- **Koch Chemical Technology Group**, with seven subsidiaries including **Koch-Glitsch**, which makes equipment for refineries and chemical plants; **Koch Membrane Systems**, which makes filtration equipment; **Optimized Process Designs**, which provides engineering and construction services for the gas processing and petrochemical industries; **John Zink Hamworthy Combustion**, which owns 12 brands of products used by the oil and gas industry; **Koch Knight**, which produces corrosion-proof ceramic and plastic materials for industry; **Koch Heat Transfer Company**; and **Koch Specialty Plant Services**.

- **Koch Minerals**, one of the world's largest dry commodity handlers for petroleum coke, coal, sulfur, wood pulp, paper products, and other substances. Its subsidiaries include **Koch Carbon**, which stores, transports, and trades products such as petcoke; **Koch Exploration Company**, which acquires, develops and trades petroleum and natural gas properties; and **Koch Oil Sands Operating**, which holds more than 1 million acres of leases in the tar sands of Alberta, Canada.

- **Koch Pipeline**, which operates 4,000 miles of pipelines that transport crude oil, refined petroleum products, liquefied natural gas, and chemicals.

- **Koch Supply & Trading**, which trades physical commodities such as crude oil; refined products such as heating oil, gasoline, and diesel; petrochemicals such as benzene, toluene and styrene; natural gas and liquefied natural gas; metals such as aluminum, copper, zinc, lead and steel. It also transports these

---

[35] Forbes, "#2 Koch Industries," America's Largest Private Companies. https://www.forbes.com/companies/koch-industries/

[36] Greenpeace, "Koch Industries," Company Facts. http://www.greenpeace.org/usa/what-we-do/koch-industries-company-facts/

[37] More information is on the Flint Hills Resources archived website at https://web.archive.org/web/20120119104652/http://www.fhr.com/about/default.aspx

[38] Dembicki, Geoff, "The Kochs: Oil Sands Billionaires Bankrolling the US Right," The Tyee, March 22, 2011. https://thetyee.ca/News/2011/03/22/KochBrothers/index.html

products worldwide through **Koch Shipping**, and trades in their financial derivatives through **Koch Derivatives Trading and Metals**. [39]

Koch Industries and its subsidiaries have a long track record of environmental violations and corrupt practices. Some notable cases include[40]:

- In 1989, a Senate committee investigating Koch Industries' business with Native Americans found that Koch had misrepresented the amount of oil it purchased from tribal oil fields. Documents showed Koch got 803,874 more barrels of oil than it paid for in 1986; 671,144 more barrels in 1987, and 474,281 more barrels in 1988[41]. The value of all the extra oil was estimated at $31.2 million, accounting for almost one-fourth of the company's crude oil profits[42]. A federal grand jury was called to hear the case, but declined to issue indictments after Kansas Sen. Bob Dole cautioned against a "rush to judgment" on the Senate floor[43]. Instead, Bill Koch, as part of a longstanding feud with his brothers Charles and David, sued Koch Industries under the False Claims Act. The jury found the company had committed more than 24,000 false claims worth $214 million. The Kochs settled for $25 million in 2001[44].

- In 1994, a leak at a pipeline-to-barge facility for anhydrous ammonia fertilizer near St. Louis killed worker Rodger McGlothlin, 36, and hospitalized worker Karl Lycke, 30. Workers were not required to wear breathing gear during barge loadings[45].

- In 1995, the Department of Justice, Environmental Protection Agency, and Coast Guard filed a civil suit against Koch Industries for more than 300 spills that discharged 11.6 million gallons of oil into the waters of six states[46]. In 2000 Koch Industries agreed to pay $30 million in civil penalties and $5 million on

---

[39] More information is on the Koch Ag & Energy Solutions website at http://www.kochagenergy.com/

More information is on the Koch Fertilizer website at http://www.kochfertilizer.com/

More information is on the Koch Energy Services website at http://www.kochenergyservices.com/

More information is on the Koch Methanol website at http://www.kochmethanol.com/

More information is on the Koch Agronomic Services website at http://www.kochagronomicservices.com/

More information is on the Koch Chemical Technology Group website at http://www.kochind.com/chemtech/

More information is on the Koch-Glitsch website at http://www.koch-glitsch.com/default.aspx

More information is on the Koch Membrane Systems website at http://www.kochmembrane.com/

More information is on the Optimized Process Designs website at http://www.opdepc.com/

More information is on the John Zink Hamworthy Combustion website at https://www.johnzink.com/

More information is on the Koch Knight website at http://www.kochknight.com/index.html

More information is on the Koch Heat Transfer Company website at http://www.kochheattransfer.com/

More information is on the Koch specialty Plant Services website at http://www.kochservices.com/

More information is on the Koch Industries website at http://www.kochind.com/minerals/

More information is on the Koch Carbon website at http://kochcarbon.com/

Find out more about petcoke in Oil Change International, Petroleum Coke: The Coal Hiding in the Tar Sands, January 2013. http://priceofoil.org/2013/01/17/petroleum-coke-the-coal-hiding-in-the-tar-sands/

More information is on the Koch Exploration Company archived website at https://web.archive.org/web/20170417140711/http://kochexploration.com/default.aspxv

More information is on the Koch Oil Sands Operations archived website at https://web.archive.org/web/20160318024605/http://kochexploration.com/canada/Default.aspx

More information is on the Koch Pipeline website at https://www.kochpipeline.com/

More information is on the Koch Supply & Trading website at http://www.ksandt.com/

[40] For a list of government actions and court cases against Koch Industries and its subsidiaries, see Corporate Research Project, "Koch Industries: Corporate Rap Sheet." https://www.corp-research.org/koch_industries

[41] Associated Press, "Oil Firm named in Probe of Unpaid Indian Royalties," *Los Angeles Times,* May 10, 1989. http://articles.latimes.com/1989-05-10/news/mn-3059_1_koch-industries-senate-panel-extra-oil

[42] Dickinson, "Inside the Koch Brothers' Toxic Empire."

[43] Dickinson, "Inside the Koch Brothers' Toxic Empire."

[44] Duggan, Ed, "Koch whistle-blower case settled for $25 million," *South Florida Business Journal,* June 4, 2001. https://www.bizjournals.com/southflorida/stories/2001/06/04/story6.html?page=all

[45] Boyd, John, "Pipeline-Barge Facility Remains Sit in Wake of Deadly Ammonia Leak," *Journal of Commerce,* February 23, 1994.

[46] U.S. Department of Justice, "Koch Industries and Affiliates Charged for Hundreds of Oil Spills in Six states," Press Releaase, April 17, 1995. https://www.justice.gov/archive/opa/pr/Pre_96/April95/213.txt.html

environmental projects, the largest civil fine ever imposed on a company under federal environmental law[47].

- In 1996, a Koch pipeline carrying butane – also known as lighter fluid – ruptured under a subdivision in Lively, Texas, filling the area with gas fumes. Danielle Smalley and Jason Stone, both 17, got in her father's pickup to go alert authorities. The truck stalled over a dry creek bed, and when she tried to restart the car, the spark from the ignition ignited the fumes, engulfing the truck in a fireball and killing the teenagers instantly[48]. An investigation by the National Transportation Safety Board in 1998 blamed the rupture on "failure to adequately protect a pipeline from corrosion" at the rupture site as well as hundreds of other sites. NTSB also found Koch Industries had not conveyed important safety information to the public, as the Smalleys did not know the pipeline was running through their subdivision[49]. A Texas jury awarded Danielle's father, Danny, $296 million in his wrongful death lawsuit against Koch Industries, then the largest such judgment in American history[50].

- In 2000, the DOJ and EPA announced another settlement with Koch Industries, this time in connection with violations of the Clean Air Act at its Pine Bend refinery in Rosemount, Minn., and two refineries in Corpus Christi, Texas. Koch agreed to pay $4.5 million in fines and $80 million to upgrade pollution-control technologies[51].

- In a separate case in 2000, a federal grand jury in Corpus Christi, Texas, returned a 97-count indictment against Koch Industries, Koch Petroleum Group, and four corporate managers for releasing 91 metric tons of cancer-causing benzene, 15 times more than the legal limit, then trying to cover it up. The company faced $350 million in fines and the managers faced stiff prison sentences. In 2001 Koch Industries pleaded guilty to concealing violations of air quality laws and paid $10 million in criminal fines and $10 million on environmental projects in Corpus Christi[52].

- In 2009, subsidiary Invista agreed to pay $1.7 in civil penalties and spend up to $500 million to correct more than 680 violations of air, water, hazardous waste, emergency planning, and pesticide regulations at 12 facilities in seven states[53].

- In 2010, the DOJ sued Georgia-Pacific and nine other companies over PCB contamination in the Fox River in Wisconsin. Georgia-Pacific settled for $7 million in what Appleton Papers, which had already spent more than $140 million to clean up the river, called a "sweetheart deal[54]."

In his 2014 *Rolling Stone* article, "Inside the Koch Brothers' Toxic Empire," Tim Dickinson documents how the brothers used their subsidiary Koch Supply & Trading to speculate on oil derivatives, volatility swaps, and proprietary trading, in an industry entirely exempt from regulations. Their wealth quintupled from 2007 to 2012

---

[47] U.S. Environmental Protection Agency, "Koch Industries to Pay Record Fine for Oil Spills in Six States," News Release, January 13, 2000. https://www.justice.gov/archive/opa/pr/2000/January/019enrd.htm

[48] Dickinson, "Inside the Koch Brothers' Toxic Empire."

[49] National Transportation Safety Board, "NTSB Determines that Inadequate Corrosion Protection Caused Fatal Texas Pipeline Rupture," November 3, 1998. https://www.ntsb.gov/news/press-releases/Pages/ NTSB_Determines_that_Inadequate_Corrosion_Protection_Caused_Fatal_Texas_Pipeline_Rupture.aspx

[50] Drake, Witham, "Jury Awards Damages in Case against Wichita, Kan.-Based Oil Company," Knight Ridder/Tribune Business News, October 27, 1999.

[51] Environmental Protection Agency, "EPA and DOJ Announce Record Clean Air Agreement with Major Petroleum Refiners," News Release, July 25, 2000. https://yosemite.epa.gov/opa/admpress.nsf/ 905a0f1800315fd385257359003d4808/cd3a52c26b10e67a85256927005e5ee7!OpenDocument

[52] U.S. Department of Justice, "Koch Pleads Guilty to Covering Up Environmental Violations at Texas Oil Refinery," News Release, April 9, 2001. https://www.justice.gov/archive/opa/pr/2001/April/153enrd.htm

[53] U.S. Environmental Protection Agency, "United States Announces Largest Settlement Under Environmental Protection Agency's Audit Policy," News Release, April 13, 2009. https://yosemite.epa.gov/opa/admpress.nsf/d0cf6618525a9efb85257359003fb69d/1e9d18f061b4da818525759700632926!OpenDocument

[54] "Suit filed over Fox River pollution cleanup," *Milwaukee Journal Sentinel,* October 24, 2010. http://archive.jsonline.com/news/wisconsin/104937079.html/

and multiplied more than 10 times from about $4 billion in 2002 to over $40 billion each by 2014[55]. From 1988, to 2015, the Kochs' net worth grew 3,855 percent, compared to a growth of 713 percent for the S&P 500[56]. Most of this growth occurred after 2007 when they started trading in oil derivatives; between 2011 and 2012 alone, their net worth increased 24 percent[57]. The Kochs used that money to buy Georgia-Pacific for $21 billion in 2015 and Molex for $7 billion in 2013. They also used it to fund a full-spectrum dominance system of political candidates, think tanks, media manipulators, front groups, academic agents, and more in what the International Forum on Globalization calls the "Kochtopus[58]."

The Kochtopus is a map of the money, structure and scale of the network that the Kochs are using to influence democracy. Besides funding climate denial, the Kochs are attacking labor unions, women's rights, voting rights, health care, immigrants, clean energy, and clean water. They run all of this influence on democracy through the Kochtopus system that includes[59]:

- Media manipulators, or press professionals such as Rush Limbaugh and Glenn Beck who create positive media coverage favorable to the Kochs' libertarian ideology[60].

- Think tanks such as the Cato Institute, Mercatus Center, American Enterprise Institute, and Heartland Institute that promote policy proposals for less government to protect people and planet but more rights for corporations and investors.

- Astroturf agents, or front groups such as Americans for Prosperity that project the appearance of popular support for ideas and policies that benefit large corporations.

- Wealth warriors, or lobbyists, accountants, and tax attorneys who work to keep wealth out of the government by creating and exploiting tax loopholes and lowering taxes for billionaires and corporations. These include the American Legislative Exchange Council, KochPAC, and the U.S. Chamber of Commerce.

- Congressional collaborators, or like-minded candidates that Koch campaign contributions help elect so they can pass laws favorable to Koch profits. Koch campaign contributions in 2016 totaled more than $1.8 million and went almost exclusively to Republicans. Top recipients included Mike Pompeo (R-Kan.), Roy Blunt (R-Mo.), Speaker of the House Paul Ryan (R-Wis.), Marco Rubio (R-Fla.), and Rob Portman (R-Ohio)[61].

- Courtroom collaborators, or like-minded judges appointed or elected to rule in the Kochs' favor. For example, Supreme Court Justices Antonin Scalia and Clarence Thomas attended retreats hosted by the Koch brothers before taking part in the Citizens United decision that opened up floodgates for political spending in 2010[62].

[55] Dickinson, Tim, "Inside the Koch Brothers' Toxic Empire," *Rolling Stone*, September 24, 2014. https://www.rollingstone.com/politics/news/inside-the-koch-brothers-toxic-empire-20140924

[56] Kiersz, Andy, "Here's how Donald Trump's net worth since 1988 stacks up against other successful businessmen," *Business Insider*, August 22, 2015. http://www.businessinsider.com/donald-trump-vs-other-billionaires-2015-8

[57] Koch Cash, "Kochs' net worth jumps 24% from last year; IFG and allies launch emergency effort on Koch Cash," International Forum on Globalization. http://koch-cash.org/kochs-net-worth-jumps-24-from-last-year-ifg-and-allies-launch-emergency-effort-on-koch-cash-2/

[58] Koch Cash, "The Kochtopus," International Forum on Globalization. http://kochcash.org/the-kochtopus/

[59] Menotti, Victor, IFG's KOCHTOPUS: The Kochs' Influence Network explained," International Forum on Globalization, 2012. https://www.youtube.com/watch?v=zaqh4p41lpQ

[60] The Daily Caller News Foundation has received more than $825,000 from Koch Foundations. See Greenpeace, "Koch Industries: Secretly Funding the Climate Denial Machine." http://www.greenpeace.org/usa/global-warming/climate-deniers/koch-industries/

[61] Open Secrets, "Koch Industries: Money to congressional candidates." https://www.opensecrets.org/orgs/toprecips.php?id=D000000186&type=&sort=&cycle=2016

[62] Lichtblau, Eric, "Advocacy Group Says Justices May Have Conflict in Campaign Finance Cases," The New York Times, January 19, 2011. http://www.nytimes.com/2011/01/20/us/politics/20koch.html

- Academic agents funded by private contracts with both public and private universities to hire like-minded faculty to teach their ideologies[63].

- Physical force by police, military and private contractors to maintain security at Koch meetings and industrial operations, and overall if civil unrest breaks out.

Although the Kochs are active in several issue areas, their highest priority is derailing legislation that would regulate carbon emissions. Why? Because Koch Industries and its subsidiaries emit a lot of carbon. Although the exact carbon footprint of Koch Industries is unknown because the United States does not mandate carbon reporting, *Think Progress* has estimated it at about 300 million tons per year, or about 5 percent of the entire carbon footprint of the United States[64].

If the Kochs had to pay for the damage their carbon emissions do to the climate and environment, it would significantly cut into their profits. The Environmental Protection Agency calculated the social cost of carbon at an average of $36 per metric ton of carbon dioxide (CO_2), $1,000 per ton of methane (CH_4), and $13,000 per ton of nitrous oxide (N_2O) in 2015, increasing steadily through 2050[65]. Even if all Koch Industries emissions were made up of relatively cheap carbon dioxide, they would have to pay $10.8 billion a year in damages to the climate and environment if the external costs of their business were accounted for. By comparison, the $100 million they have spent on climate denial over the last 20 years is a relative bargain.

Kochs in action

Perhaps the prime example of the Koch climate denial machine swinging into action occurred in 2009, with the introduction of the American Clean Energy and Security Act, also called the Waxman-Markey Bill after its authors Rep. Henry Waxman of California and Ed Markey of Massachusetts, both Democrats. The bill would have established an ambitious cap and trade scheme to reduce American greenhouse gas emissions to 20 percent below 2005 levels by 2020, increasing to 83 percent below 2005 levels by 2050. It also mandated that 25 percent of the nation's energy be produced from renewable sources by 2025, created new energy efficiency programs, put limits on the carbon content of motor fuels, and required greenhouse gas standards for new heavy duty vehicles and engines[66]. Introduced on May 15, 2009, Waxman-Markey was made possible by two previous landmark events: Massachusetts v. EPA (2007), in which the Supreme Court ruled that the EPA must regulate greenhouse gases unless it finds a valid scientific reason not to[67]; and the endangerment finding of 2009 in which the EPA declared that "greenhouse gases contribute to air pollution that may endanger public health or welfare.[68]"

As Govtrack notes, Waxman-Markey "met with a mixture of concern, measured praise, and outright criticism from business interests, environmentalists, and Democratic lawmakers." Environmental groups were concerned that it gave away most of the initial carbon credits and used the proceeds to subsidize industries including coal. Some opposed the idea of trading carbon credits altogether. Republicans were almost all opposed. Still, a diverse group of stakeholders supported the bill, including Defenders of Wildlife, Alliance for Climate Protection,

[63] For more information, see Greenpeace, "Koch Pollution on Campus." http://www.greenpeace.org/usa/global-warming/climate-deniers/koch-pollution-on-campus/
[64] Johnson, Brad, "Koch Industries: The 100-Million Ton Carbon Gorilla," *Think Progress*, January 30, 2011. https://thinkprogress.org/koch-industries-the-100-million-ton-carbon-gorilla-ab5c1551eb2b/
[65] Environmental Protection Agency, "The Social Cost of Carbon," January 19, 2017. https://19january2017snapshot.epa.gov/climatechange/social-cost-carbon_.html
[66] Tutwiler, Patrick, "Climate Change Legislation: Where Does It Stand?" GovTrack Insider, April 27, 2010. http://www.tmacog.org/Air%20Quality/Climate%20Change%20Legislative%20Update.pdf
[67] Greenhouse, Linda, "Justices Say E.P.A. Has Power to Act on Harmful Gases," The New York Times, April 3, 2007. https://www.nytimes.com/2007/04/03/washington/03scotus.html
[68] Milbourn, Cathy, "EPA Finds Greenhouse Gases Pose Threat to Public Health, Welfare / Proposed Finding Comes in Response to 2007 Supreme Court Ruling," U.S. Environmental Protection Agency News Release, April 17, 2009. Cached version at https://webcache.googleusercontent.com/search?q=cache:jKANaTmLC0sJ:https://yosemite.epa.gov/opa/admpress.nsf/0/0ef7df675805295d8525759b00566924+&cd=1&hl=en&ct=clnk&gl=us

Environmental Defense Fund, National Wildlife Federation, The Nature Conservancy, Audubon Society, Natural Resources Defense Council Sierra Club, League of Conservation Voters, United Auto Workers, Exelon, General Electric, Dow Chemical Company, Pacific Gas and Electric Company, DuPont, Dow Chemical, and Ford. After a long debate, the American Clean Energy and Security Act passed the House, 219-212[69], on June 26, 2009[70].

As Congress adjourned for its July 4 recess, the Koch-funded Americans for Prosperity took a three-pronged strategy to target the Senate, which was slated to take up Waxman-Markey later in July. First, they made sure plenty of constituents lobbied their senators at home to oppose to cap-and-trade. "We launched something we called Hot Air," Tim Phillips, president of Americans for Prosperity, told Frontline in 2012. "We got up a hot-air balloon, put a banner on the side of it that said, 'Cap-and-trade means higher taxes, lost jobs, less freedom.' And we went all over the country doing events and stirring up grass roots anger and frustration, concern[71]." These events featured appearances by paid climate deniers such as Christopher Monckton, a British consultant, policy adviser, writer, columnist, and hereditary peer supported by Competitive Enterprise Institute, which has taken almost $750,000 from the Koch brothers[72]. Also present were "carbon cops" who went through the crowds claiming to be overreaching EPA regulators ready to shut down backyard barbecues and lawn mowers[73].

Second, Americans for Prosperity took out a lot of expensive television ads, such as one that featured "Carlton," a wealthy eco-hypocrite who "wants Congress to spend billions on programs in the name of global warming and green energy, even if it causes massive unemployment, higher energy bills, and digs people like you even deeper into the recession.[74]" Finally, Americans for Prosperity also circulated the "No Climate Tax" pledge[75], requiring elected officials to oppose new spending to fight climate change. By 2013 the pledge had 411 signatories including the entire Republican leadership in the House, a third of the members of the House as a whole, and a quarter of U.S. senators[76].

Meanwhile, President Obama was preparing for his first international climate conference in Copenhagen in December 2009. There were high hopes for his participation, as Bush had withdrawn from the Kyoto Treaty. But on November 17, 2009, everything changed when thousands of internal emails between scientists from the University of East Anglia Climatic Research Unit were hacked and made public. The emails, going back to 1996, showed unguarded discussion of whether to release scientific data, how best to combat climate denial, and in some cases derisive comments about specific deniers. In one message, climate researcher Phil Jones said he had used a "trick" borrowed from scientist Michael Mann to "hide the decline.[77]" Describing the controversy as "Climategate," the Koch network plunged into action. "The blue dress moment may have arrived," said Chris Horner of the Competitive Enterprise Institute. At the Cato Institute, one scholar alone gave 20 interviews denouncing the emails. Tim Phillips of Americans for Prosperity told conservative bloggers at the Heritage Foundation that the controversy was "a crucial tipping point" in the fight against climate science[78].

---

[69] Clerk, U.S. House, "Final Vote Results for Roll Call 477." American Clean Energy and Security Act passed the House 219-212, with 44 Democrats opposed, 8 Republicans supporting. http://clerk.house.gov/evs/2009/roll477.xml

[70] Broder, John, "House Passes Bill to Address Threat of Climate Change," The New York Times, June 26, 2009. https://www.nytimes.com/2009/06/27/us/politics/27climate.html?_r=1&hp

[71] Hockenberry, John, and Catherine Upin, "Climate of Doubt," Frontline, October 23, 2012. https://www.pbs.org/wgbh/frontline/film/climate-of-doubt/

[72] Greenpeace, "Competitive Enterprise Institute (CEI)," Koch Industries Climate Denial Front Group. https://www.greenpeace.org/usa/global-warming/climate-deniers/front-groups/competitive-enterprise-institute-cei/

[73] Mayer, Dark Money, Random House Large Print, p. 346.

[74] Mayer, Dark Money, Random House Large Print, p. 343

[75] Americans for Prosperity, "No Cl!mate Tax." Archived at https://web.archive.org/web/20101104233321/http://www.noclimatetax.com/

[76] Holmberg, Eric, and Alexia Fernandez Campbell, "Koch: Climate pledge strategy continues to grow," Investigative Reporting Workshop, American University School of Communication, July 1, 2013. http://www.investigativereportingworkshop.org/investigations/the_koch_club/story/Koch_climate_pledge_strategy/ See also Mayer, Jane, "Koch Pledge Tied to Congressional Climate Inaction," The New Yorker, June 30, 2013. https://www.newyorker.com/news/news-desk/koch-pledge-tied-to-congressional-climate-inaction

[77] Revkin, Andrew C., "Hacked E-Mail Is New Fodder for Climate Dispute," The New York Times, November 20, 2009. https://www.nytimes.com/2009/11/21/science/earth/21climate.html?mcubz=0&mtrref=undefined

[78] Mayer, Dark Money, Random House Large Print, pp. 350-52.

Since the email hack, eight separate investigations have cleared the scientists involved of any wrongdoing[79]. In particular, Mann said, the word "trick" was shorthand in science for finding a good way to solve a problem, not trying to commit fraud. At issue were two sets of climate data over the past 100 years, one from measured temperatures, the other from tree rings. For some reason, the tree rings stopped indicated a rise in temperature starting in about 1960, while the measured temperatures kept going up. After finding indications the tree ring data was unreliable, along with other indirect data consistent with measured temperatures, the scientists had agreed to drop the tree ring data – a move discussed in the scientific literature for a decade.

Yet the scientists were targeted personally over a two-year period by the climate denial machine. Sen. James Inhofe (R-Okla.) and other recipients of Koch money sent letters to Pennsylvania State University, where Mann worked, demanding an investigation. Virginia Attorney General Ken Cuccinelli demanded all records relating to Mann's previous employer, University of Virginia. A former CIA officer contacted Mann's colleagues offering a $10,000 reward for dirt about Mann. Americans for Prosperity hired right-wing talk show host Mark Levin to accuse Mann of inventing global warming to justify a government takeover. Another front group, the Commonwealth Foundation for Public Policy Alternatives, part of the Koch-funded State Policy Network, ran attack ads against Mann in the campus newspaper, held anti-Mann campus protests, and lobbied Republican legislators to withhold funding for Penn State unless it fired Mann. The university agreed to investigate. Death threats arrived, including an envelope full of white powder[80].

The East Anglia email hack cast a shadow over the Copenhagen Climate Summit, widely considered a failure. The conference was chaotic. On the last day, China, India, Brazil, and South Africa were holding secret talks when Obama and Secretary of State Hillary Clinton got word and crashed the meeting[81]. The countries did agree to keep temperature rise to no more than 2°C, but set no targets for emissions reductions. All previous references to 1.5°C were dropped, as was the previous goal of reducing carbon emissions 80 percent by 2050[82].

Meanwhile, in the Senate, Democrats John Kerry and Joseph Lieberman and Republican Lindsay Graham worked for months to negotiate a deal on American Clean Energy and Security Act. They wrote language into the bill that gave away permits for offshore drilling in the Gulf of Mexico, increased the production of natural gas, provided subsidies for nuclear power, exempted utilities for three years, and even revoked the authority of the EPA to regulate greenhouse gas. The oil companies wanted to pay a set fee rather than trade on the open market, an idea that got translated in the media as a "gas tax." Graham in particular came under attack, with one Tea Party activist questioning Graham's sexuality at an event in South Carolina. Eventually, Graham withdrew from the coalition, and the bill died when Senate Majority Leader Harry Reid dropped it to work on immigration instead[83]. As cap and trade fell apart in Congress, a methane explosion killed 29 miners at a Massey Coal mine in West Virginia, while another methane explosion on BP's Deepwater Horizon oil rig killed 11 workers and sent 60,000 barrels of oil a day for 87 days into the Gulf of Mexico, the worst environmental disaster in American history.

[79] The eight major investigations include: House of Commons Science and Technology Committee (UK); Independent Climate Change Review (UK); International Science Assessment Panel (UK); Pennsylvania State University, first panel 2010, second panel 2012 (US); United States Environmental Protection Agency (US); Department of Commerce (US); and National Science Foundation (US).
[80] Mayer, *Dark Money*, Random House Large Print, pp. 353-58. See also Mann, Michael E., The Hockey Stick and the Climate Wars: Dispatches from the Front Lines, Columbia University Press, 2012.
[81] Lerer, Lisa, "Obama's dramatic climate meet," Politico, December 18, 2009. https://www.politico.com/story/2009/12/obamas-dramatic-climate-meet-030801
[82] Vidal, John, Allegra Stratton, and Suzanne Goldenberg, "Low targets, goals dropped: Copenhagen ends in failure," The Guardian, December 18, 2009. https://www.theguardian.com/environment/2009/dec/18/copenhagen-deal
[83] Lizza, Ryan, "As the World Burns," The New Yorker, October 11, 2010. https://www.newyorker.com/magazine/2010/10/11/as-the-world-burns

Fossil fuel politics

The next few years saw more back and forth between fossil fuel interests and defenders of climate science. Although climate science won some significant victories in the court of public opinion, they continued to lose elections as fossil fuel barons flooded the political system with cash. The problem of corporate spending on elections became markedly worse after 2010 with *Citizens United v. Federal Election Commission*[84], in which the Supreme Court ruled that free speech clause of the First Amendment prohibits the government from restricting political spending by corporations, nonprofits, labor unions, and other associations. While corporations and unions could not give money directly to campaigns, the court ruled, they could spend unlimited money to persuade the public through other means, including ads. The decision stemmed from a case in which a conservative group called Citizens United wanted to air ads for a film critical of Hillary Clinton during the 2008 Democratic primary. The 2002 Bipartisan Campaign Reform Act, also known as the McCain-Feingold Act, prohibited communications that mentioned a candidate within 60 days of a general election and 30 days of a primary. The Supreme Court struck this law down, finding that the First Amendment protects not only individual speech but also corporate speech. Because disseminating speech costs money[85], the court said, limiting a corporation's ability to spend money on political speech is unconstitutional.

Two months later, the U.S. Court of Appeals for the District of Columbia applied the precedent set by *Citizens United* to *SpeechNOW.org v. Federal Election Commission*, overturning all contribution limits to Political Action Committees. The resulting "Super PACs" cannot contribute directly to political parties or coordinate directly with political campaigns, but can spend whatever they want on speech to support or oppose political candidates. This opened the door to enormous political spending by wealthy individuals and corporations. Whereas in 2008, independent expenditures through PACs made up about half of outside election spending of $71.6 million, by 2012, Super PACs made up 95 percent of outside election spending of $188.6 million – an increase of 515 percent. Super PACs have continued to make up the vast majority of outside spending in elections ever since[86].

Although both liberals and conservatives have markedly increased political spending, conservatives have been much better organized and financed[87]. No one better exemplifies this than the Koch brothers. From 1998 to 2008, the three Koch foundations, Koch Industries, and KochPAC spent a total of $254 million on politics[88]. By 2012, the Kochs were using a network of 17 tax-exempt groups and limited-liability companies to spend at least $407 million to turn out conservative voters and run ads against Obama and Democrats in Congress[89]. This network was centered around Freedom Partners, Center to Protect Patient Rights, and TC4 Trust[90], and its spending dwarfed union spending by a factor of 2.6 to 1[91]. During the 2014 midterms, the Kochs were expected to spend $290 million, the equivalent of the annual incomes for 5,270 households[92]. By 2016, the Kochs had an

[84] For more information, see "Citizens United v. Federal Election Commission," SCOTUSblog. http://www.scotusblog.com/case-files/cases/citizens-united-v-federal-election-commission/

[85] The precedent was set in Buckley v. Valeo (1976), in which the Supreme Court ruled that limits on election spending in the Federal Election Campaign Act of 1971 were unconstitutional because a restriction on spending for political communication necessarily reduces the quantity of expression.

[86] "Total Outside Spending by Election Cycle, Excluding Party Committees," Open Secrets, Center for Responsive Politics. https://www.opensecrets.org/outsidespending/cycle_tots.php

[87] Cillizza, Chris, "How Citizens United changed politics, in 7 charts," Washington Post, January 22, 2014. Archived at https://web.archive.org/web/20170124174556/https://www.washingtonpost.com/news/the-fix/wp/2014/01/21/how-citizens-united-changed-politics-in-6-charts/

[88] Mayer, "Covert Operations."

[89] Gold, Matea, "Koch-backed political network, built to shield donors, raised $400 million in 2012 elections," Washington Post, January 5, 2014. https://www.washingtonpost.com/politics/koch-backed-political-network-built-to-shield-donors-raised-400-million-in-2012-elections/2014/01/05/9e7cfd9a-719b-11e3-9389-09ef9944065e_story.html?utm_term=.cc39030bcca0

[90] Maguire, Robert, "Inside the $400-million political network backed by the Kochs," Washington Post, January 5, 2014. https://www.washingtonpost.com/politics/inside-the-koch-backed-political-donor-network/2014/01/05/94719296-7661-11e3-b1c5-739e63e9c9a7_graphic.html?utm_term=.bd89f71663ad

[91] Fang, Lee, "Chart: Koch Spends More Than Double Top Ten Unions Combined," Republic Report, March 7, 2014. https://www.republicreport.org/2014/unions-koch/

[92] Bump, Philip, "The Koch brothers may spend $290 million on this election. That's how much 5,270 American households make in a year," Washington Post, June 17, 2014. https://www.washingtonpost.com/news/the-fix/wp/2014/06/17/the-koch-brothers-may-spend-290-million-on-this-election-thats-how-much-5270-american-households-make-in-a-year/?utm_term=.693d14bba8d0

election spending budget of $889 million – about the same as each of the two major political parties[93]; however, once Trump won the Republican Party nomination, they decided to sit out the presidential race and scaled back spending to $750 million – still a record sum[94].

Predictably, all this political spending had an impact. In 2010, Republicans swept the midterm elections, capturing the House of Representatives and making gains in the Senate. Of the 85 new Republicans in the House, 76 had signed the Kochs' "No Climate Tax" pledge, and of those 76, 57 had received campaign contributions from the Kochs[95]. One freshman House Republican was Trey Gowdy, who knocked off what is now an anomaly in politics – a Republican who believes in human-caused climate change. Bob Inglis had represented the 4th District of South Carolina from 1992 to 1998, then again from 2004 to 2010. "You know, I'm pretty conservative fellow," Inglis told Frontline. "I got a 93 American Conservative Union rating, 100 percent Christian Coalition, 100 percent National Right to Life, A with the NRA, zero with the ADA, Americans for Democratic Action, a liberal group, and 23 by some mistake with the AFL-CIO. I demand a recount. I wanted a zero[96]." But Inglis did accept the scientific consensus on climate change and supported a revenue-neutral carbon tax to deal with it. He came to this conclusion after a Congressional trip to Antarctica while serving on the House Science Committee, and at the urging of his five children. As a result, he was derided by conservative talk radio hosts as "off the reservation, somewhere with Al Gore." Inglis lost his primary election, 29 to 70 percent, in what Myron Ebell of the Competitive Enterprise Institute called an overwhelming victory[97].

Inglis's fate became an example for other Republicans – and even Democrats — who would now think twice before talking about the scientific consensus on climate change. Besides electing more Republican climate deniers, the effect of the Koch money was to polarize the political debate over climate change. In 2008, both presidential candidates, Barack Obama and John McCain, acknowledged that climate change is real and needs to be addressed. That year Democratic Speaker of the House Nancy Pelosi and previous Republican Speaker of the House Newt Gingrich filmed a commercial together for former Vice President Al Gore's Alliance for Climate Protection. "We don't always see eye to eye, do we Newt?" Pelosi says. "No," Gingrich replies. "But we do agree our country must take action to address climate change[98]." By 2011, after taking fire in the Republican presidential primary, Gingrich disavowed the ad as "the dumbest thing I've done in the last four years.[99]" Mitt Romney, the 2012 Republican candidate for president, also flipped on climate change. In his 2010 book, *No Apology*, Romney said he believes climate change is occurring and human activity is a factor[100]. By 2011, when asked about climate change at a campaign event, he said: "My view is that we don't know what's causing climate change on this planet. And the idea of spending trillions and trillions of dollars to try to reduce CO_2 emissions is not the right course for us[101]." Even Democrats were affected by this shift in the Overton window on the climate debate. Whereas President Obama had campaigned in 2008 on action to address climate change, after the defeat of cap and trade he chose a strategy of silence as climate change was considered "not a winning message[102]." During the 2012

[93] Confessore, Nicholas, "Koch Brothers' Budget of $889 Million for 2016 Is on Par With Both Parties' Spending," The New York Times, January 26, 2015. https://www.nytimes.com/2015/01/27/us/politics/kochs-plan-to-spend-900-million-on-2016-campaign.html

[94] Vogel, Kenneth, "Behind the retreat of the Koch brothers' operation," Politico, October 27, 2016. https://www.politico.com/story/2016/10/koch-brothers-campaign-struggles-230325

[95] Holmberg and Campbell, "Koch: Climate pledge strategy continues to grow."

[96] Hockenberry and Upin, "Climate of Doubt."

[97] Hockenberry and Upin, "Climate of Doubt."

[98] Allen, Nate, "Nancy Pelosi and Newt Gingrich Commercial on Climate Change," Alliance for Climate Protection, April 17, 2008. https://www.youtube.com/watch?v=qi6n_-wB154

[99] O'Brien, Michael, "Newt Gingrich: Nancy Pelosi Climate Change Ad Is 'Dumbest Thing I've Done In The Last 4 Years," Huffington Post, December 27, 2011. https://www.huffingtonpost.com/2011/12/27/newt-gingrich-nancy-pelosi_n_1171530.html

[100] Jacobson, Louis, "On Mitt Romney and whether humans are causing climate change," Politifact, May 15, 2012. http://www.politifact.com/truth-o-meter/statements/2012/may/15/mitt-romney/mitt-romney-and-whether-humans-are-causing-climate/

[101] Johnson, Brad, "Romney Flips To Denial: 'We Don't Know What's Causing Climate Change,'" Think Progress Green, October 28, 2011. Archived at https://web.archive.org/web/20111130020720/https://thinkprogress.org/green/2011/10/28/355736/romney-flips-to-denial-we-dont-know-whats-causing-climate-change/

[102] Goldenberg, Suzanne, "Revealed: the day Obama chose a strategy of silence on climate change," The Guardian, November 1, 2012. https://www.theguardian.com/environment/2012/nov/01/obama-strategy-silence-climate-change

election, there was no mention of climate change in six hours of televised debates, and neither candidate chose to bring up the issue[103].

Although ascendant, climate denial did not go unchallenged. In 2012, a leak of documents from the Heartland Institute showed $8.6 million in funding for "climate change projects" from an anonymous donor; plans to raise $7.7 million from the Charles G. Koch Foundation, among others; and development of a climate denial curriculum for K-12 classrooms[104]. Later it emerged that the leaker was Peter Gleick, president of the Pacific Institute, who had posed as a board member online to get the documents[105]. That same year, a study by climate skeptic Richard Muller funded by the Charles G. Koch Foundation surprisingly found that climate change is real and caused by humans. Muller, a professor of physics at University of California-Berkeley, changed his mind about climate change, calling himself a "converted skeptic[106]." Then in 2015, documents released by Greenpeace showed that scientist Wei-Hock Soon, whose work attributing climate change to the sun was widely touted in the climate denial sphere, was deeply tied to corporate interests. Soon had accepted $1.2 million from the fossil-fuel industry over the previous decade but had failed to note the conflict of interest in most of his scientific papers. Email correspondence with funders like ExxonMobil, the Koch foundation, and DonorTrust, showed him describing the papers as "deliverables.[107]"

After winning re-election in 2012, President Obama continued to address climate change, but through a new strategy: executive action. The centerpiece of his policy was the EPA Clean Power Plan, introduced in June 2014 to regulate carbon emissions from power plants. With a goal of cutting carbon pollution from power plants 32 percent from 2005 levels by 2030, the policy allowed states to develop their own plans to reach state-specific goals from a suite of possible actions[108]. In response, 24 states sued the federal government, accusing the EPA of exceeding its authority in requiring them to transform their energy systems[109]. Incredibly, most of these states were already on their way to meeting the goals of the Clean Power Plan[110], which was supported by the majority of each state's residents[111].

The Clean Power Plan was the basis for Obama's breakthrough deal with China's President Xi Jinping, announced November 2014, in which the United States agreed to cut greenhouse gas emissions 26 to 28 percent below 2000 levels by 2025 while China agreed to peak carbon emissions and increase renewable energy to 20 percent by 2030[112]. The U.S.-China agreement provided great momentum going into the Paris Climate Conference in 2015. No one knew whether every country in the world would commit to meaningfully reduce carbon emissions, but they did, in part because the world's two largest emitters of carbon had already made commitments. Like the Clean Power Plan, the Paris Agreement was also signed through executive action. While the agreement's rules and procedures were legally binding, its emissions reductions targets were not, which meant it did not have to be ratified by the Senate. In interviews, Secretary of State John Kerry dismissed concerns

[103] Broder, John, "Both Romney and Obama Avoid Talk of Climate Change," The New York Times, October 25, 2012. https://www.nytimes.com/2012/10/26/us/politics/climate-change-nearly-absent-in-the-campaign.html

[104] DeMelle, Brendan, "Heartland Institute Exposed: Internal Documents Unmask Heart of Climate Denial Machine," DeSmog Blog, February 14, 2012. https://www.desmogblog.com/heartland-institute-exposed-internal-documents-unmask-heart-climate-denial-machine

[105] Gleick, Peter, "The Origin of the Heartland Documents," Huffington Post, February 20, 2012. https://www.huffingtonpost.com/entry/heartland-institute-documents_b_1289669.html

[106] Muller, Richard, "The Conversion of a Climate-Change Skeptic," The New York Times, July 28, 2012. https://www.nytimes.com/2012/07/30/opinion/the-conversion-of-a-climate-change-skeptic.html

[107] Gillis and Schwartz, "Deeper Ties to Corporate Cash for Doubtful Climate Researcher."

[108] .S. Environmental Protection Agency, "FACT SHEET: Overview of the Clean Power Plan," August 3, 2015. https://archive.epa.gov/epa/cleanpowerplan/fact-sheet-overview-clean-power-plan.html

[109] Cama, Timothy, "Two dozen states sue Obama over coal plant emissions rule," The Hill, October 23, 2015. http://thehill.com/policy/energy-environment/257856-24-states-coal-company-sue-obama-over-climate-rule

[110] Groom, Nichola, and Valeria Volcovici, "Most states on track to meet emissions targets they call burden," Reuters, September 18, 2016. https://www.reuters.com/article/us-usa-climatechange-lawsuit-insight/most-states-on-track-to-meet-emissions-targets-they-call-burden-idUSKCN11O0E1

[111] Yale Program on Climate Change Communication, "Majority of Citizens in States Suing to Stop the Clean Power Plan Actually Support the Policy," November 2, 2015. http://climatecommunication.yale.edu/publications/61-of-the-public-in-the-states-suing-to-stop-the-clean-power-plan-actually/

[112] The White House, "FACT SHEET: U.S.-China Joint Announcement on Climate Change and Clean Energy Cooperation," Office of the Press Secretary, November 11, 2014. https://obamawhitehouse.archives.gov/the-press-office/2014/11/11/fact-sheet-us-china-joint-announcement-climate-change-and-clean-energy-c

a future Republican president could simply walk away from the deal. Republicans, he said, had "eliminated themselves from contention in the general election" because of their approach to issues such as climate change[113].

Sadly, that is exactly what happened after the election of 2016, again with a lot of influence from the Kochs. Interestingly, Trump was not the choice of the Koch brothers during the Republican primary. They invited five of the primary candidates – Jeb Bush, Ted Cruz, Rand Paul, Marco Rubio, and Scott Walker – to an annual gathering in California informally referred to as the "Koch primary.[114]" But before the Kochs could choose one, Trump surged ahead of the pack, powered by a strain of anti-immigrant protectionism that repulsed the Kochs so much they threatened to back the Democrat Hillary Clinton[115]. Ironically, it was the Kochs' funding of anti-establishment Tea Party protests in 2009 and 2010 that helped lay the groundwork for Trump. "We are partly responsible," said one former network staffer. "We invested a lot in training and arming a grassroots army that was not controllable, and some of these people have used it in ways that are not consistent with our principles, with our goal of advancing a free society, and instead they have furthered the alt-right." After Trump won the Republican primary, the Kochs decided to stay out of the presidential election altogether[116].

Even sitting out the presidential election, the Kochs still wielded a great deal of influence in 2016. First, they concentrated on Congress, funding 18 candidates in the Senate and 82 in the House. Among the top Senate candidate recipients were Roy Blunt (R-Mo.), Ron Johnson (R-Wisc.), Jerry Moran (R-Kan.), James Lankford (R-Okla.), Marco Rubio (R-Fla.), Tim Scott (R-S.C.), Rob Portman (R-Ohio), Kelly Ayotte (R-N.H.), Pat Toomey (R-Penn.), and Mike Lee (R-Utah). Of those, only Ayotte lost. Once Trump was elected, the Kochs oversaw much of the transition team. Initial cabinet members with Koch ties included Vice President Mike Pence, Attorney General Jeff Sessions, Education Secretary Betsy DeVos, Chief of Staff Reince Priebus, White House Advisor Kellyanne Conway, and CIA Director Mike Pompeo[117], who as a representative from Kansas took $375,000 from Koch Industries from 2009 to 2018[118].

Heading up Trump's EPA transition team was none other than longtime climate denier Myron Ebell, director of the Koch-funded Competitive Enterprise Institute. Ebell chose Oklahoma Attorney General Scott Pruitt – who had sued the EPA 14 times including four lawsuits over the Clean Power Plan[119] – to be EPA administrator. Ebell was also the main driver behind Trump's withdrawal from the Paris Agreement. After hearing that Trump was waffling on whether to uphold his campaign promise to leave the pact, Ebell's CEI put out a TV ad with video montages of Trump promising to do just that during the election. "There was a debate, a real debate in the White House and in the Cabinet," Ebell told CNN. "So we just wanted to remind the president which side he's on[120]." It worked.

[113] Sevastopulo, Demetri, and Pilita Clark, "Paris climate deal will not be a legally binding treaty," Financial Times, November 11, 2015. https://www.ft.com/content/79daf872-8894-11e5-90de-f44762bf9896

[114] Reston, Maeve, "'Koch primary' heads to California," CNN, August 1, 2015. https://www.cnn.com/2015/07/31/politics/koch-brothers-presidential-2016/index.html

[115] Isenstadt, Alex, "What the Kochs think about Trump now," Politico, May 5, 2016. https://www.politico.com/story/2016/05/trump-megadonors-koch-222825

[116] Vogel, Kenneth, "Behind the retreat of the Koch brothers' operation" Politico, October 27, 2016. https://www.politico.com/story/2016/10/koch-brothers-campaign-struggles-230325

[117] The Real News Network, "Trump, The Koch Brothers and Their War on Climate Science," Narrated by Danny Glover, May 23, 2018. https://therealnews.com/stories/trump-the-koch-brothers-and-their-war-on-climate-science

[118] Open Secrets, "Mike Pompeo." https://www.opensecrets.org/members-of-congress/summary?cid=N00030744&cycle=CAREER&type=I

[119] The New York Times, "Pruitt v. EPA: 14 Challenges of EPA Rules by the Oklahoma Attorney General," January 14, 2017. https://www.nytimes.com/interactive/2017/01/14/us/politics/document-Pruitt-v-EPA-a-Compilation-of-Oklahoma-14.html

[120] Griffin, Drew, and Miranda Green, "The man behind the decision to pull out of the Paris agreement," CNN, June 3, 2017. https://www.cnn.com/2017/06/03/politics/myron-ebell-paris-agreement/index.html

Chapter 12

Climate Crisis & Selling Doubt
Part III: What can we do?

by

Cathy Cowan Becker, M.A.

"The Greatest Threat To Our Planet Is The Belief That Someone Else Will Save It" **– Robert Swan**

Climate Crisis "Keep/Stop/Start" Actions

Below for your consideration, are several author-suggested action items. For a further explanation of the Keep/Stop/Start organizational tool, please see How to Read This Book.

KEEP:

1) *Keep learning.* You are reading this book, which means you want to learn about climate change. Good! Now check out these 12 classic books about climate change, climate denial, and climate solutions:

   - An Inconvenient Truth: The Crisis of Global Warming, by Al Gore

   - This Changes Everything: Capitalism vs. the Climate, by Naomi Klein

   - The End of Nature, by Bill McKibben

   - Storms of My Grandchildren: The Truth About the Coming Climate Catastrophe and Our Last Chance to Save Humanity, by James Hansen

   - The Sixth Extinction: An Unnatural History, by Elizabeth Kolbert

   - Six Degrees: Our Future on a Hotter Planet Paperback, by Mark Lynas

   - Climate Changed: A Personal Journey through the Science, by Philippe Squarzoni

   - The Madhouse Effect: How Climate Change Denial Is Threatening Our Planet, Destroying Our Politics, and Driving Us Crazy, by Michael Mann and Tom Toles

   - Merchants of Doubt: How a Handful of Scientists Obscured the Truth on Issues from Tobacco Smoke to Global Warming, by Naomi Oreskes and Erik M. Conway

   - Don't Even Think About It: Why Our Brains Are Wired to Ignore Climate Change Paperback, by George Marshall

   - Reinventing Fire: Bold Business Solutions for the New Energy Era, by Amory B. Lovins and the Rocky Mountain Institute

- Drawdown: The Most Comprehensive Plan Ever Proposed to Reverse Global Warming, edited by Paul Hawken

Also check out these 12 publications and websites that do great reporting about climate change:

- The Guardian
- The Washington Post
- The New York Times
- EcoWatch
- InsideClimate News
- DeSmog Blog

- Grist
- Think Progress
- The Daily Climate
- Climate Desk
- Climate Central
- Real Climate

Finally, check out these 12 organizations where you can join with like-minded people who care about and take action to protect the climate and environment:

- Climate Reality
- Sierra Club
- Citizens Climate Lobby
- Greenpeace
- Friends of the Earth
- Food and Water Watch

- Center for Biological Diversity
- Union of Concerned Scientists
- National Wildlife Federation
- Natural Resources Defense Council
- Rainforest Action Network
- World Wildlife Fund

STOP:

1) *Stop arguing with climate deniers.* It is a waste of your time and energy. Remember that a lot of these people are getting paid by the comment, so it is in their interest to suck up as much of your time and energy as possible. Of course, we want to set the record straight when we see misinformation about climate change, so it's perfectly fine to make a comment or two pointing people who might be reading the exchange in the right direction. But when deniers try to suck you into further engagement, don't take the bait. Most onlookers won't read beyond the first few comments anyway.

2) *Conversely, if you are in despair* over climate change and the lack of action by the United States, stop. Despair is not an option. Yes, as this chapter has shown, the fossil fuel industry has successfully delayed climate action for decades. But that doesn't mean they will be able to delay it indefinitely. Once people caught on to delays in regulating smoking, acid rain, the ozone hole, and other issues on which corporate interests have spread misinformation, regulation came swiftly and decisively. Our society has now caught onto misinformation from climate denial.

   Often people say "It's too late, we've passed the point of no return." But this is folly, too. Although our burning of fossil fuels has already done damage to the climate, scientists don't know where the tipping point is – or even if there is one. We do know that the longer we burn fossil fuels, the worse it will get, which is reason enough to stop. The transition to renewable energy will be a process, and the sooner we get started, the more damage we can avert. We don't know for sure that we can save an inhabitable planet, but we do know we can't if we don't try.

START:

1) *Start visiting the Skeptical Science website.* One indispensable tool for dealing with climate denial is the website Skeptical Science at https://www.skepticalscience.com. This site explains climate change science and rebuts global warming misinformation. One great feature is the list of "Most Used Climate Myths," each of which the website rebuts in basic, intermediate, and advanced reading levels with citations to the actual science. If you run across climate misinformation that you are not sure how to rebut, Skeptical Science is your go-to resource to show why denier talking points don't work.

2) *Start Visiting Climate Change Communication website.* Another invaluable resource is the Yale Program on Climate Change Communication, which conducts scientific research on public climate change knowledge, attitudes, policy preferences, and behavior at the global, national, and local levels. The Yale Program is known for its "Six Americas" research sorting climate attitudes into Alarmed, Concerned, Cautious, Disengaged, Doubtful, and Dismissive. Its Climate Opinion Maps identify opinion about various aspects of climate change down to the county and Congressional district level. Finally, its Climate Change in the American Mind report and associated reports let you see how attitudes about climate change have changed over time. Find out more at http://climatecommunication.yale.edu.

3) *Start Countering climate denial* and here are six things you can do:

- Talk about climate change. Social science research about climate change shows that people simply don't talk about it. That allows us to push it to the back of our minds. Simply talking about climate change helps keep it at the front of the agenda.

- Write a letter to the editor. Respond to a story in the paper, keep it short, follow instructions for submittal, and call to make sure they got it.

- Contact your local, state, and national representatives about climate legislation. Calls are more effective than emails, and emails are more effective than signing petitions. Pro tip: Program their phone numbers and emails into your phone for quick and easy reference.

- Sign up to get action alerts from environmental groups, and fill them out regularly. If you are on their mailing list, often they will have your information pre-filled to save time. Add a few personal thoughts to letters and comments to make them more effective.

- When you see people spreading climate misinformation – deliberately or not – call it out. Let them know they have been duped and give them the real information.

- Learn about solutions to climate change and work to make them happen. One example is Citizens Climate Lobby's carbon fee and dividend that would tax fossil fuels at the point of extraction and distribute the proceeds in equal shares to all American households.

~ ~ ~

Cathy Cowan Becker, M.A. can boast a twenty-year career in journalism, communications, and public relations. She is currently communications coordinator for the Mershon Center for International Security Studies at Ohio State University, where she is also pursuing a dual master's in public administration and environment and natural resources. Cathy journeyed on environmental study-abroad trips in Costa Rica in 2015 and Iceland in 2016. She also traveled to Paris for the 2015 Climate Conference and to Oceti Sakowin camp at Standing Rock in 2016. She marched with 400,000 people for climate in New York City in 2014, as well as with 150,000 for science and 250,000 for climate in Washington, DC in 2017. Cathy lives in Grover City, Ohio. Read more at: www. climateabandoned.com.

Chapter 13
Climate Crisis & the Media

by

Betsy Rosenberg

(To my daughter, Jenna, and the upcoming generation.)

Someday the media will learn that the environment is not an intermittent news story, not a special interest, not a win-lose sports event, not a luxury, not a fad, not a movement, not discredited, not faltering, and not something to pay token attention to one day a year. It is a beat far more important than Wall Street or Washington. Its laws are stronger than Newton's laws, it's moves are more important than the Federal Reserve's, its impact overwhelms that of NAFTA or the stock market or the next election. The environment is not one player on the field; it IS the field. It holds up, or fails to hold up, the whole economy and all of life, whether the spotlight is on it or not. — **Donella "Dana" Meadows,** Pioneering Environmental Scientist, Professor, Writer

I first heard this quote more than a decade ago at an environmental conference. Meadows' words so resonated with me that I decided to reach out to her; convinced she and I would become fast friends over our shared frustration with mainstream media's short-sightedness and bias against news about our environment. Soon after that I was stunned and deeply saddened to learn that Meadows had died a few years prior at the age of 59 in 2001.

Two decades have passed since Meadows wrote those poignant and prescient words, but little has changed when it comes to mainstream coverage of our imperiled environment—especially in broadcasting which is my background. I believe that when it comes to coverage about our climate crisis—or lack thereof—the sin of omission is the crucial missing piece in the awareness and action challenge.

Brief Background

From television and radio news networks to newspapers and magazines, outlets that specialize in covering news have marketed their programming as information and analysis that can help us understand our world and hold leaders accountable. Sadly, they have fallen woefully short, especially when it comes to what's happening to nature at the hands of humans. In that regard, it has been a case of Mission Unaccomplished. Mainstream news providers, especially the television networks, have abandoned their responsibility to serve the public's right and need to know the truth about our endangered climate, compromised oceans, disappearing forests, species extinction, and other signs of ecosystem decline.

I left the security of an on-air position with CBS Radio 15 years ago to become an independent environmental "news and views" content creator. The plan was and still is, to bring my acquired eco-expertise back to a mainstream news platform for maximum public reach and impact. I never dreamt it would take this long to fill what is still a glaring, inexplicable, and inexcusable void and weakness in corporate media's coverage of all things ecological— especially in light of recent climate-fueled weather events that have left thousands of American citizens homeless, joining the rising global populations of climate refugees.

I am concluding this chapter at the end of 2018 which will go down as the year climate change reached a tipping point. A succession of catastrophic weather events, bookended by dire government reports, made it all but impossible to ignore—or deny—that something has gone terribly wrong with our weather system. The regular patterns of fire season, the ferocity of storms, depths of flooding and length of droughts are no longer staying within recognizable parameters—something I grew up taking for granted. I have been using the term "global weirding" for more than a decade, and sadly, it seems to be increasingly appropriate.

Weather on steroids has become the new normal, or abnormal. Wildfires in California left 90 people dead and destroyed an estimated 20,000 homes and buildings. Hurricanes Michael and Florence took some 120 lives and flooded or flattened entire towns. The alarm has been sounded: Code Red, with 2018 going down as the year climate change "came out of the closet" and was finally on the public radar, suddenly being mentioned in news media, even on network television. What took so long, and the cost of that lapse to our atmosphere and society is the focus of this chapter.

I wasn't always obsessed with climate change; growing up I did not identify with being an environmentalist, though even as a child I hated waste. I recall in 3rd grade feeling unhappy to see kids throwing away their uneaten lunches—early signs of what I jokingly call my "enviro-mental illness." The first half of my career was spent covering breaking news for KCBS in San Francisco and my 20 years with CBS included two years at the network in New York City. In 2007 I left news to focus exclusively on environmental issues as a result of the dearth of such coverage; what I refer to as "the green gap." Beginning with waste prevention minutes on Earth Day in 1997 (on KCBS), and later breaking new ground with the nation's first daily environment-themed interview show (on Air America Radio), the beat became my calling.

As global warming grew to be a more significant concern than the garbage piling up in our landfills, my focus and fear for our future expanded. So did my anger and anxiety as the scientific findings grew more dire while news coverage remained scant. Newspapers were, and are, better than the radio and TV news channels, with more coverage now —relatively—than five to ten years ago, but still too little too late considering the existential threat.

I have spent the past decades actively in search of a robust mainstream media platform from which to disseminate environmental news and views. At this writing, I have yet to break through the "green ceiling" in my industry, which is twice as thick as the glass ceiling. While there are many women in radio and television news, both on-air and in management, where are the environmentally-focused personalities and programmers? Missing in action! Or put another way there is color discrimination against Greens. Turn on CNN or MSNBC and you'll see shades of Black, Brown (ethnic diversity), Red and Blue-hued hosts (political opponents) and commentators. But where are the Greens: the eco-advocates, environmental activists and climate commentators? Only recently has CNN begun to bring scientists and academics on to discuss climate change but more often they've reverted to the usual political pundits, including Republican "deny-o-saurs" like Rick Santorum, Tom Delay, Joni Ernst, and Stephen Moore.

All of those mentioned above appeared on either CNN or MSNBC to cast doubt on the alarming findings from both the IPCC report released in October, and the National Climate Assessment published in late November. In the days that followed CNN continued to populate panels with known climate doubters, if not deniers, but never the same time as proponents for climate action. Surely the climate experts would have immediately shot down the misstatements, but CNN's anchors are not well steeped enough in the science to do that, perpetuating confusion and doubt, all the while running their ad campaign about "Facts First." The slogan for those promos was, "If repeated often enough, lies can become truth," referring to Donald Trump, while failing to note the irony in their contribution to climate silence as well as Trump being president! I have long maintained that more comprehensive coverage of the climate crisis would better equip voters to elect leaders who are more eco-literate and climate savvy. There is a cost, and there are serious consequences, for that negligence. This was especially irksome since I had been trying to pitch CNN execs on climate solution programming and was met with resistance, being told: "our viewers are not interested." True then or not, that interest is growing by the day.

The story of my efforts to find a network that supports such content reveals a great deal about how, and why, the news media has abandoned its duty to cover environmental news, our looming climate crisis and attendant collapse of our oceans, marine, and wildlife, forests, biodiversity. The abysmal and abject failure on the part of entities tasked with informing the public is both the cause and effect of a more significant problem: Americans' inability, or unwillingness, to come to grips with sobering realities that should otherwise warrant action and demand lifestyle changes on the part of citizens, communities, and country.

The Problem

Part of the problem is cultural: too often in the U.S. we have treated "the environment" as an afterthought issue and viewed protecting it is as an altruistic or aspirational endeavor. Until very recently climate change has appeared at, or near the bottom of national concerns in public opinion polls. That detached attitude is mirrored in the media, though which came first is a chicken and egg conundrum. Whether it's Americans' love of freedom and independence or the rights of individuals being a societal priority, conservation and stewardship have not been prominent values in modern day America and consequences of that negligence will be felt for generations.

Another reason we are losing the battle for a healthy and safe climate has been the assumption, on the part of news managers, that the public has a low interest in ecological issues and there has been catering, or pandering, to give viewers what they supposedly want to watch, (high-interest stories) vs. what they need to be informed of (high importance stories). Seemingly, there have been no grown-ups in the newsroom when decisions were made—or avoided—regarding environmental coverage. It is primarily our children and grandkids who will pay dearly for this lapse. At a time when cable TV news networks are obsessed with politics and Trump, ad nauseam, whether it's the more centrist CNN or the so-called "slanted networks— Fox on the right and MSNBC on the left—as of this writing there is not one hour of ongoing programming dedicated to our climate and related crises, and that is shameful. I cannot fathom a better utilization of mass media outlets than to mobilize collective action to preserve life on this planet since responding, with urgency and proper attention make or break any other human endeavors.

As mentioned above, there has been little media progress since I began covering environmental news in 1997. Most of my focus has been on the "broadcast" side, mainly commercial radio and television networks. On the newspaper side, the Guardian, New York Times, and Washington Post have done better over the last few years, impressing me with their leadership on climate coverage. I can recall when, in 2013, the NYT closed their environment desk, arguing—rather unconvincingly—that they would better serve readers by integrating stories about our changing planet into other sections of the paper. A few years later the powers that be at the Times seemingly woke up to "smell the carbon," hiring a half dozen reporters and editors to cover climate. I don't know the inspiration behind the Guardian's relatively early dedication to covering climate, but I attribute at least some of the paper's prescience to being a UK-based publication. Western European countries have traditionally been smarter about environmental issues and resource use than we've been in the U.S.

As for magazine coverage of climate and environmental news I have been impressed with Bloomberg, Business Week and The Economist over the past few years but I've been paying attention to these stories for two decades and wonder why it took them so long to catch on. While news magazines like Time and Newsweek did a decent job of explaining the problems, there was little ink devoted to solutions.

Another standout in environment and climate coverage has been public radio and television. NPR (National Public Radio) and PBS (Public Broadcasting System) have long fed their audience's appetite for intelligent interviews and program content on substantive issues, much more so than their commercial counterparts. For as long as I can recall NPR has offered a weekly green-themed show, Living On Earth (LOE), which pre-dated my program on Air America, and has endured. As excellent as LOE is, it is pre-produced and recorded before airing which takes away from the air of immediacy I included in my show. Once a news gal, always a news gal!

As for talk radio, long dominant on the AM dial, it has been populated mainly by conservative white male (and a few female) hosts. When I read, a decade ago, that 91% of all talk radio/call-in shows were hosted by conservatives, it partially explained why I could not break through. The people in charge of hiring those hotheads are more interested in creating heat (and too often hate) then they are in shedding, and spreading, light. From Rush Limbaugh to Sean Hannity, and previously Bill O'Reilly to Glenn Beck, these "talkers" are notoriously loose with the facts and deliberate in their attempts to cast doubt on established science and downplay the genuine threats of a climate in flux due to human activity. Don't they have kids, nieces, nephews and grandchildren they love? Having been a guest on Hannity's radio and television shows more than a dozen times, I can attest to the willful stubbornness I experienced on the part of these deny-o-saurs. Whether out of ignorance or arrogance – likely both – these highly paid hosts are not interested in the facts and routinely perpetuate myths and attempt to "kill the messenger" with red meat tactics while ignoring the meat, or substance, of the message. The fact that they're not held accountable for their highly irresponsible positions—on environmental issues alone—blows my mind... and my top...angering me no end. I have long said, after several rebuffed attempts to give Hannity books on climate change after my appearances, that someone should start a campaign to get him "off set."

Many people excuse the yelling and name calling of guests saying the hosts are just being "entertaining" but I cut them no slack, not when there is so much on the line. In the mean time they are laughing all the way to the bank. Rather, they should be escorted to jail given the severe and irreversible effects of climate collapse. It should be a crime to deliberately spread lies about something so critical to humanity's survival— and on a supposed news channel no less! We need more sanity and less inanity from Hannity!

The fact that Fox fans and Breitbart readers absorb the lies— and repeat them as memes if ever in a confrontation about climate change— shows the power of persuasion to the uninformed, even if "not believing" goes against their own interest (as if destroying our life protecting atmosphere would be in anyone's interest!). Given how large their audience, and how long the right-wing talkers have dominated radio (still representing 90% of all shows), I am convinced that right-wing media—along with the Russians—have had a hand in the election of Donald Trump. If more voters were aware of what is truly at stake for their families and future, they should not be voting for climate doubters to lead our country.

In addition to the selective ignorance, we also have a genuine literacy problem when it comes to the environment. Most of my non-activist friends have no clue who Bill McKibben is or that we're in the midst of the sixth mass extinction event, the first one caused by humans. Similarly, in my spontaneous eco pop-quizzes, almost no one knows the significance of the number 350, even when I give a clue such as "in the context of climate change" to otherwise well-educated people. That is not to say if someone fails my little quiz they are not aware, even if only vaguely, that the climate is changing. But in my view it shows a low level of interaction with these matters if a leading author and activist on the climate front – Bill McKibben – does not ring a bell, nor does his organization 350.org, purposely named to sound the alarm that we are already well past the safe level of 350 parts per million and thus, feeling the impacts (we're at or near 410ppm).

This eco-illiteracy problem is not necessarily the fault of citizens. Whenever I ask people who is responsible for educating American adults about ecological issues, without fail, I get blank stares. The fact is that no one, and no government agency, is tasked with informing citizens about what's happening to our planet at our own hands and what we must do to avert disaster. Whether by design or default, we are a nation that's pretty clueless about the state and complexity of ecosystems and their rate of decline across the board.

The Solution

If I am depressing you, hold on for the hopeful news. Since having zero TV programs designed for mainstream consumption on all, or any, green topics, has gotten us to this low level of understanding, imagine the potential impact of daily programming that both connects the dots on what is happening to nature/us, and what we can do to reverse course! Call me nuts, but I imagine that potential every day, which is why I've been fighting to fill this void for 20 years.

So why the dearth of content about saving life on Earth, our only home, on which all things are possible—or were before we started messing with Mother Nature? While there is no proof, I have had enough anecdotal experiences to draw some conclusions. The people in charge of programming news networks – both radio and TV – are not the types to break new ground or experiment with new genres. Prior to my encounters with program execs, I would have guessed that compelling new content areas, and fresh advertising streams would be an irresistible draw. However, resist they do, twisting themselves into pretzels making excuses for their non-interest. Instead, they go with the tried and true program line-ups which for the past few years, mainly fall into the political category. The path of least resistance is more comfortable, or certainly a safer bet, for an executive who wants to keep his or her job. But at what cost to society or their own families?

The aim of news programs is to attract the highest ratings, not always to report the most important stories, trends, and events of the day. Since the 2016 election, CNN's most common format is for one of their hosts to lead a discussion with a panel of pundits on the latest Trump administration misstep, scandal, or Russia investigation, punctuated only by a government shutdown, horrendous crime, or extreme weather coverage (which is mostly devoid of any mention of what's fueling these destructive events). Such programming is cheap and easy to produce. Bring on the political commentators, let them argue with each other over every little tweet, then rinse and repeat. While such easy formats keep the needle moving, it leaves little room for other "real news" and so the public remains under-informed on key issues like a broken climate and warmed-over, plastic-polluted, oceans "on acid" where coral reefs could be wiped out in 30 years.

There can be no strong democracy with a weakly informed populace and for that matter no good politics—or anything— on a bad planet. My perpetual question is: are the network program execs on news channels actually ignorant or do they deliberately minimize coverage on our many eco-challenges for their own purposes? Likely some of both since my encounters with them leave me with the impression that "they don't know what they don't know." In my experience, they don't want to know how rapidly the interconnected web of life is unraveling and even if they are aware, they have assumptions about their audiences that may or may not be accurate. However, they have an obligation to connect dots and inform the public about solutions. They have abandoned that responsibility.

The predominant presumption is that viewers or listeners are not interested in environmental content because it's either "too complicated, too depressing, too boring," or…fill in the blank. However, if essential stories are not presented, and dots left unconnected, then the public remains largely unconcerned. The less people know the less likely they are to express interest in these topics leading news programmers to conclude there is no imperative to make a shift in what gets covered. Can you see how these circular forces keep us stuck?

The other assumption is that there's little we can do about climate change so why bother talking about it? Well, there's less we can do about Trump's irrational behavior or lies so why talk about that night and day? The reality is there are no legitimate excuses, and there has been a bias against news about our natural world. According to Media Matters in 2016 evening newscasts and Sunday shows on ABC, CBS, and NBC and FOX collectively decreased their total coverage of climate change by 66 percent compared to 2015, even though there were a host of critically important climate-related stories, including the announcement of 2015 as the hottest year on record, the signing of the Paris climate agreement, and numerous extreme weather events.

You could say news executives are color-blind in that they can't, or won't, see the green elephant in the room, hiding in plain sight. Moreover, where are the climate savvy commentators and eco-advocates? MIA and I call that the green bias! When CNN's Original Series programming staffers turned down a proposal for a solutions-based environmental series one of the reasons cited for their decision to pass was they did a Town Hall with Al Gore the month prior and a Morgan Spurlock episode a few years earlier on waste. But that barely scratched the surface, and without ongoing programming where context and complexities can be presented, there is little lasting import or impact. The takeaway is they see such programming as something to be avoided rather than as an opportunity to be game-changing and provide a critical public service.

One assumption is there's concern about alienating advertisers, particularly commercials from the American Petroleum Institute, promoting oil and gas exploration as if those energy sources had a rosy future with no consequences to society. I wish I knew how heavily money from the fossil fuel industry influences decisions about content, but I can only speculate.

One of the more glaring examples of this neglect has been the nearly total lack of focus on climate and other environment or energy matters during the past few national elections. When one is as obsessed with the climate crisis as some—but not enough of us—are, it's surreal to watch question after predictable question in the debates about the topics du jour while the future of life hangs in the balance and doesn't even warrant a single question from moderators. And from the very professionals whose job it is to inform the American public! This is much more than an annoying lapse. As long as news outlets treat these existential threats as an afterthought, the public will continue to perceive a lack of urgency. It's what I call, the "crime of omission about our emissions" but it is not a victimless crime if one surveys the impacts of extreme weather on lives and livelihoods.

Although the topic of climate change was brought up by a few of the leading Democratic presidential candidates in 2016, most often and vociferously by Bernie Sanders, it was never "exploited" by the top contenders in a way that should, and perhaps could, have sunk the Republican candidates, certainly the one who called it a "Chinese hoax." Let's hope Democrats don't make the same mistake in 2020, especially when the GOP is so weak on this issue. Sanders was on my radio show more than a decade ago talking about legislation he'd introduced to cap emissions, so he has been an early and steadfast climate hawk. The fact that he was the only leading candidate to take the issue seriously—bringing it up regularly in debates and interviews— speaks volumes about what is wrong with this picture, though there are already signs that things will be different next time.

When I pressed a few high-level aides with the Hillary Clinton campaign on why she wasn't using climate as a leverage issue against Trump, I was told it didn't rate as one of the top concerns, even for Democrats. Given how close the election ended up, and the dire climate consequences of a Trump presidency, I would argue this was a significant tactical error.

I have been focused in this chapter on network news, primarily television and radio. While I have touched on a few standout newspapers, there was one episode that was revealing involving a major California paper. The San Jose Mercury News, which covers Silicon Valley and the greater Bay Area, has a strong journalistic reputation so when its longtime natural resources and environment writer, Paul Rogers, was speaking on a panel about media and the environment a few years ago, I attended. After the discussion, someone in the audience asked why the paper did not devote much if any, space to solutions. Rogers, somewhat dismissively, responded by saying "It's not our job to educate about solutions, that's for Greenpeace and The Sierra Club." His candid comment revealed at least part of the problem; if it's up to the green NGO's to inform the public about eco threats and solutions—but those groups can only reach their members— then who is talking to the rest of America?

It's not the EPA. In a 2017 campaign rolled out to "Save the EPA" (under Scott Pruitt's terrible tenure) a list of all EPA services was publicized to show what we cannot afford to lose under Trump's anti-environment appointees. In looking at the top ten duties tasked to the EPA I could not help but notice that "educating the public" is not among them. So, again, I ask, who or what institution, is responsible for educating the public about environmental issues critical to preserving human life? When I pose this question to people, they often say it's going to be up to our kids to solve. While the upcoming generation is learning more about environmental science in school than we Baby Boomers did, there isn't time to wait for them to come into power as the climate window closes.

It is the duty of we citizens, parents, and politicians to set a good example and not leave behind an environment that is in worse shape than what we inherited. I lay much of the blame on our news media, though the failure (until late 2018) to cover and treat climate news with proper scale and scope is partly a reflection of our society's tendency to marginalize ecological issues and environmentalists.

What will it take to get America to wake up and smell the carbon so we can start to take seriously the need to get off our gasses? For years, friends who knew of my efforts to find a mainstream news platform to help bring us from critical mess to critical mass would say, "It will take a crisis to get people to pay attention." To which I would reply: "What crisis do you want to talk about, there are so many happening all around us (whether in the news or not)?

Talk of a climate connection rose (slightly) in the wake of 2017's trio of terror-inducing "horror-canes" which leveled Caribbean islands and parts of Houston, Florida Keys and nearly all of Puerto Rico. Hurricanes Harvey, Irma, and Maria – which a friend blended to call "Harmia" – did spark some press and public talk about climate change but the coverage stopped before ever getting to the "what do we do about it?" part.

That's where the kind of program pitched to CNN would come in. "Meet the Solutionaries" is a 6-8 episode series which would, as its title implies, showcase exemplary individuals and groups making a difference in turning the eco-tide. Although I cannot think of more timely or compelling content now, and CNN would certainly look like a programming innovator, the network turned me down. The official reason? "We are not ready to pull the trigger on an environmental series and don't think our viewers would sustain interest." And yet... the typical 24-hour news day on CNN and MSNBC is filled with different hosts talking about the same thing...Trump and politics. If there is any discussion of climate, it is usually done with political commentators, but there is also the latest science, economic and health impacts, consequences for national security with a growing world refugee crisis, and the connection between our addiction to oil and rise in terrorism. Why can't supposedly well-informed news executives see that these stories are not so much "about the environment" as they are about us—and our future survival?

In recent years, as the popularity of online news platforms has grown and there is more climate content many have asked why I'm so fixated on traditional, or corporate, media. The answer is multifaceted. First off, even though the audiences may be on the decline for cable TV, rumors of the industry's demise are greatly exaggerated. CNN, MSNBC, and Fox still draw millions of viewers, and the glaciers are melting faster than these audiences are shrinking! It is on these networks that you can access the broadest possible reach of mainstream viewers, including the bulk of yet-to-be-converted climate concerned citizens. Contrast that with self-selecting internet news. Conservatives read Breitbart and the Daily Caller while progressives flock to Axios, the Daily Kos and Huffington Post. When I did my program on the internet, at a site called the Progressive Radio Network, I reached green-leaning liberals, predominantly, "the choir." That has never been my interest because unless we grow the choir, we won't move the needle enough in time to avert the worst consequences to planet and people.

Secondly, I am not a millennial; I'm a baby boomer and my fellow middle-aged boomers still get their news from television. That's whom I am trying to reach because those of us in our 50s, 60s and 70s still have several decades ahead of us (or so we hope!) and make consumer choices each day that can have either a positive or negative impact. We also vote every few years, and that may be where we can have the most significant influence.

When faced with the facts, it's hard to turn away. That's where the Fourth Estate has abandoned its obligations and I contend that the missed opportunity has cost us precious time. Every day that we are not aggressively attacking this enemy of our own making means more lives lost in future "freak" weather events.

In Fall of 2018, the back to back government reports issuing alarming warnings about warming oceans, rising seas and emissions levels that must be cut in half over the next decade did finally get the mainstream media talking. The networks covered the reports and even mentioned climate in conjunction with the record hurricanes and fires. However, they quickly reverted to the usual far less consequential news programming.

The one exception was NBC's Meet the Press which, in late December, devoted its entire hour to climate change. The show was hailed as courageous (and long overdue!) but was primarily the result of pressure after host Chuck Todd put a climate denier on a previous segment on climate. Because it was in the wake of the stark IPCC and National Assessment on Climate reports, critics pointed out that there's no longer any legitimate debate on global warming, therefore there is no room for deniers who create a sense of "false balance" and confuse the public, allowing for continued complacency.

While it was gratifying to see an entire program discussing the climate crisis, the show was short on solutions and may well have been a one-off. The question I have is why is such critically important programming—never dull or depressing if appropriately handled since there is SO much to talk about—considered such a radical thing? What better use of our airwaves now in response to this earth emergency? In light of the scary reports, and continuing weird, wacky and wicked weather, I will be pitching anew in 2019. Wish me luck and stay tuned!

Climate Crisis "Keep/Stop/Start" Actions

Below for your consideration, are several author-suggested thought, behavior, and action items. Action nourishes inspiration! For a further explanation of the Keep/Stop/Start organizational tool, please see How to Read This Book.

KEEP:

Keep caring by continuing to educate yourself. There are numerous places to learn more, mostly online, including Eco-Watch, Climate Progress, Inside Climate News, Grist and the Daily Climate. I hope by the time you read this I will have succeeded in getting a television program on environmental issues onto a mainstream platform that is connecting all the dots and tapping into the wisdom of so many solutionaries. In the meantime, you can visit my website at www.betsyrosenberg.com and under broadcast archives you will find hundreds of audio segments on all shades of green topics, featuring interviews with scientists, politicians, activists, authors and sustainable business leaders.

STOP:

Stop feeling discouraged about the slow state of progress on the climate front. Although there are reasons aplenty to be depressed or even despairing about what we have done to our precious planet, the truth is people are finally waking up to some scary facts, and soon the deny-o-saurs will not dare to roar or rear their ugly heads, ever again in public!

START:

1) *Start talking about the climate crisis* whenever possible, as loudly and frequently as people and places will allow.

2) *Start to communicate what you know* via op-eds and speaking in your community, at your kids' school, your church or synagogue, your workplace. If people object, saying it's "too political" or "too controversial" correct them politely by acknowledging that yes, the issue has been politicized by special interests invested in continuing the status quo, but it should never have been since our one thin atmosphere being compromised by our emissions is a practical matter! Too often this has been used as an excuse not to discuss the biggest issue of our time. I say let's stop complaining about the weather and consider what's fueling it and what we can do about it! I support the usual recommended behaviors: to fly and drive less, eat a plant-based diet and have fewer children. If you've already had your family raise your kids to be good stewards of our only home. Soon it will be theirs to tend. Let's hope they do a better job but let's help give them a chance.

3) *Start to call and/or write the* television stations you watch— whether local or national—and let management know that you'd like to see more coverage on climate change and other eco-challenges, including on solutions!

~ ~ ~

Betsy Rosenberg is an award-winning national broadcast journalist and a green media trailblazer. After two decades as an on-air anchor and reporter with the CBS Radio Network, she shifted her full-time focus to environmental news. Beginning in 1997, with one-minute waste prevention features on KCBS in San Francisco (TrashTalk), Betsy went on to produce and host the nation's first daily interview show dedicated exclusively on ecology issues, with an emphasis on human impacts. EcoTalk was heard in forty U.S. markets on the Air America Radio Network from 2004 to 2007 and explored topics across the green spectrum, with a heavy emphasis on the emerging climate crisis. Betsy has earned the distinction of being a world-class, eco-journalism pioneer, interviewing leading scientists, politicians, business innovators, activists, and artists in the environment, climate, and energy sectors. She has appeared as a commentator on local and national television programs, including more than a dozen appearances on Fox's Sean Hannity Show. Betsy lives in Austin, Texas Read more at: www.climateabandoned.com.

Chapter 14
Climate Crisis & Stress

by

G. Elizabeth Kretchmer, M.F.A.

"Yet few think…of the healing power of nature." **– John Muir**

Planet Earth is chronically ill.

As with many chronic illnesses, a variety of symptoms plague our planet, and with every passing day, the situation feels increasingly dire. Way back in November 2016, I attended a film festival in Friday Harbor, Washington. At the time, I thought I already understood climate change, and prior to that weekend, I had been moderately worried about it. But the powerful documentaries shown over the course of just a few days opened my eyes wider. Then, later that week, Donald Trump won the presidential election.

Ever since, it felt as if I was shouldering an immense burden, now that I more fully understood the gravity of our global situation, and the October 2018 report by the United Nations Intergovernmental Panel on Climate Change made me feel even worse[1]. I worry about my children. I fear for the future of humanity and all sentient beings. I suffer from guilt as I shop and drive and live my first-world life, releasing carbon into our thinning atmosphere every day, even as I try to minimize my carbon footprint. The pressure of climate change is wearing on me, and I know I'm not alone. As citizens of a planet suffering from a serious chronic illness, many of us are wrestling with an insidious undercurrent of climate crisis stress.

A Refresher About Stress

Stress levels rise when we feel physically threatened, when our core needs aren't being met, when our values feel compromised, or when we feel like we're losing control.

Physical symptoms can include racing heartbeat, fatigue and insomnia, and gastrointestinal upset, but chronic stress can also lead to a proliferation of more severe physical ailments. As if the physical manifestations aren't bad enough, stress can also affect cognitive function. We might obsess about some things while totally forgetting about others. We have trouble concentrating. We struggle to make sound, rational decisions. And when it comes to our emotions, stress can be entirely debilitating. We become irritable and angry, get easily frustrated, and suffer from depression, despair, panic attacks, and guilt. We have difficulty handling otherwise trivial problems and we might even lash out at those we love most.

Although conventional wisdom has claimed that stress can be helpful, it's well known that it can also be dangerous for us, individually and collectively, as our world becomes more complex. Health guru Deepak Chopra called modern life a "battleground…of stress[2]," and our battles are coalescing into a full-blown war we have to fight every day.

---

[1] "Do What Science Demands 'Before It Is Too Late,' Secretary-General Stresses in Statement on Special Global Warming Report." United Nations Meetings Coverage and Press Releases. October 8, 2018. https://www.un.org/press/en/2018/sgsm19282.doc.htm.
[2] Deepak Chopra. "Two Inescapable Facts About Your Stress." Deepak Chopra, May 8, 2015.

Climate Crisis Stress

The atomic bomb became a worldwide source of stress in the mid-twentieth century, inspiring William Stafford's "At the Bomb Testing Site" in 1953:

> *At noon in the desert a panting lizard*
>
> *waited for history, its elbows tense,*
>
> *watching the curve of a particular road*
>
> *as if something might happen.*
>
> *It was looking at something farther off*
>
> *than people could see, an important scene*
>
> *acted in stone for little selves*
>
> *at the flute end of consequences.*
>
> *There was just a continent without much on it*
>
> *under a sky that never cared less.*
>
> *Ready for a change, the elbows waited.*
>
> *The hands gripped hard on the desert*[3].

James Hansen, the former Director of NASA's Goddard Institute for Space Studies, described the energy trapped by manmade global warming pollution as far more lethal, "equivalent to exploding *400,000* Hiroshima atomic bombs per day, 365 days per year.[4]" It's no wonder, then, that so many of us feel like Stafford's lizard, our hands gripping hard as we wait for the unthinkable.

Stress on the Frontline

The millions who have been assaulted by catastrophic storms, historic droughts, severe floods, and deadly wildfires are on the climate crisis frontline. Clearly, they've been confronting stress in the midst of so many losses, from the loss of power to loss of loved ones. They, more than anyone else, are keenly aware of the economic and physical trauma caused by climate change events, as well as the emotional impact.

Ross Martin, a Manhattan landscape architect, described his experience with Hurricane Sandy, aka Frankenstorm, in 2012. As the storm hit, he hunkered down in his apartment in Zone B, where evacuation was recommended but not mandatory. "The winds sounded like a jet plane and shook the buildings. The East River breached a nearby seawall, and the road became a river of saltwater with torrents several feet high. The Con Ed Plant, five blocks north, blew several transformers, sending up fireworks into the sky, rattling our windows, knocking out power. It looked like the end of the world." Like many of Sandy's victims, he ran out of food, and all the local stores and restaurants were flooded and closed. He went hungry for days while also being cut off from cell service and the Internet and, therefore, from loved ones and the world at large.

More recently, Hurricane Harvey wreaked emotional havoc in Houston, Texas. Called one of the worst natural disasters in United States history and a 1-in-1,000-year flood event, the August 2017 tropical storm epitomized the ever-increasing weather severity climate scientists have been warning about, while also demonstrating how meteorological events inflict significant fear and worry around the world, not only for those in the midst of the

---

[3] Copyright ©, by William Stafford from Ask Me: 100 Essential Poems. Reprinted by permission of Graywolf Press.
[4] Joe Romm. "Earth's Rate Of Global Warming Is 400,000 Hiroshima Bombs A Day," ThinkProgress, December 22, 2013. http://thinkprogress.org/climate/2013/12/22/3089711/global-warming-hiroshima-bombs/

storm but also for their loved ones in distant places. Facebook was flooded with posts after Harvey hit, as friends and relatives desperately searched for information.

Dr. Howard Frumkin, Professor of Environmental and Occupational Health Sciences at the University of Washington, said that mental health issues are one of the greatest public health impacts from storms like these. After Hurricane Katrina assaulted New Orleans and displaced hundreds of thousands, the rates of domestic violence, substance abuse, and depression rose substantially[5].

Climate Crisis and PTSD

The initial shock and trauma of a severe weather event can clearly be damaging to the psyche, but its effects often continue long after the storm has passed. Victims, as well as first responders, frequently experience post-traumatic stress disorder (PTSD) from climate change events, living through the disaster over and over via flashbacks, nightmares, and obsessive thought patterns. They often suffer from debilitating physical effects as well. The combination of physical and mental malaise can ultimately lead to behavioral changes, including violent behavior. One study after Hurricane Floyd hit the eastern seaboard in 1999 showed an increase in inflicted traumatic brain injury on children "well past the immediate disaster period." The report's authors suggested the stress of natural disasters of this magnitude can contribute to an increase in psychiatric symptoms as well as financial hardship and loss of social ties, all of which can lead to a rise in abusive situations[6].

PTSD is especially likely to occur when individuals experience more than one major traumatic event or a continuing pattern of events. Residents and aid workers in North Africa and the Middle East have been battling ongoing food shortages for years. In British Columbia, firefighters have been fighting devastating wildfires for the last several summers. California citizens have long been dealing with drought, while Miami residents have to withstand frequent flooding of their streets from sea level rise. As climate change contributes to an ever-rising frequency of severe weather events, the likelihood increases that more and more citizens and first responders will suffer the effects of PTSD from chronic or repeated exposure.

Women, already vulnerable to domestic abuse and economic hardship, fare especially poorly in traumatic experiences, according to Psychiatrist Allen Frances and Psychologist Katie Cherry[7]. Other disadvantaged populations are also at greater risk. The Inuit communities in Canada, for example, who were once known as the People of the Sea, are reporting more and more cases of depression and substance abuse and greater deterioration in overall self-worth as the polar ice cap disappears and their cultural identity dissipates. A 2015 paper published by the *Indian Journal of Occupational and Environmental Medicine* confirms that stress from ongoing climate events, such as rising ambient temperatures, contributes to a corresponding rise in mood disorders, aggressive behavior, and acculturation issues for the population at large. Farmer suicide is one frightening consequence as prolonged drought wipes out crops and destroys soil fertility. Interestingly, these suicides have far-reaching effects, as they impact the availability of food for many others[8].

When it comes to climate crisis-triggered stress and PTSD, one oft-overlooked group is the community of climate scientists and activists. The nature of their work puts them squarely in the face of doom and gloom, day in and day out. "To be a climate scientist is to be an active participant in a slow-motion horror story," wrote Kate Marvel, associate research scientist at NASA's Goddard Institute for Space Studies and Columbia University's

---

[5] Stephen Miller. "Why Climate Change Belongs in the Health Care Debate." Yes! Magazine, June 29, 2017. http://www.yesmagazine.org/planet/why-climate-change-belongs-in-the-health-care-debate-20170629

[6] H. Keenan, S. Marhsall, M.A. Nocera, D. Runyan. "Increased Incidence Of Inflicted Traumatic Brain Injury In Children After a Natural Disaster." American Journal of Preventive Medicine. April 2004, Volume 26, Issue 3, p. 189-193.

[7] Jayne O'Donnell. "Harvey Can Be Hazardous To Your Mental Health, Even If You Aren't In the Path of the Storm." USA Today, August 26, 2017. https://www.usatoday.com/story/news/2017/08/26/harvey-mental-health-path-storm/605042001/

[8] S.K. Padhy, S. Sarkar, M. Panigrahi, and S. Paul. "Mental Health Effects of Climate Change." Indian Journal of Occupational and Environmental Medicine, January-April 2015.

Department of Applied Physics and Applied Mathematics[9]. She added that she dreads the day she'll have to tell her son what we've done to his planet. This overwhelming stress weighing on climate scientists and activists is further compounded when they're confronted by climate deniers. Even former Vice-President Al Gore, climate activist extraordinaire, has acknowledged occasional disheartenment; one of his darkest days was when he heard the news that President Trump vowed to pull the United States from the Paris Climate Accord. "Despair can be paralyzing," Gore warned[10].

The Shadow of Apocalypse

Beyond those who are directly impacted by severe weather incidents are the millions who comprehend the severity of the issue at hand and what the future will bring for all. News reports abound regarding the latest species facing extinction or the spread of another water-borne infectious disease. Debates drag on about the plight of starving refugees and rising sea levels in our own back yards. Predictions by renowned scientists, including speculation that our Earth could become uninhabitable by the end of this century, terrorize nearly every parent and grandparent on the planet. It's no surprise when blood pressures rise with the planet's temperature and emotions destabilize as polar ice melts. The planet's future viability seems to be slipping farther and farther from our grasp, primarily as a result of economic greed, and we're seeing more mental health issues as we realize the end of the world isn't just fodder for a Hollywood movie.

In 2012, the National Wildlife Federation (NWF) issued a report, co-authored by Forensic Psychiatrist Lise van Susteren and the Federation's Vice President for Education Kevin J. Coyle. That report should be required reading for everyone, as it predicted that climate change would become a "top-of-mind worry" for many Americans. The report foresaw a steep rise in social disorders in the population at large, stemming from chronic worry and manifesting as "depressive and anxiety disorders, post-traumatic stress disorders, substance abuse, suicides, and widespread outbreaks of violence.[11]" Unfortunately, this already sounds all too familiar.

An increasing percentage of our population already reports feeling the effects of guilt, which compounds stress. On a personal level, we feel like we're failing Earth every time we fill up our fuel tanks or buy something shrink-wrapped in plastic. On a larger scale, many of us, particularly citizens of the United States, feel guilty by association. The NWF report said the United States is perceived worldwide as a "global warming villain," and that was back in 2012; surely, that reputation has worsened since the election of Donald Trump. The concept of good guys vs. bad guys has escalated since the 2016 election, and many of my own friends have confessed they've begun to take a closer look at friends and neighbors who refuse to accept climate change as a serious matter. It's possible that climate change is the largest socially divisive matter since the Civil War. How unfortunate, given that we'll need to come together if we are to have any chance of solving this universal problem.

Kim Rice, a retired radiologist, became an overnight environmental activist after the 2016 election threw her into a state of guilt and despair. Her home in Washington State's San Juan Islands once overlooked a sea abundant with orcas and other marine life. Those sea animal populations have dwindled dramatically, and she now looks out her panoramic window at a parade of barges and container ships transporting fossil fuels and plastic gadgets. Rice grieves for her grandchildren, dreading they may never know the joys of the natural environment where she has spent much of her life, and she also fears for their future. "They'll be spending their lives witnessing or participating in endless wars over disappearing resources," she says. "It's an existential problem, and those types of problems are, perhaps, the most stressful of all." She, too, feels the effects of the divisiveness that has arisen

[9] Kate Marvel. "We Should Never Have Called It Earth." On Being with Krista Tippett, August 1, 2017. https://onbeing.org/blog/kate-marvel-we-should-never-have-called-it-earth/?utm_source=On+Being+Newsletter&utm_campaign=1ab6bc282d-20170812_cloud_cult_newsletter&utm_medium=email&utm_term=0_1c-66543c2f-1ab6bc282d-69961737&goal=0_1c66543c2f-1ab6bc282d-699617
[10] An Inconvenient Sequel: Truth to Power, dir. Bonni Cohen and Jon Shenk, Paramount, 2017.
[11] Kevin J. Coyle and Lise Van Susteren. "The Psychological Effects of Global Warming on the United States and Why the U.S. Mental Health Care System Is Not Adequately Prepared." National Forum and Research Report February 2012. National Wildlife Federation, p. v.

among us about the climate crisis and how mean-spirited some of the climate deniers have become. "This crisis has put us all in a cauldron that's going to exacerbate how poorly we treat one another," she said.

Frumkin called Rice's perspective "living under the shadow of possible apocalyptic scenarios." Van Susteren referred to it as "pre-traumatic stress disorder[12]." Regardless of what we call it, the idea that we're carrying the weight of the world on our shoulders is no longer metaphorical. Some days, we naturally feel we just can't really make a difference no matter how hard we try. And the grim outlook on so-called social media only heightens our anguish. Many specialists worry that today's children are at the greatest risk for stress complications because they're painfully aware that their generation and subsequent ones will bear the brunt of climate mistakes, and neglectful decisions, made in the past.

For those of us who love Earth, watching her demise is like watching a loved one suffer from a relentless disease over which we have no control. We must do whatever we can to help, but as many caregivers say, we also need to nurture ourselves and take care of each other.

The serenity prayer, notably adopted by Alcoholics Anonymous in the 1940s, has helped thousands of alcoholics find strength and hope. Regardless of one's religious position or drinking proclivities, it can offer solace for environmental enthusiasts too:

> *God, grant me the serenity*
> *to accept the things I cannot change;*
> *the courage to change the things I can;*
> *and the wisdom to know the difference.*

I am learning to accept the disheartening fact that our species is destroying the environment at an alarming rate. I cannot change that. I am also beginning to recognize how, like many others in the developed world, I've become addicted to convenience. This is something I *can* change. I can cut back on unnecessary travel, for example. I can make more foods and other products at home from scratch. I can endure other inconveniences, one Ziploc bag at a time.

On the other hand, I also know that I'll never be perfect. I have to remember and accept that managing stress in the face of climate change isn't about being perfect. It's not about solving the entire world's problem, in the same way that managing a chronic illness doesn't mean trying to find the cure. Managing it involves changing our thoughts, actions, and expectations. By recognizing how the climate crisis is impacting our mental health and by understanding the choices we have, we can make decisions that allow us to enjoy the only lives we have, on the only planet we have, while staying true to our core values and fulfilling our needs.

Coping with Climate Change Stress

As many patients with chronic diseases such as arthritis or fibromyalgia can attest, pain is often invisible and intangible. People who don't feel it don't understand or believe the magnitude of the symptoms or disease. As a result, false assumptions are made, and inadequate solutions are proposed. Sometimes, the illness is completely denied, and the patient's concerns are dismissed with a callous, "It's all in your head."

The situation surrounding the climate crisis is similar. The standard recommendations for dealing with stress, such as meditation, healthy eating, adequate sleep, and exercise, apply to climate crisis stress too. These sorts of activities are soothing and help us feel better, but there are other things we can do to mitigate our climate misery, both as individuals and as a society.

---

[12] Daniel Oberhaus. "Climate Change Is Giving Us Pre-Traumatic Stress." Motherboard: It's not Just You, February 4 2017. https://motherboard.vice.com/en_us/article/vvzzam/climate-change-is-giving-us-pre-traumatic-stress

Climate Crisis "Keep/Stop/Start" Actions

Below for your consideration, are several author-suggested action items. For a further explanation of the Keep/Stop/Start organizational tool, please see How to Read This Book.

What *You* Can Do, as an Individual

It's easy to become disheartened with each new report we read. As the character Tyrion Lannister said in the epic *Game of Thrones* television series, our "minds aren't made for problems [this] large[13]." However, rather than crawling under a rock, as many of us are often tempted to do, the better choice is to face the dragon.

KEEP:

1) *Living!* First and foremost, continue to experience the planet you love. Great thinkers, dating all the way back to Cyrus the Great, have known that nature can calm the mind and heal the soul. Contemporary studies confirm this, reflecting reductions in the stress hormone cortisol, blood pressure, and heart rates, when we get out and enjoy nature. Unfortunately, many of us still spend too much time trying to solve our problems from our indoor desks and laptops. So, take a break and get outdoors as often as possible, whether that means soaking in a natural hot spring or strolling around the block and listening to the sassy squirrels in the trees.

2) *Learning!* Of the many available resources about our environment and the climate crisis, one of the most user-friendly is Drawdown: The Most Comprehensive Plan Ever Proposed to Reverse Global Warming, edited by Paul Hawken. This book lists 100 substantive actions to save the planet, and while it contains enough information to be inspiring, it's not overwhelming[14].

3) *Loving!* Frequent reminders about why you love the environment, through mediums that speak to the heart like art, film, or literature, can soothe the soul. One of the best uplifting films I've seen lately is *Tomorrow: Take Concrete Steps to a Sustainable Tomorrow*. Its filmmakers, Melanie Laurent and Cyril Dion, present numerous alternative solutions to our global agricultural, economic, energy, educational, and political problems.

Poetry can be especially powerful, and there's no one like Wendell Berry to trigger an internal sense of joy about nature and its importance. One of my all-time favorite poems of his is "The Peace of Wild Things":

When despair for the world grows in me
and I wake in the night at the least sound
in fear of what my life and my children's lives may be,
I go and lie down where the wood drake
rests in his beauty on the water, and the great heron feeds.
I come into the peace of wild things
who do not tax their lives with forethought
of grief. I come into the presence of still water.

[13] 13 Game of Thrones, Episode 3: "The Queen's Justice," writ. David Benioff and D.B. Weiss, dir. Mark Mylod, HBO 2017.
[14] Paul Hawken, ed., Drawdown: The Most Comprehensive Plan Ever Proposed to Reverse Global Warming (New York: Penguin Books, 2017).

And I feel above me the day-blind stars
waiting with their light. For a time
I rest in the grace of the world, and am free[15].

Homo sapiens are a social species, so *living, learning, and loving* also mean connecting with others. Seek out those who share your concerns and exchange ideas with them. And seek out those who are despondent, too, and give them reason to hope.

STOP:

Wasting! Waste is a major cause of Earth's impending demise, Goodall said. According to the Natural Resources Defense Council, America wastes up to 40 percent of its food, not to mention 18 percent of farming fertilizer and 21 percent of irrigation water used to grow food that ultimately winds up in the landfill[16]. But it's not just food we waste. Environment Victoria, an advocacy organization in Australia, reports that almost 99 percent of everything we buy becomes waste within six weeks of purchase[17].

We experience stress when our actions conflict with our values, so it makes sense that we can mitigate stress by changing our buying habits. Beth Henderson, an environmental consultant, has found that waste reduction improves her self-esteem, generates a sense of empowerment, and supports her personal desire to live a virtuous lifestyle. In short, it makes her happy[18].

START:

1) *Joining!* Attend rallies and other community events or organize your own neighborhood discussion groups. Revolutionary naturalist Dr. Jane Goodall said that meeting with people is how you can most successfully create change. "If we think locally, get together with other likeminded people, [and] take action, we realize there is something we can do," Goodall said[19].

2) *Writing!* John Steinbeck said that writing helped him find happiness again. William Carlos Williams believed it relieved distress. Virginia Woolf said that when she put her pain into words, it could no longer hurt her. Numerous studies have demonstrated that writing can be therapeutic, even healing. Sherry Reiter, a pioneer in poetry therapy, said expressing the self is a matter of "creatively clearing one's way through the brushfire and discovering a path forward[20]." And perhaps best of all, it helps show how your story links to our collective story in this turbulent time.

3) What *We* Can Do, as a Society

If climate change contributes to societal unrest, it stands to reason that if we come together in unity, we can create a powerful force to counteract its negative effects.

---

[15] Credit: Copyright © 1999, by Wendell Berry, from The Selected Poems of Wendell Berry. Reprinted by permission of Counterpoint.

[16] Dana Gunders. "Wasted: How America Is Losing 40 Percent of Its Food From Farm to Fork To Landfill." Natural Resources Defense Council, August 16, 2017. https://www.nrdc.org/resources/wasted-how-america-losing-40-percent-its-food-farm-fork-landfill

[17] "Why Waste Matters." Environment Victoria, June 16, 2016. https://environmentvictoria.org.au/resource/waste-matters/

[18] Beth Henderson. "Preventing Waste: It Just Feels Good." WeHateToWaste, January 19, 2015. http://www.wehatetowaste.com/preventing-waste-feels-good/

[19] Dr. Jane Goodall Teaches Conservation: Official Trailer, creative dir. Ankush Jindal, MasterClass, 2017.

[20] Sherry Reiter. Writing Away the Demons: Stories of Creative Coping Through Transformative Writing (St. Cloud, MN: North Star Press, 2009).

KEEP:

Swimming! The Chinook salmon is a keystone species in the ecosystem of the Pacific Northwest, and the health of these cold-water fish is in great peril thanks to changes in weather patterns, warmer waters, the corresponding proliferation of pests and viruses, and ocean acidification. Many streams and rivers will likely become uninhabitable for salmon during this century in the same way other parts of the Earth will become uninhabitable for humans.

Although the salmon undoubtedly sense they're at risk, they don't give up. They keep fighting their way upstream by the thousands each year, sometimes for hundreds of miles, against strong currents and up waterfalls, to return to their birthplaces so they can spawn before they die. Why? For the sake of future generations.

Climate scientists and activists are frequently asked if it's too late. United Nations Secretary-General Antonio Guterres urged the world to "rise to the challenge of climate action and do what science demands before it's too late.[21]" It's hard not to give up, but Al Gore said in his June 2017 climate reality leadership workshop that, so long as we keep making changes in our homes, schools, workplaces, and communities, we can still influence the *pace* of climate change and avoid complete catastrophe[22].

STOP:

Letting big corporations and climate deniers control our emotional wellbeing! This is easier said than done, but if we collectively chip away at the power of big corporations and climate deniers bit by bit we can also chip away at our stress. Here are five steps we can take:

1) Put our collective dollars where our mouths are. Corporations and politicians respond to economics. When families, employers, churches, book clubs, and other economic or social organizations band together to support those companies and governmental leaders that take climate change seriously, and take business away from those that don't, we can make the difference we long to make.

2) Join together in grassroots protests against offending corporations and politicians. By sending messages as a group, our voices are louder. A side benefit is elevated dopamine, the feel-good hormone, derived from the camaraderie and sense of belonging.

3) Educate others. If each of us gives one lunch-hour presentation per month to a local business or governmental agency about the impact of climate change and what these organizations can do about it, we could make a big difference! Sharing our knowledge with others makes us feel better, alleviates stress, and allows us to be more in control, so it's a win-win for speaker and audience alike.

4) Use the power of social media. Nobody likes bad press, and sometimes that's exactly what's needed to precipitate change. After Greenpeace posted its "Everything is NOT Awesome" video on YouTube in 2014, LEGO announced that it wouldn't renew its contract with Shell Oil Company.

5) Turn them off. Group boycotts have been used in the past to make a statement and have successfully garnered attention for the issues at hand.

---

[21] "Do What Science Demands 'Before It Is Too Late,' Secretary-General Stresses in Statement on Special Global Warming Report." United Nations Meetings Coverage and Press Releases. October 8, 2018. https://www.un.org/press/en/2018/sgsm19282.doc.htm.
[22] Al Gore. Climate Reality Project Leadership Training Forum. Bellevue, WA, June 2017.

START:

Recognizing our stress! In a 2017 paper co-authored by the American Psychological Association (APA) and ecoAmerica, the concept of "ecoanxiety" was introduced[23]. That report, as well as the one published by the NWF, emphasized that the impact of environmental risk on mental health must be addressed. Academic curricula for sociology and psychology students should include modules about climate crisis-induced stress. Mental health professionals should be proactively involved in community disaster preparedness and other strategic crisis discussions to ensure that vulnerable communities are adequately identified and considered. Finally, policymakers should strive for greater social cohesion through accessible workshops, open discussion forums, up-to-date communication channels, and opportunities for citizens to connect with one another in as they anticipate continued issues that will affect their families, homes, and livelihoods.

~ ~ ~

G. Elizabeth Kretchmer. M.F.A. is a writer of fiction, blogs, and essays, a wellness workshop facilitator, an environmental advocate, and an organic farmer. Our natural landscape and the wellbeing of our individual and collective psyches play important roles in all aspects of her work, from her novels to the solar panels she installed on her farm outbuildings. Trained to be a Climate Reality Project leader in 2017, she also holds an MFA in writing. G. Elizabeth lives in Washington State. Read more at: www.climateabandoned.com.

---

[23] 23 S. Clayton, C.M. Manning, K. Krygsman, and M. Speiser. Mental Health and Our Changing Climate: Impacts, Implications, and Guidance. (Washington DC: American Psychological Association and ecoAmerica, 2017), p. 27.

Chapter 15
Climate Crisis & Zombie Myths

by

William McPherson, Ph.D.

"Manmade global warming is the greatest hoax ever perpetrated on the American people." **– Senator James Inhofe,** in a 2003 Senate speech

"Ice storm rolls from Texas to Tennessee. I'm in Los Angeles, and it's freezing. Global warming is a total and very expensive, hoax!" **– President Donald Trump,** in a 2013 tweet

Zombie Myths and Climate Politics

Zombie myths are fallacies that infuse climate politics, and they never die; they just keep coming back over and over again despite any science and facts to the contrary because they serve an ideology. As we have seen, climate myths derive from denial ideology and perpetuate inaction on climate. They are particularly useful to politicians who want to paralyze the public and avoid climate action. They provide alleged rationales by which one can avoid climate policies that are viewed as costly or unpopular, and they serve fossil fuel billionaires and those who take their campaign donations.

Trump's Mythical Rationale for Withdrawing from Paris Agreement

President Donald Trump relied on a number of climate myths when he announced the withdrawal of the U.S. from the Paris Agreement on June 1, 2017:

Jobs and Economic Growth Zombies

Trump stated that the Paris Agreement would result in "lost jobs, lower wages, shuttered factories, and vastly diminished economic production." He cited a number of negative statistics about the predicted economic impact from the climate deal, including a $3 trillion drop in gross domestic product, 6.5 million industrial sector jobs lost, and 86 percent reduction in coal production, all by 2040.

All of these were derived from a March 2017 study prepared by NERA Economic Consulting, with an acronym not spelled out. NERA analyzed the potential impact of *hypothetical* regulatory actions necessary to meet the goals of the Paris Agreement. Their analyses have been questioned by a number of leading economists and statisticians. NERA is known as a captive of the denial wing of conservative organizations[1]. The NERA study makes worst-case assumptions that may inflate the cost of meeting U.S. targets under the Paris Accord, while largely ignoring the economic benefits to U.S. businesses from building and operating renewable energy projects[2].

The zombie myth of economic damage from climate action pervades in denial ideology. It is a rationalization based on fear of change, that a low-carbon economy will somehow be a poorer economy with fewer jobs and a meager lifestyle. In fact, jobs increase with expansion of renewable energy. There are already more jobs (374,000)

[1] Politifact, June 1, 2017
[2] CBS, June 2, 2017

in renewable energy fields than in coal and oil combined (187,000)[3]. In addition to fewer jobs, climate inaction actually means a decline in economic growth, particularly in out years[4].

Reducing Carbon Doesn't Affect Temperatures

Trump said, "Even if the Paris Agreement were implemented in full, with total compliance from all nations, it is estimated it would only produce a two-tenths of one degree... Think of that, this much...Celsius reduction in global temperature by the year 2100. Tiny, tiny amount.[5]"

The co-founder of the MIT program on climate change, Jake Jacoby, says the Trump administration cited an outdated report, taken out of context. Jacoby said the actual global impact of meeting targets under the Paris Accord would be to curb rising temperatures by 1 degree Celsius, or 1.8 degrees Fahrenheit[6].

This myth embodies much of the rationale found in denial ideology. For example, the Heartland Institute asserts that climate change is natural, not human-induced. Another version of this myth is that climate change already existed before human development, a truism that echoes the theme that it is entirely natural. It also embodies the assertion that there has been a hiatus in global warming during the past two decades, since 1998. In other words, Trump and believers in denial ideology assert that carbon emissions have no effect on global temperatures and that reducing them will have little to no effect on temperature rises. They cite the fact that carbon emissions have been steadily rising during the past two decades, while temperatures remain stable.

This zombie myth is based on cherry-picking data. Beginning with 1998 is a means of disproving temperature trends by using an especially high temperature record year as a baseline (1998 was an El Nino year, with the highest recorded global average), then picking subsequent years with lower global averages for comparison. It is also based on the use of defective satellite measurements. Data cited by John Christy and associates at the Earth System Science Center at the University of Alabama in Huntsville is corrupted by the deterioration of orbits of the satellites. Comprehensive data on temperatures have shown a definite upward trend, as indicated in the data from NOAA:

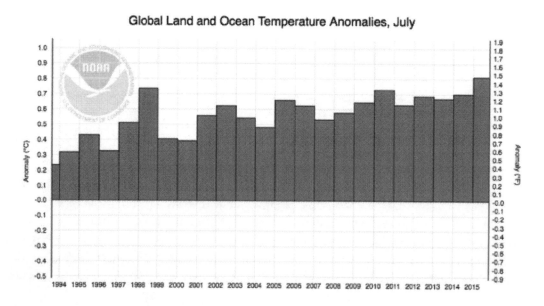

Global Land and Ocean Temperature Anomalies, July

---

[3] Forbes, January 27, 2017; insideclimatenews.org, May 27, 2017
[4] Nicholas Stern, The Economics of Climate Change
[5] Based on MIT Study
[6] CBS, June 3, 2017

The data clearly indicate an upward trend since 1994, with only 1998 showing much deviation from that trend. Nevertheless, perpetrators of the climate myth that temperature increases have paused use 1998 as a baseline to claim there is no warming, despite increasing concentrations of carbon dioxide. This is an unethical manipulation of data and an immoral manipulation of facts.

Another factor overlooked by perpetrators of the myth is that 90 percent of warming goes into the ocean[7]. Because water is more able to absorb heat, it retains a lot of the temperature increases from global warming. There is a downside to this factor in climate change, however: As the ocean warms, it expands, raising sea levels and impacting coastal communities. Warm water is also less likely to absorb and hold carbon dioxide, so it returns some CO_2 to the atmosphere, accelerating the buildup of carbon concentrations.

Saving Coal Jobs

President Donald Trump said one reason he withdrew his country from the Paris Agreement was to save coal jobs, but the move was unlikely to revive the U.S. coal industry. Coal-fired power generation in the United States is expected to fall by 51 percent by 2040, with a 169 percent increase in renewable power helping to fill the void[8]. Nearly all coal station shutdowns are due to economic factors such as the price of natural gas and declining cost of renewable, not simply regulations.

Relying on the myth of saving coal jobs is a tactic of sectorial politics; that is, it appeals to a segment of the population who is impacted by social change by blaming some other sector or elite group. Although Trump did not directly blame the natural gas industry for loss of coal jobs, he did imply that changes compelled by the Paris Agreement are responsible. Even if coal production revives somewhat, employment will not increase much because of automation. Nevertheless, climate policies are convenient whipping boys for politicians who want to elicit support from coal miners out of work.

Other Climate Myths are Zombies that Keep Coming Back

Temperature Records Are Biased Toward Heat

One myth that resurfaces frequently is that ground temperature stations give biased readings because they are sited near heat sources such as air conditioners (with hot external coils) or asphalt surfaces that retain heat. This myth, perpetuated by Anthony Watts at wattsupwiththat.com, has been debunked by climate scientists[9]. Many sources of temperature measurement are used, including ocean buoys, radiosondes (balloons with instruments), and satellites; any biases are corrected with cross-checking.

Climate Scientists Are Getting Rich on Grants

This myth uses distortion of the grant process with the assertion that climate scientists have to report temperature increases to create alarm among funding agencies. In other words, climate scientists are applying for grants strictly for acquisitive reasons, not to do real research on climate. An example is a statement by James Taylor of Heartland Institute about Michael Mann, the Penn State scientist who devised the so-called hockey stick graph:

> Mann's career is dependent on taxpayer handouts. He draws his salary from a public university, Penn State. He solicits government grant money to carry out special research projects. Government officials, who are accountable to voters for how they spend the tax money they collect, occasionally follow up on

---

[7] https://www.climate.gov/news-features/understanding-climate/climate-change-ocean-heat-content
[8] Bloomberg New Energy, June, 2017
[9] https://www.epa.gov/heat-islands/measuring-heat-islands

taxpayer concerns that government-funding recipients appear to be misusing government funds. This, apparently, makes Mann furious[10].

Mann has been vilified in climate denial blogs, and his research, confirmed many times, has been attacked. He has even received death threats, but Mann has been cleared of wrongdoing[11]. The greedy climate scientists myth flies in the face of the reality that grants are difficult to get and have declined in recent years. As one climate analyst said, "Despite sporadic accusations otherwise, climate researchers are scrambling for a piece of a smaller piece of the government-funded pie, and the resources of the private sector are far, far more likely to go to groups that oppose their conclusions.[12]"

It Can't Be True Because Al Gore Has a Giant House

Ad hominem arguments, reactions directed against a person rather than the position they are maintaining, are common in denial ideology; if you can denigrate the messenger, you can ignore the message. Many ideologues accuse climate activists of hypocrisy because they use air transportation to attend conferences or have large residences, cars, etc. Al Gore is a particular target because of his production of the *An Inconvenient Truth* documentaries and Climate Reality Project training. In fact, Gore's large residence is solar powered.

A more revealing aspect of *ad hominem* arguments is that they indicate the weakness of climate denial assertions. As the previous quotation from James Taylor demonstrates, when you cannot disprove the facts underlying such data analyses as the hockey stick, you attack the scientists. Attacks such as Taylor's rely on distracting people from the facts so they do not accept climate science.

Record Levels of New Ice Every Year

Fortunately, this is one of the easiest myths to dismiss. One only has to look at the research to see how this argument fails[13]:

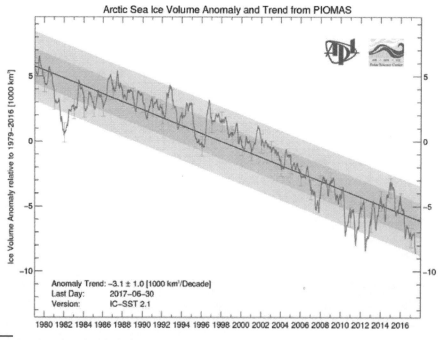

Arctic Sea Ice Volume Anomaly and Trend from PIOMAS

---

[10] https://www.forbes.com/sites/jamestaylor/2013/08/09/global-warming-activist-michael-mann-wages-war-on-government-transparency/#18d6df227e1a

[11] https://www.skepticalscience.com/broken-hockey-stick.htm

[12] https://arstechnica.com/science/2016/05/if-climate-scientists-push-the-consensus-its-not-for-the-money/

[13] http://psc.apl.uw.edu/research/projects/arctic-sea-ice-volume-anomaly/

Arctic Sea ice is not the only measure of climate change, of course, but it is significant because of the change in albedo, the reflection of heat from the sun, which occurs when arctic ice melts. Dark water absorbs much more heat than the white ice, a positive feedback in climate change that accelerates global warming.

Other measures of ice volume also confirm the loss of ice. Greenland's ice sheet is melting more and more, and its glaciers have started moving more rapidly. This is because summer thaw produces water crevasses, moulins, which pour melt water down to the bedrock, lubricating the movement of the glaciers to the sea. This phenomenon will contribute in a major way to sea level rise. In addition, Greenland is experiencing seven times more glacial earthquakes than it did in the 1990s. Other land-based glaciers, such as those in the Alps and the Andes, are melting rapidly, threatening water sources for populations in the valleys below. Glacier National Park will soon have no glaciers.

One area of contention is the melting of Antarctic ice. While scientists have not yet determined the effect of climate change on the Antarctic, large ice shelves have calved off the Antarctic Peninsula. The calving of the Larson ice shelf is releasing floating sea ice, which does not contribute to higher sea levels. The crucial fact that involves climate change is that the ice shelves have bottled up the land glaciers, and with calving, land glaciers flow faster. This will increase sea levels as land ice calves off the glaciers. There is also the question of air and water temperatures in the Antarctic region; research indicates increasing water temperatures particularly in the area around the Antarctic Peninsula stretching toward South America[14].

Volcanoes Make Way More Pollution Than We Do!

Volcanoes contribute much less carbon pollution than carbon dioxide from human sources. "Human activities emit sixty or more times the amount of carbon dioxide released by volcanoes each year.[15]" Volcanoes do produce sulfur dioxide aerosols, another form of pollution that ironically helps cool the climate. These aerosols helped reduce temperatures about one degree Celsius in 1993, from the eruption of Mount Pinatubo in the Philippines. This has been a hot topic of discussion in geoengineering, technical fixes for climate change: Should humans reproduce the effects of volcanoes by cooling the planet with aerosols spread by aircraft? Since there are many possible unintended consequences of geoengineering, most climate analysts tend to oppose large-scale geoengineering projects while research continues.

God Will Save Us

This is probably the most egregious climate myth, and it takes two forms: We do not control the climate, only God does; or, climate science is a form of devil worship, so we can counter it with devotion to God. Each of these requires further thought and discussion.

Perhaps the most explicit statements of God's role in climate were by two members of Congress: "The Earth will end only when God declares it's time to be over. Man will not destroy this Earth," said Representative John Shimkus (R-IL)[16]; and Senator James Inofe (R-OK) said, "My point is, God's still up there. The arrogance of people to think that we human beings would be able to change what He is doing in the climate is, to me, outrageous[17]."

Both politicians are known as climate denial leaders, and it is not surprising that they look to non-empirical sources for their assertions, which have no scientific basis. Some of the divinity myth stems from apocalyptic beliefs based on the *Bible's* Book of Revelations, which claims God will destroy the Earth when the time comes, and we do not have to worry about the climate crisis because that time is imminent.

---

[14] https://phys.org/news/2014-12-antarctic-seawater-temperatures.html
[15] https://www.climate.gov/news-features/climate-qa/which-emits-more-carbon-dioxide-volcanoes-or-human-activities
[16] *New York Times*, November 16, 2010
[17] Rockford Register Star, March 10, 2012

A more egregious form of religious argument about climate change is the assertion that climate science is nature worship or, *in extremis,* devil worship. One author portrayed climate policy as a religion: "Climate change policy is almost religious in the way it is being rolled out. It is, and that's because it is central to the Gaia Earth-worship and Lucifer 'consciousness-raising' that the UN is trying to engender[18]."

While this is far out on the curve of denial ideology, it nevertheless illustrates a tendency to view climate science as religion. It is understandable that denial ideologues would see climate science in their own terms, as a belief system rather than scientific findings. It is in their interest to reduce science to ideology, but scientists do not respond at this level. Most, understandably so, will have nothing to do with the ideology of denial that equates scientific work with religion.

Climate Crisis "Keep/Stop/Start" Actions

Below for your consideration, are several author-suggested action items. For a further explanation of the Keep/Stop/Start organizational tool, please see How to Read This Book.

KEEP:

1) *Keep an open mind.* Openness may allow zombie myths into your psyche, but it also gives way to facts and reality. Zombie myths are meant to browbeat you into closed submission.

2) *Keep being curious.* Skeptical Science is a nonprofit organization dedicated to cutting off the head of a zombie myth. They have collected and debunked hundreds of climate myths, and they offer a website and app to present their findings. The fossil fuel industry and the billionaires who own them have funded numerous think tanks such as the Heritage Foundation that have paid staff whose job it is to develop climate lies to preserve the industry's financial position. Industry leaders and their paid staff know they only have to create doubt to paralyze and confuse the public; they don't have to scientifically prove anything, unless you keep being curious and open enough to stop them.

STOP:

1) *Stop spreading zombie myths* by confronting them at every opportunity. Visit skepticalscience.org to find the truth; it is quite educational to review the 197 climate myths they have identified to date. Opinion pieces or letters to newspaper editors may repeat one or more myths, so these should be countered immediately with rebuttals. Speeches such as Trump's abandonment of the Paris Agreement should elicit discussion and responses that point out his reliance on climate myths. I use a PowerPoint presentation that lists the Trump myths and sources that debunk them.

While it is satisfying to confront the myths with facts, we need to look behind people's willingness to accept myths and ask why they find them attractive. Ideology, as noted elsewhere in this book, has appeal to people who face a major dilemma: continue to enjoy a lifestyle they like and face disaster, or prepare for a different lifestyle. It is reasonable to suggest gently we do not have to fear a carbon-free lifestyle. Renewable energy can support our secure way of life, so we will not become cave dwellers again!

---

[18] Wishart, Ian. 2013. Totalitaria: What If the Enemy Is The State? Howling at The Moon Publishing Ltd.

START:

1) *Start to resist zombie climate myths* and dampen their effect on the debate. Consider the ideology that underlies them. Denial ideology relies on the premise that there is no human effect on the climate and that they can ignore the findings of climate science that indicate that effect. Once people accept this premise, they will cling to myths that confirm their bias: myths such as the pause or hiatus in temperature increases or that climate scientists and Al Gore are promoting climate change to line their pockets. It is sometimes futile to argue the fallacy of the myths, even with the arsenal of evidence cited above. It is better to look at the underlying premise and ask why people believe what they believe. What is their motivation in perpetrating the zombie myth? Who benefits from the myth's survival? A neighbor once asked, "Do you really believe in climate change?" I replied that I want to know what's happening with the climate, regardless of what I do or do not believe.

2) *Start changing beliefs.* After confronting the problem that climate denial is a belief system, the issue is how to change those beliefs. Some people will never change their beliefs about climate, but others may be open to persuasion. I have found it is useful to ask, "How do you know that?" Then, after they sternly cite some pseudoscience source, ask about the credentials of the scientists. There is a risk that this line of argument will lead to a contest of my scientists versus yours, but it more often leads people to think about where they are getting their information. I find it is better to keep asking questions than to try to assert something contradictory to your interlocutor's own beliefs.

3) *Start avoiding passivity.* Seldom do we get into these arguments because most people do not have the fervor nor stamina to argue for climate denial. A lot of denial is passive, and we say, "I haven't thought much about it," or, "I don't have time to worry about that now." It is difficult to get people to address remote concerns about climate consequences, remote in both time and space, as most have more immediate concerns about the economy, education, and other impending issues. In effect, many have abandoned the cause of climate action because it seems less salient to them, and their reticence to act is bolstered by belief in climate myths. In order to overcome this inertia, it may be necessary to point to some of the effects of climate change that are already evident, such as flooding in Miami from sea level rise, where fish have been seen swimming up the boulevards.

~ ~ ~

William McPherson, Ph.D. is a retired environmental diplomat with 21 years' service in the U.S. Foreign Service, including assignments in Tokyo and Geneva. He continued his work on international environmental issues after retirement, working with Earth Negotiations Bulletin, a newsletter on environmental negotiations. He has served in assignments in Europe, Japan, Korea and the Philippines as well as other countries. He is an activist with the Sierra Club, working on issues of climate change and coal exports. His work on climate includes promoting carbon pricing and oil transport safety in Washington State. He is the author of author of three books: *Ideology versus Science, Climate Weather and Ideology* and *Sabotaging the Planet*. Read more at: www.climateabandoned.com

Chapter 16:

Climate Crisis & Ideology Versus Science

by

William McPherson, Ph.D.

"The good thing about science is that it is true whether or not you believe it." – **Neil deGrasse Tyson**

"The challenge that we're taking up in this Energy and Enterprise Initiative that I'm working on is basically creating a safe space for conservatives to pay attention to science, because right now it doesn't seem safe to pay attention to the science because the ideology says no, the science is wrong." – **Bob Inglis,** Frontline, PBS, October 23, 2012

Climate Change and Ideology

Perhaps besides the ongoing conflict in the Middle East, gun control, and abortion, climate change is one of the most polarizing subjects in our current political sphere. Unsurprisingly, it engenders some of the worst of human instincts; when people are confronted with something they do not wish to face, they tend to devise methods to avoid it. One reaction to the social, political, and physical pressures generated by climate change is denial, an ideology that provides something for people to cling to when reality or grim projections of a future reality becomes difficult to grasp.

The climate crisis presents us with an ostensible dilemma: Do we continue enjoying our current lifestyle, even at the cost of future disaster, or do we prepare for and embark upon an entirely different lifestyle? For many, facing climate change insinuates that we must give up luxuries or even some of the basic necessities of life. While this is not necessarily true, because renewable energies really *can* maintain most modern-day conveniences, many mistakenly believe or assume there are no alternatives to the current energy system that supports our lifestyles. Nevertheless, change is inevitable. Either a low-carbon lifestyle or a severely diminished lifestyle will occur due to the consequences of climate change.

Ideology, a belief system that relies upon assumptions and premises about reality to frame perception, comes to the rescue of those who face this dilemma and enables people to wall off the implications of climate change. Ideology also guides thinking toward conclusions that are compatible with one's desires; it supports confirmation bias in choosing which facts to comprehend and which to ignore. Think of ideology as the tattered, gross, old, ratty, relatively useless blanket from your childhood. When things in the world get too scary and you are bombarded with realities that don't agree with your set of beliefs, you can run and hide under that blanket and feel safe again, ignoring those dissenting opinions and finding solace in blissful ignorance, no matter how full of holes that blanket is.

Ideology Versus Science

Denial ideology is not partisan. Climate change denial is not Republicans versus Democrats, conservatives versus liberals, or business versus environmentalism. Some conservative Republican businesspeople, like Bob Inglis, actually promote climate action. Bob's Energy and Enterprise Institute, RepublicEn, takes the following approach to climate action: "Climate change is real, and we believe it's our duty and our opportunity to reduce the risks. But to make a difference, we have to fight climate change with free enterprise instead of ineffective subsidies and regulations."[1]

While not everyone agrees with this approach, it is a sign that denial ideology is nonpartisan.

Former Republican cabinet members Hank Paulson and George Schulz proposed climate action in their manifesto, "The Conservative Case for Carbon Dividends." Businesses, even Exxon-Mobil and other oil companies, have recently acknowledged the findings of climate science. While many Republicans still deny it, particularly members of the U.S. Congress and the Trump Administration, others do not. For this reason, climate denial cannot and should not be identified as a merely Republican stance, nor should all Republicans or conservatives be accused of or presumed to be in denial. Thus, if party and political stance have nothing to do with it, we must consider other reasons for climate denial.

Once we strip away the partisan factor, we are left with one possibility: Climate denial is ideology versus science. To understand climate denial, we must examine the difference between ideology and science. Science is a testable set of findings and conclusions about reality, while ideology is an untestable set of premises and assumptions about reality.

In brief, ideology is comprised of fixed beliefs about a changing reality. By making assertions about reality that have no basis in fact, climate denial ideology promotes a set of comforting myths to assuage anxiety about the future. (Note that climate myths are addressed in another chapter of this book.) If we believe humans have no effect on the climate, we will continue our present practices of consumption and energy use, with no qualms about our future. Ideology, which constructs a complex belief system, enables us to ignore science when the scientific conclusions are inconvenient. In other words, ideology can serve as either the reins that guide the horse in a specific direction or, contrarily, as the blinders that keep the horse on track, refusing to acknowledge any other direction than the way it is already headed. Science can be considered the driver of the wagon, using facts and research to lead the horse in the best possible direction.

Heartland Institute

Nowhere is the distinction between ideology and science clearer than in the programs of the Heartland Institute. In its periodic reports, "Climate Change Reconsidered," Heartland provides some rationale for denial ideology:

> ['Climate Change Reconsidered'] summarizes the research of a growing number of scientists who say variations in solar activity, not greenhouse gases, are the true drivers of climate change. We describe the evidence of a solar climate link and how these scientists have grappled with the problem of finding a specific mechanism that translates small changes in solar activity into larger climate effects. We summarize how they may have found the answer in the relationships between the sun, cosmic rays and reflecting clouds[2].

In other words, we can ignore findings of climate science that humans are, in fact, driving climate change through greenhouse gas emissions because climate change is driven only by natural forces such as solar activity. Anyone who asserts this claim not only fails to understand basic climatic and environmental science, but they are also just plain wrong. While global temperatures have increased by an average of 1°C (1.8°F) since 1880, solar irradiance (the power of sunrays) has remained relatively constant over this same time period[3].

[1] http://www.republicen.org/
[2] https://www.heartland.org/_template-assets/documents/CCR/CCR-I-Executive-Summary.pdf
[3] https://www.skepticalscience.com/solar-activity-sunspots-global-warming.htm

This is a constant theme in denial ideology: We are not changing the climate, and only natural forces are involved. Another version of this myth is this truism: Climate change already existed before human development." This absolves humans of all responsibility for climate change and provides a biased confirmation of the belief that we can ignore climate science. It is completely false, as scientists have shown, since solar activity has actually diminished as temperature has risen in recent decades. Nevertheless, Heartland maintains its egregious stance of natural causes in the face of a completely different reality: that temperature increases while solar activity does not and that those increases are largely due to human activity. Perhaps they would be better referred to as the Heart*less* Institute!

Heartland has made its reputation as a source of denial ideology by contradicting basic findings of the UN Intergovernmental Panel on Climate Change (IPCC), known by many as the world's most trusted, least partisan, best-funded international entity in climate research. Specifically, Heartland states, "On the most important issue, the IPCC claim that most of the observed increase in global average temperatures since the mid-twentieth century is very likely due to the observed increase in anthropogenic greenhouse gas concentrations. We once again reach the opposite conclusion, that natural causes are very likely to be dominant[4]."

The myth that "natural causes [are] dominant" has been debunked by scientific research that shows that natural changes cannot explain most of the temperature increases in the last fifty years. "Natural drivers of climate cannot explain the recent observed warming. Over the last five decades, natural factors (solar forcing and volcanoes) alone would actually have led to a slight cooling[5]".

In Heartland's alternative world, the need to confront the implications of energy use is averted by denying that humans are responsible. If natural forces are the only causes of climate change: (1) There is little or nothing we can do about them; and (2) They are likely to be cyclical, so recently increasing temperatures are likely to taper off. This may be false reassurance to many who are rightfully concerned about global warming, but it is not science; on the contrary, it is nothing but ideology.

Evidence of how far Heartland Institute will go in promoting its ideology comes from one of its blunders. In the run up to its 2012 annual conference, Heartland generated controversy about a billboard featuring Unabomber Ted Kaczynski, saying, "I still believe in global warming. Do you?" When asked about this, James Taylor of Heartland (not *that* James Taylor, so rock 'n' roll fans can rest easy) claimed so-called alarmists use the same tactics. This suggests that he considers climate science a kind of ideology, a polar opposite to the denial ideology promoted by Heartland. This stance, coupled with the bad optics of the Kaczynski billboard, led a number of former Heartland supports to withdraw their funding. Trying to fight climate science with ideology can be fruitless, and Heartland's inability to maintain credibility has been frustrating and costly for its leaders. Heartland Director Joseph Bast admitted, "We've won the public opinion debate, and we've won the political debate as well...but the scientific debate is a source of enormous frustration.[6]" Apparently, no matter how hard you want science to agree with your convoluted ideology, it just won't. Go figure!

Based on these quotes alone, it is evident that Heartland Institute has abandoned climate science and chooses instead to promote denial ideology as its equivalent. Equating ideology and science is fraught with risk, as the untestable nature of ideology makes it brittle. When reality intrudes on comfortable denial myths, their tenuous nature will instigate doubts about the credibility of denial ideology.

Koch-Funded Groups

Heartland Institute is only one of several groups that abandon climate science and promote denial ideology. Many of these are funded partially or fully by Koch Industries, a petroleum conglomerate and the second-largest

---

[4] https://www.heartland.org/_template-assets/documents/CCR/CCR-Interim/Front Matter.pdf
[5] http://nca2014.globalchange.gov/report/our-changing-climate/observed-change
[6] Nature, July 28, 2011.

private corporation in the U.S., right behind Cargill. Charles and David Koch fund a number of denial groups, including: Heartland Institute, Americans for Prosperity (AFP), American Legislative Exchange Council (ALEC), and some manifestations of the Tea Party. The significance of these groups in terms of climate change is that they are purportedly grass-roots organizations that influence policy decisions. In fact, they are AstroTurf groups, portrayed as grass roots but actually dominated and funded mainly by the Koch brothers.

AFP has sponsored a number of climate-denial events and has openly lobbied against climate action in legislatures and Congress. It has called on President Trump to "withdraw fully from the Paris Climate Treaty and stop all taxpayer funding of UN global warming programs." Unfortunately for them, our President can't be so easily persuaded! Or can he? On June 1, 2017, AFP got its wish when Trump abandoned that Paris agreement and the UN Green Climate Fund. AFP lauded the action with the following statement:

> Americans for Prosperity has been an outspoken opponent of this damaging international accord since it was signed by President Obama in 2015. And this year, AFP called for the administration to withdraw from the U.S. from the Paris agreement in our Reform America 2017 agenda, our set of top priorities for the year. As President Trump said in the Rose Garden Thursday, the non-binding Paris accord imposed a "draconian financial and economic burden" on the country. The deal put hardworking American families in the crosshairs by imposing hundreds of job-killing regulations on American businesses and discouraging investment, innovation and economic growth, all in the name of subsidizing failing green energy projects abroad[7].

As a percentage of the federal budget, U.S. contributions to the UN Green Climate Fund has been about 0.035 percent; former President Obama had pledged another 0.07 percent but this pledge was revoked by Trump, hardly the suggested "draconian financial and economic burden." To put this in perspective, one of the Trump administration's early budget proposals contained a $2 trillion accounting error, 1,000 times larger than the $2 billion commitment to the Green Climate Fund[8]. Of course, Trump and the AFP can argue that the Paris Agreement is "job-killing" but that argument flies in the face of the fact that renewable energy creates twice as many jobs as the fossil fuel industry. Coal jobs in particular are declining for economic reasons that have nothing to do with the Paris Agreement.

AFP claims members in more than thirty states where it lobbies legislatures to defeat climate actions. In conjunction with the ALEC, an organization comprised of business representatives and state legislators, AFP has succeeded in putting a burden on renewable energy by penalizing installers of solar panels. ALEC drafted a model bill for thirty-seven state legislatures. Several of these, including Arizona and Indiana, have proposed fees on solar installations for connection to the grid. ALEC has drafted the following resolution on solar energy:

> **WHEREAS**, the subsidy rate for onsite solar power equipment is significantly higher than the subsidy rate for utility-scale solar power; and
>
> **WHEREAS**, onsite solar power is substantially more costly to produce than conventional power and utility-scale solar power; and
>
> **WHEREAS**, solar power receives substantially more subsidies than conventional power sources;
>
> **THEREFORE, BE IT RESOLVED** that the American Legislative Exchange Council encourages state policymakers to encourage free markets and affordable energy by *refraining from granting*

---

7 https://americansforprosperity.org/americas-withdrawal-paris-climate-agreement-major-victory-american-taxpayers
8 https://www.cnbc.com/2017/06/02/why-trump-is-seeing-red-about-the-green-climate-fund.html

special privileges to the solar power industry to sell electricity directly to consumers. (Emphasis added)[9]

While citing costs as a rationale for opposing solar subsidies, ALEC has a broader reason for abandoning most government regulations. ALEC uses a libertarian rationale when it comes to its energy policy:

Because energy is so ubiquitous to our lives, it is critical that lawmakers implement market-oriented energy policies that allow energy to be produced more efficiently at lower costs with fewer economic disruptions and lower environmental impacts. *If a state or federal government imposes overly onerous regulations or adopts policies that drive up energy costs, the effects will soon be felt throughout the entire economy.* Food, medicine and other household goods will become more expensive almost immediately, disproportionately affecting those with low and fixed incomes. (Emphasis added)[10]

ALEC relies on this rationale to bolster their opposition to a number of climate policies. With regard to the Obama Clean Power Rule, ALEC proposed the following resolution:

> **NOW, THEREFORE, BE IT RESOLVED** that the American Legislative Exchange Council urges EPA to analyze the combined potential impacts of its future regulations for coal-fired power plants and to share such analysis with all states and Congress to require the Congressional Budget Office (CBO) to provide an independent cost/benefit analysis of all major EPA rules transmitted to Congress; and

> **BE IT FURTHER RESOLVED** that EPA should minimize such impacts on the nation and each state as it develops each individual regulation; and

> **BE IT FURTHER RESOLVED** that Congress should take all necessary steps to ensure that EPA analyzes the combined potential impacts of such future regulations on the U.S. electricity sector, jobs, energy prices, individual states, and regions of the country and to share such analysis with all states[11].

For those of you who skipped the previous section of legal jargon, allow me to offer a quick summary: Essentially, ALEC chooses to take an extremely libertarian stance in regard to solar power and pushes to keep that industry from selling directly to consumers. They also believe the federal government should basically avoid any sort of regulation on coal-fired power plants at all costs. It is a bit ironic that they call for no regulations on coal and oil yet demand restrictions on the solar industry. It may be safe to say ALEC probably prefers Koch over Pepsi.

The Trump administration has proposed to abandon the Clean Power Rule, and resolutions like the one above may have some influence on that decision. The rule is currently being litigated in the courts. If it is abandoned, the U.S. will only have to achieve carbon reduction of approximately half the 26 percent pledged by the U.S. in Paris. This would not bother Trump or ALEC, as they have abandoned the entire U.S. pledge.

Ideology and Cognitive Dissonance

While it is understandable that some fossil fuel interests such as Koch Industries want to forestall climate action, it is necessary to look at other reasons why Republican politicians and groups such as Heartland Institute profess climate denial ideology. Heartland President Bast has admitted that the "scientific debate" is "frustrating."

---

[9] https://www.alec.org/model-policy/resolution-concerning-special-markets-for-direct-solar-power-sales/
[10] https://www.alec.org/issue/energy/
[11] https://www.alec.org/model-policy/resolution-concerning-the-combined-impacts-of-future-epa-regulations-for-coal-fired-power-plants/

It is only a debate, by definition, in Heartland's eyes because they seem to think they have science on their side, yet this renders their ideology groundless.

In fact, there is no scientific debate about climate change. Heartland and its followers would love to have scientific confirmation of their premise that there is no anthropogenic climate change, that only natural forces affect the climate. Unfortunately, that is impossible, because scientific evidence obliterates this premise.

It would be rather logical to expect that when the premise of the argument is obliterated, the argument will change; however, this flies in the face of social science research. Leon Festinger and associates studied a religious cult in a book titled, *When Prophecy Fails*. In that study, they found that when a prophecy does not pan out, some members will retain their belief regardless. He labeled this tendency *cognitive dissonance* and defined it as the ability to hold a thought contradictory to reality, despite evidence that refutes it.

It is clear from the statements and actions of groups such as Heartland Institute and AFP/ALEC that the premise that there is no anthropogenic climate change underlies most of their programs. They resolve cognitive dissonance for their followers by using unfounded statements about science to dismiss climate findings. Once you abandon climate science, it is possible to argue that there is no threat or problem of climate change, and any proposed remedies will be deemed unnecessary.

When they couple denial ideology with their libertarian views, Koch-funded organizations consider it mandatory to oppose regulations and policies that work to address climate change. Heartland lauded Trump for withdrawing the U.S. from the Paris Agreement because it frees the market from onerous regulations. They also applaud Trump's very open rejection of climate science. AFP and ALEC have influenced legislators to put restrictions on renewable energy because they view any subsidies or support for solar power as violations of the free market ethos.

What is more sinister is that they also argue that renewable energy is unnecessary. Why? Because they presume there is no need for an energy transition. The basic premise of denial ideology, that there is no need to face the climate crisis, is the most dangerous aspect of the denial movement, and it will result in repercussions throughout the political and economic system as leaders in both spheres delay needed action. Climate change will eventually catch up with us and compel action, but the longer we wait, the higher the cost will be.

Perhaps this is the most pernicious aspect of climate denial ideology: Politicians and business executives use it to simply avoid difficult decisions. As you have read in other chapters, the inaction at the federal level, coupled with bad business decisions, has exacerbated the climate crisis, and these will continue to delay needed action on the climate crisis.

If climate scientists and Al Gore are wrong and we act on climate change, we can only create a better world. If denial ideologues are wrong, we will leave the world in much worse shape than we found it.

Climate Crisis "Keep/Stop/Start" Actions

Below for your consideration, are several author-suggested action items. For a further explanation of the Keep/Stop/Start organizational tool, please see How to Read This Book.

KEEP:

1) *Keep resisting.* Resist the false arguments of denial ideologues who are driving our political agenda. Refute the *alternative facts* used by the Trump Administration to justify climate inaction and the withdrawal from the Paris Agreement.

2) *Keep questioning the ideology,* and understand why people adhere to it. When the truth is inconvenient, people will find reasons to avoid it. Unquestioned premises lead to false conclusions such as the earth is not warming, or that climate change is completely natural. Learn the facts and the effects of climate change through sources such as the Climate Reality Project (https://www.climaterealityproject.org/climate-101). Use these facts to refute the ideology when you are confronted with climate denial.

STOP:

Stop emphasizing the downside of climate consequences. This may seem untimely or unusual to suggest, but negativity often feeds denial ideologues. If you argue that floods, drought, and other effects of the climate crisis require immediate change, you may find yourself being laughed at. Scientists are currently engaged in efforts to attribute storms and drought to climate change, but clear links have not been established, and mere probability will only water down your argument. It is likely that extreme weather is made more severe by the increase in atmospheric energy generated by carbon concentrations, but no single storm or wildfire can be directly linked to climate change as a singular event without any other causes. Denial ideologues will use this complexity, and the uncertainty about climate science that it engenders, to counter arguments that climate change is dangerous.

Rather, emphasize that continued fossil fuel use means we are abandoning our obligations to future generations, in a selfish, mindless way that will leave our children and grandchildren with a massive problem that will require massive solutions. As one scientific study noted, "continued high fossil fuel emissions today place a burden on young people to undertake massive technological CO_2 extraction if they are to limit climate change and its consequences... Continued high fossil fuel emissions unarguably sentences young people to either a massive, implausible cleanup or growing deleterious climate impacts or both[12]."

START:

Start viewing arguments by denial protagonists as ideology rather than science. Look at the premises of all arguments and question their validity. Ask questions: "What is the science behind that argument? Where does that argument come from? Who really benefits if I believe this?"

It is not useful to tumble angrily into arguments about scientists. For example, one could question the credentials of Heartland-endorsed scientists such as Fred Singer or Craig Idso, principal authors of the "Climate Change Reconsidered" reports. Unfortunately, such arguments often devolve into *ad hominem* attacks rather than reviews of actual scientific findings. Interlocutors who believe in denial ideology may try to attack scientists rather than discuss the actual science.

It is better to focus on facts, such as the increase in world temperature. For example, 2014 to 2016 were the hottest three years on record, and 2017 is likely to come in as hot, second only to 2016. The so-called hiatus in temperature during the first two decades of the twenty-first century, a zombie myth discussed elsewhere in this book, has been invalidated by these temperature records and should be returned to the grave. Discussions of natural causes such as solar activity should rely on facts about actual decline of sun activity during increases in temperature.

~ ~ ~

William McPherson, Ph.D. is a retired environmental diplomat with 21 years' service in the U.S. Foreign Service, including assignments in Tokyo and Geneva. He continued his work on international environmental

[12] https://www.earth-syst-dynam.net/8/577/2017/

issues after retirement, working with Earth Negotiations Bulletin, a newsletter on environmental negotiations. He has served in assignments in Europe, Japan, Korea and the Philippines as well as other countries. He is an activist with the Sierra Club, working on issues of climate change and coal exports. His work on climate includes promoting carbon pricing and oil transport safety in Washington State. He is the author of author of three books: *Ideology versus Science, Climate Weather and Ideology* and *Sabotaging the Planet*. Read more at: www.climateabandoned.com.

Chapter 17

Climate Crisis & the Failure of Political Leadership

by

William McPherson, Ph.D.

"Strong international support for the Paris Agreement entering into force is testament to the urgency for action, and reflects the consensus of governments that robust global cooperation is essential to meet the climate challenge." – **Ban Ki-Moon,** former UN Secretary-General, October 5, 2016

The Politics of Denial and the Consensus of Governments

Although history is written by the winners, it doesn't tend to forget the losers. In fact, those who have stood on the wrong side of history are typically used as examples of how brutally uninformed, ideologically unsound, and obstructionist humans can be. Until recently, these antagonists have only disagreed with the masses on matters of ethics, politics, and governance. These are all subjective points; however, we are now living in a time where politicians deny and refute fundamentally proven scientific facts. This intellectual resistance does not allow for the continued environmental destruction of our planet. Rather, it severely impairs our country's standing on the international scale.

President Donald Trump has withdrawn the U.S. from the Paris Agreement. Scott Pruitt, former Administrator of the Environmental Protection Agency (EPA), questioned climate science openly: "I think that measuring with precision human activity on the climate is something very challenging to do and there's tremendous disagreement about the degree of impact, so no, I would not agree that [carbon dioxide is] a primary contributor to the global warming that we see."

With climate denial such as this, it is difficult for U.S. diplomats to maintain credibility in international fora. In fact, it has left the U.S. isolated on the world stage. Most assuredly, the individual who replaced Scott Pruitt, Andrew Wheeler, will be more insidious to protection of the environment since he is known to be more of a work-behind-the-scenes person and won't be as easily caught in his corruption.

That isolation was evident during the G-7 meeting in May 2017. The G-7 is comprised of several nations: Canada, France, Germany, Italy, Japan, the UK, the U.S., and the EU, and all nations other than the U.S. endorsed the Paris Agreement. The joint communiqué said: "Expressing understanding for this process, the heads of state and of government of Canada, France, Germany, Italy, Japan and the United Kingdom, and the presidents of the European Council and of the European Commission reaffirm their strong commitment to swiftly implement the Paris Agreement."

[1]Notably absent from the list was the United States.

The G-7 dispatch, a separate statement from the meeting, tried to finesse U.S. isolation with the following text: "[The U.S.] is in the process of reviewing its policies on climate change and on the Paris Agreement and thus is not in a position to join the consensus on these topics.[2]" Is it any wonder that political pundits are starting to refer to the G-7 as the G-6? Perhaps it's simple math: We're no longer a part of it.

Although other countries use a light touch in reacting to U.S. intransigence, it is clear that international cooperation is damaged by U.S. actions. A group of nations called the G-20 drafted their own communiqué that

---

[1] The New York Times, May 27, 2017.
[2] CNN, May 27, 2017.

states, "We take note of the decision of the United States of America to withdraw from the Paris agreement," adding, "The leaders of the other G-20 members state that the Paris Agreement is irreversible," and, "we reaffirm our strong commitment to the Paris agreement[3]."

The G-20 includes a number of large economies such as China, India, Brazil and Russia, as well as oil producers such as Saudi Arabia, so it is clear that the U.S. is isolated from all of the major players in international climate cooperation. It is also clear that with this isolation comes a number of costs.

It would be one thing if we were like Chile, one of our small, quiet, friendly allies in the southern hemisphere that contributes a mere 0.22 percent of the world's CO2 emissions and 0.3 percent of the world's GDP. However, with no offense intended toward Chile, the U.S. is undeniably a bit more prominent on the international scale than that. We are the second-largest emitter and the largest economy in the world, and our continued use of fossil fuels imperils climate stability throughout the world.

Even worse, our lack of cooperation may weaken the resolve of other countries to participate in worldwide action on climate change. Much of the U.S. reticence to support international climate action comes from political opposition within the U.S., fueled by climate denial. It is in this regard that the ideology of climate denial, pervasive in U.S. politics, has an impact on international negotiations.

Domestic Politics and International Negotiations

In 1992, the UN Framework Convention on Climate Change (UNFCCC) was signed at the Rio Summit (UN Conference on the Environment and Development), and the U.S. Congress ratified that agreement unanimously. Among the senators in Congress that year was Mitch McConnell, current Republican leader of the Senate. The follow-up Kyoto Protocol (1997), however, was not ratified by the U.S. Congress, and the U.S. is not a party to it. Kyoto became a political football in U.S. politics, viewed as unfair because it only applied to developed countries.

Similar dynamics played out in the lead-up to the Paris Agreement. Recognizing that U.S. opposition would be strong unless developing countries were included, former President Obama visited China, India, and Brazil to discuss climate change, and leaders of those countries signed on to the efforts to frame an international agreement that would include all countries. Eventually, 197 countries agreed, and only Syria and Nicaragua were going to hold out because Syria's leaders wanted to exclude themselves due to dealing with a civil war, while Nicaragua was going to refuse because they didn't think the accord was strong enough! Yet, both countries did sign and, sadly, that left the US as the only country not willing to sign. Thus the U.S. has ceded its leadership as the only country that does not support the Paris Agreement.

In November of 2014, President Obama visited China and signed an agreement with Xi Jinping on climate. Chinese President Xi said, "We agreed to make sure that international climate change negotiations will reach an agreement as scheduled at the Paris conference in 2015, and we agreed to deepen practical cooperation on clean energy, environment protection, and other areas[4]. climatedev.com, November 18, 2014." The statement released by the White House in regard to this matter reads, in part:

Today, the Presidents of the United States and China announced their respective post-2020 actions on climate change, recognizing that these actions are part of the longer-range effort to transition to low-carbon economies, mindful of the global temperature goal of two degrees C. The United States intends to achieve an economy-wide target of reducing its emissions by 26 percent to 28 percent below its 2005 level in 2025 and to make best efforts to reduce its emissions by 28 percent. China intends to achieve the peaking of CO2 emissions around 2030 and to make best efforts to peak early[5].

[3] Guardian, July 8, 2017.
[4] climatedev.com, November 18, 2014.
[5] whitehouse.gov, November 12, 2014.

Politicians use climate myths to undermine international efforts to address climate change. As we saw in the chapter on denial myths, the Trump administration not only withdrew the U.S. from the Paris Agreement but also revoked the commitment to reduce emissions by 25 to 28 percent by 2025. Where did this reversal come from?

Some politicians, such as Senator Mitch McConnell, contend that the agreement gave China a fifteen-year free ride to increase emissions, but the rate has to start declining now for the emissions to peak in 2030. Other politicians, such as then-candidate Donald Trump, pooh-poohed the Chinese climate program: "If you look at China, they're doing nothing about it[6]." Apparently, Trump was not aware of the Chinese cap-and-trade program already underway in a number of Chinese cities and the nationwide economy in 2017; then again, to be fair, I'm sure there are quite a number of things of which Donald Trump is not aware.

Sabotaging the Paris Agreement: Abandoning Climate Action

In the run-up to the Paris Agreement in 2015, U.S. politicians made a number of statements indicating opposition to former President Obama's negotiation of the Agreement. In fact, because U.S. Congress was controlled by Republicans, the Obama administration insisted that negotiators frame the agreement as non-binding, so it would not be considered an official treaty, which requires ratification by two-thirds of the Senate. Other parties, especially the EU, disliked this approach but conceded to the U.S. position in order to achieve the agreement. Undoubtedly, these are now having second thoughts about their concession, given the Trump administration withdrawal.

Opposition by domestic politicians, while not impinging directly on negotiations of the agreement, does reduce the credibility of U.S negotiators. Also, as we have seen, it raises legitimate doubts about U.S. commitment, doubts that have borne out. Other parties of the Paris Agreement feel a sense of abandonment in regard to U.S. policies, since the U.S. previously often took the lead in both climate science and climate action. Research on climate has received major funding from Congress before, and renewable energy has often originated in U.S. research, such as solar photovoltaic energy as just one example.

It comes as a shock to many nations that U.S. politicians have turned their backs on all this effort. Senator Mitch McConnell (R-KY), Senate Majority Leader, said, "This unrealistic plan that the president [Obama] would dump on his successor would ensure higher utility rates and far fewer jobs[7]." He also claimed, "I read the agreement... Requires the Chinese to do nothing at all for sixteen years while these carbon emission regulations are creating havoc in my state and other states around the country.[8]" To our foreign readers, let me just tell you that you are not the only ones who are shocked by our conservative politicians' views on climate change; we're just as dumbfounded as you are.

Senator James Inhofe (R-OK), Chairman of the Senate Environment Committee, complained that China will increase its emissions until 2030, then argued that the agreement would impose a $300 billion tax on the U.S. economy. Inhofe also said, "Even if they did agree to reducing emissions, we wouldn't believe them. They don't end up doing what they say they're going to do in these agreements.[9]" Essentially, James Inhofe suggests that nothing China can do or say will ever convince him of their intentions to cut emissions and transition to a clean energy economy. Talk about one of the worst episodes of *According to Jim* ever!

Furthermore, Inhofe considered Obama administration policies as "reckless" and misleading to the world: "The president and his State Department officials are recklessly leading the world to believe we will live up to emission reductions the administration can't substantiate and won't even defend before congressional committees[10]." This type of language leads other countries to view U.S. statements with suspicion and to doubt

[6] Business Insider, September 25, 2015.
[7] Agence France Press, November 12, 2014.
[8] CNN November 12, 2014.
[9] Huffington Post, November 12, 2014.
[10] The Wall Street Journal, November 27, 2015.

the ability of negotiators to carry through U.S. commitments, and they are right to do so. It's likely those left standing with the U.S. are quickly whittling down to a few.

Plenty of other statements on China by U.S. politicians are also based on misinformation. Regarding the U.S.-China Agreement, Senator John Barasso (R-WY) said, "To me, this is an agreement that's terrible for the United States and terrific for the Chinese government and for the politicians there, because it allows China to continue to raise their emissions over the next sixteen years.[11]" Barasso, like McConnell and Inhofe, misused terms of the agreement to imply that the Chinese are not going to do anything for sixteen years. Contrary to that, China has already bent the curve of increasing emissions and reduced the growth rate. In addition, China has canceled more than 100 coal plants, accelerated solar and wind power installations, and instituted cap-and-trade. Those are some interesting moves for a country that is said to have created the hoax of climate change to take advantage of U.S. trade.

Yes, believe it or not our president actually said that.

Will Other Countries Abandon Climate Action?

According to the above statements, neither the G-7 nor the G-20 have indicated backsliding in the immediate wake of the U.S. withdrawal from the Paris Agreement. Some questions remain, however, about just how firm these commitments will be if the U.S. completely ignores the need for climate action and proceeds with plans for fossil fuel development and use. Trump policies are resonant of a Reagan-era push for increased extraction and use of shale oil and gas, and coal:

> The Trump administration will embrace the shale oil and gas revolution to bring jobs and prosperity to millions of Americans. We must take advantage of the estimated $50 trillion in untapped shale, oil, and natural gas reserves, especially those on federal lands that the American people own... The Trump Administration is also committed to clean coal technology, and to reviving America's coal industry, which has been hurting for too long[12].

Considering the Trump administration's plans, other parties to the Paris Agreement might have second thoughts about their commitments. In the wake of Trump's withdrawal from the Paris Agreement, there are already inevitable questions about U.S. commitments.

During the administrations of both Bush and Obama, for instance, the United States pushed for all countries to adhere to a single set of transparency rules for reporting emissions, while China has long argued for a weaker set for developing countries. If the United States were to leave, it would lose its ability to shape these discussions. "If you're interested in pushing China to do more," said [former lead negotiator Todd] Stern, "then the best way to do that is to have us at the table.[13]" Now we'd be lucky to get a seat at the kids' table, let alone negotiate international climate policies with the adults.

Abandoning World Leadership

Another question that arises from the U.S. abandonment of the Paris Agreement is world leadership on climate issues.

Can China really lead the world on climate? They do have economic incentives *not* to push the world toward the kind of ambitious goals it needs to reduce emissions; therefore, with them in the driver seat, we might just see

---

[11] MSNBC, November 12, 2014.
[12] https://www.whitehouse.gov/america-first-energy
[13] The New York Times, May 2, 2017

complacency following the hard-won Paris Agreement that was just a starting point, not an endpoint to where the world needs to be on climate change[14].

In the absence of U.S. leadership, however, other countries have picked up the gavel and indicated leadership. Ironically, one of these is China, our erstwhile partner, and they have started the process of weaning the electricity sector off coal. China has actually taken steps to stop building coal plants; in January, they stopped construction of 103 new ones, a move that sidelined scores of projects where work had already begun and put 120 gigawatts of capacity on hold[15].

India also looms large in the climate crisis. With the world's second-largest population, soon to be first-largest, India may be even more significant than China in terms of carbon pollution. India is highly dependent on coal-fired electricity, so it does present a major problem for emissions reductions. "India, in ratifying the agreement on October 2, 2016, said it would follow a path of low carbon commitment in tandem with its national laws and development agenda, including eradication of poverty.

India also committed to reduce emissions 33 to 35 percent of 2005 levels by 2030."Politifact, June 1, 2017)[16] India's Prime Minister Modi knows the difficulties that lie ahead of him, but has a impressive track record with clean energy. Specifically, he was previously governor of a province that installed large arrays of solar power, and he has pledged to extend this policy to the entire country.

President Donald Trump played the China and India cards when he withdrew the U.S. from the Paris Agreement, echoing the sentiments of Senators McConnell, Inhofe, and Barasso. He said China "can do whatever they want for 13 years. Not us;" and "India will be allowed to double its coal production by 2020. Think of it. India can double their coal production. We're supposed to get rid of ours." Trump gets particularly annoyed when someone else seems to have an advantage over the U.S., but he rarely investigates the facts to see if his claims are valid, and that's putting it delicately. In both cases, they are not. China and India have both committed to carbon reductions and are on track to attain them.

Abandonment by Whom?

Clearly, the fears of Senators McConnell, Inhofe, and Barasso, and President Trump, have not panned out. Ultimately, it is the U.S., not China or India, that has abandoned international climate action. All of the critiques of U.S. politicians miss the mark because of faulty assumptions or ignorance of the facts.

It is interesting that nearly all statements by politicians are based on soft denial, a form of denial ideology that pervades U.S. politics. Many denial politicians say something like, "I am not a scientist," or, "The science is uncertain," just to avoid taking positions in favor of climate action. They excuse their lack of support for international climate agreements by using this form of denial. If they can avoid unpopular stances, they will not rile up their base, who have been manipulated through right-wing media think tanks, with an agenda to dislike the Paris Agreement. It is also worth noting that two of these three senators serve in the top ten oil-rich states. Perhaps that is mere coincidence, but I'll let you be the judge.

Groups such as Heartland Institute or Americans for Prosperity are external to Congress and the Trump administration, but they have influenced senators and the president through their influence on the denial base. These groups prefer to employ hard denial, such as claiming that climate change is only natural or that carbon emissions have little effect on the climate. Trump did use some of these arguments in his speech when withdrawing the U.S. from the Paris Agreement, but most politicians rely on secondary arguments such as the impact on jobs and economic growth. Also, by making invidious criticisms of other countries, U.S. politicians invoke a kind of negative nationalism: "*They* aren't doing anything. Why should we?" or, "*They* are going to take our jobs

[14] NPR, April 2, 2017.
[15] Politifact, June 1, 2017.
[16] Politifact, June 1, 2017.

away!" It is incumbent on us to counter these false assertions with facts about international action on climate change and reassert our responsibilities to the world.

Climate Crisis "Keep/Stop/Start" Actions

Below for your consideration, are several author-suggested action items. For a further explanation of the Keep/Stop/Start organizational tool, please see How to Read This Book.

KEEP:

Keep inquiring. Question statements from politicians that seem to contradict the facts, and look for the underlying reasons that they deny climate change.

When you hear arguments against U.S. action on climate change, ask what effects these arguments will have on the rest of the world. Politicians' assertions can be fact-checked with sites such as Politifact.com.

Will other countries follow the U.S. example and pull out of the Paris Agreement, or will they reaffirm their commitment to compensate for U.S. foot-dragging? Reporting on international negotiations on climate change is available in national newspapers such as the *New York Times* and the reporting service *Earth Negotiations Bulletin* (http://enb.iisd.org/process/climate_atm.htm). Dig into the details so that you will be better prepared to refute the misstatements of the politicians.

STOP:

1) *Stop the conflation of climate action* and threats to sovereignty, jobs, or the economy. The Paris Agreement and international cooperation are not signs of world government, as there are no legally binding mechanisms to enforce actions. There is no international body to administer the agreement. The United Nations Framework Convention on Climate Change (UNFCCC) secretariat in Bonn is authorized by members only to arrange meetings and exchange information, from which decisions are made by a Conference of Parties each year, and all members must agree on any decisions. More information can be found at http://unfccc.int/2860.php. The Paris Agreement does not threaten jobs or economic growth; in fact, renewable energy installations provide more than twice the employment of coal, oil, and gas installations (see the chapter on Zombie Myths), while unfettered climate change will severely damage economies.

2) *Stop disparagement of the Paris Agreement* by voicing to others that there is one provision particularly useful about its effectiveness:

Parties aim to reach global peaking of greenhouse gas emissions as soon as possible, recognizing that peaking will take longer for developing country Parties, and to undertake rapid reductions thereafter in accordance with best available science, so as to achieve a balance between anthropogenic emissions by sources and removals by sinks of greenhouse gases in the second half of this century, on the basis of equity[17].

---

[17] Article 4 of the Paris Agreement.

In plain English, this provision aims for zero net emissions by about 2050. In other words, the Paris Agreement has a specific target for worldwide emissions that is fair and achievable by all countries. Despite Trump and company, the U.S., Europe, China, India, and many other countries have the means to achieve zero net emissions by 2050 and could help other countries[18]. If you can only get across one point, this is the most salient argument for international cooperation. Only by working together can the world's peoples solve the climate crisis!

START:

1) *Start informing yourself about international agreements* on climate change and their role in addressing the climate crisis. You may have heard that the Paris Agreement would not reduce warming by two degrees Celsius as planned, and opponents of the agreement contend that it is not worth U.S. accession sometimes use this argument, but the Paris Agreement has built-in mechanisms for correcting any shortcomings. If you hear anyone saying that's not good enough, counter with the provisions that require parties to review their commitments every five years and increase them if necessary to reach the two-degree goal. The bottom line is that with the sheer amount of misinformation out there, *you* need to be the voice of reasoning and a preacher of fact.

2) *Start becoming involved in the support of the Paris Agreement* at the state and local level. Fourteen states, with one-third the U.S. population and also a third of its GDP, have formed the United States Climate Alliance for supporting the Paris Agreement, despite the Trump administration revocation. Find out if your state is one of the twelve and, if not, why not. Cities have also joined, so see if your state or city is involved by checking the lists at https://en.wikipedia.org/wiki/United_States_Climate_Alliance#Cities and https://en.wikipedia.org/wiki/Mayors_National_Climate_Action_Agenda.

3) *Start making presentations or speeches* in favor of U.S. participation in international climate programs at your church and political or civic group. The easiest way to get people to see the benefits of acting on climate is to make it relatable to them, and there is no better way to make it relatable than by talking to your peers. You can find information and talking points at http://earthjustice.org/features/paris-agreement; Point 8 is particularly good for framing a brief discussion of what the U.S. should do about the Paris Agreement.

~ ~ ~

William McPherson Ph.D., is a retired environmental diplomat with 21 years' service in the U.S. Foreign Service, including assignments in Tokyo and Geneva. He continued his work on international environmental issues after retirement, working with Earth Negotiations Bulletin, a newsletter on environmental negotiations. He has served in assignments in Europe, Japan, Korea and the Philippines as well as other countries. He is an activist with the Sierra Club, working on issues of climate change and coal exports. His work on climate includes promoting carbon pricing and oil transport safety in Washington State. He is the author of author of three books: *Ideology versus Science, Climate Weather and Ideology* and *Sabotaging the Planet*. Read more at: www.climateabandoned.com.

[18] https://www.beforetheflood.com/explore/the-solutions/zero-greenhouse-gas-emissions-by-2050-why-paris-is-good-news-for-the-planet/

Chapter 18

Climate Crisis & Energy

by

Jigar Shah, M.B.A.

"Climate change is one of the greatest wealth-creating opportunities of our generation."
– Sir Richard Branson, Founder, the Carbon War Room

In 2017, over 50 percent of all electricity plants added to the grid were from solar facilities. Add in other climate friendly sources like wind, hydro, biomass, and nuclear energy sources, and that raises to a whopping 80 percent. In fact, we have hit "peak coal," meaning that global coal use will now get smaller over time. We really can power the world with climate-friendly solutions; the real concern comes from figuring out how governments and private investors will react to this reality.

When I first talked about climate solutions, I was CEO of the Carbon War Room, a global nonprofit started by Sir Richard Branson to celebrate fearless change-makers pushing climate solutions around the world. Working with Bloomberg and others, we found that the world needed about $10 trillion of incremental investment above and beyond business-as-usual investing to reduce our carbon emissions to acceptable levels.

We felt immediate pushback from folks who thought we needed better technology to save the world. In the minds of many who lived through the dot-com and telecom bubble of the 1990s, modern technology was simply not mature or cheap enough to meet global lifestyle goals without significant sacrifice. Over the years, the Carbon War Room proved this to be a fallacy. While new technology and research should always be encouraged, much of what we had already was capable of deploying an additional $10 trillion of cost-saving solutions at scale.

This data was presented in an interesting way via the McKinsey Cost Curve. The 2009 report showed that approximately 50 percent of the climate change challenge could be addressed profitably by *existing technology,* revealing it as a business opportunity masked as a crisis. With the dramatic cost reductions of solar, wind, and battery technology, that cost curve now shows that 80 percent of the challenge can be profitably addressed by 2030.

If it is not a question of waiting for better innovation, what is stopping us from confronting the greatest challenge to mankind? Effective implementation! Since World War II, we have put in place extraordinary institutions designed to bring circa-1940s industrial technology to the forefront. The goal of these institutions is to reduce poverty by taking free markets and democracy to every corner of the globe. What's lost in this framework is that the carbon-intensive infrastructure they are pursuing will directly undermine their goals. More importantly, today's infrastructure will no longer be monolithic, centralized, and inefficient government-run projects; rather, it will consist of resilient, flexible, and distributed ones driven by the private sector. This approach will mean more jobs, more economic surplus, and more resilience to both foreseeable and unforeseeable risks.

It is time for institutions, governments, and financial players worldwide to acknowledge that the big-infrastructure era has ended. The new era of infrastructure doesn't look like nature-defying hydroelectric facilities or pollution-belching, coal-fired power plants. It isn't just bridges and toll roads either. This era of infrastructure is on-demand, shared transportation like Uber rather than high-speed rail that never gets permitted, much less funded; it is your neighbor's solar lantern that doubles their daily income and provides freedom from the promises of the government utility that never comes; it is the modular wastewater treatment system attached to the local microbrewery that generates clean water, clean energy, and natural fertilizer simultaneously, without requiring the costlier, more bureaucratic municipal waste authorities and utilities.

Most climate-friendly solutions come in flexible, modular infrastructure that requires less time to receive government permission at national and local levels. The sad truth is that by the time governments have a large monolithic project planned, permitted, financed, and built, they may be doing a ribbon-cutting for an already-obsolete project.

Importantly, as governments continue to cry poverty to provide the basic infrastructure their people need, private sector players are entering the space. Big infrastructure is slow, publicly expensive, and politically contentious. To build this smaller, faster, better infrastructure, we don't need government appropriations; however, policymakers and government agencies still play a critical role. Streamlining permitting processes helps. Deregulating the transportation, utility, water, and energy sectors will help, too, by opening the door for new competition and providing choice and lower costs for consumers. Accelerating approvals for small business loans from the U.S. Small Business Administration or Small Business Investment Companies can also make a difference.

Getting to $10 trillion

When making these estimates, it is always important to remind ourselves that they are just that: educated guesses. The International Energy Agency (IEA) undertakes an extensive effort annually, Energy Technology Perspectives (ETP). ETP 2012 concluded that an incremental $5 trillion of investment was needed and could be profitably invested versus the business-as-usual case across the world to stay on track to keeping global temperatures below 2 degrees (versus the 6 degrees we are track to reaching).

Cumulative additional investments in the 2DS compared to 6DS, 2010 to 2020

Given that we are starting later than expected and have to over-build infrastructure to stave off the worst impacts of climate change, the incremental investment the IEA identified of $5 trillion of renewable electricity generation investments is probably closer to $10 trillion at this point. Nevertheless, as the IEA noted, all these investments save money and fully pay for themselves in fuel, maintenance, and other ways.

Additional investment and fuel savings in the 2DS compared to 6DS, 2010 to 2020

Looking at the data, there are some broad themes that jump out. Given the need to ramp up carbon emission reductions rapidly, it is clear that renewable electricity and building efficiency have to lead the way, since many of the other categories simply do not have enough existing momentum.

Vehicle efficiency and, more importantly, a transition away from personal car ownership to electric vehicles is critical to allow for more renewable energy to be added to the electricity grid. The rest of the sectors are not left behind because of intrinsic potential but because moving this much capital requires trained personnel and institutional comfort. If efforts are made now to attain this comfort by 2020, sectors like industrial efficiency, agriculture, and deforestation could play a bigger role in the next decade. In fact, renewable electricity and energy efficiency must take such a large role that they will have to necessarily slow their growth after 2020, making room for the trained people to switch to other carbon-producing sectors to bring their knowledge and expertise to those critical sectors.

Addressing climate change effectively requires a planned, systematic approach. One important barrier to remember is perception. Even leaders such as Germany are hard-pressed to say that such climate change solutions as solar and wind power can anchor an entire economy. If we don't change this perception on a global scale, such investments will never happen.

Creating our carbon economy has been achieved piece by piece over the past seventy years. Dismantling it, while increasing our electric production, will create winners and very entrenched losers who still have considerable political power.

Hope and Change

In 2008, Al Gore made a John F. Kennedy-style proclamation that it was possible to achieve 100 percent renewable energy in just 10 years, stating that all we needed was willpower. Since that speech, every major government agency has said that this transition is possible, at least to 80 percent renewable energy, and the largest markets (India, China, Brazil) for new power plants are now aiming to meet 100 percent of their demand growth from clean energy.

Another big area of improvement has been in energy efficiency. It has long been known that buildings could operate using significantly less energy, but implementation has been lacking. Today, big data has unlocked energy efficiency and made it possible to change the way buildings use energy to integrate various forms of renewable electricity without destabilizing local electricity circuits. *Whether it's U.S. government data, McKinsey data, or electric utility data, all confirm that a complete global transition to clean energy is both technically and economically feasible.* The barriers to achieving that goal are social and political.

Many have criticized the United States for not being the de facto global leader on climate change solutions and policies. Japan, Germany, and even China are held up as examples of nations that have done a better job. Still, the United States is a diverse democracy with more than 324 million people and 2,000 electric utilities. Sure, the United States has let others lead the way, but in 2017, they were the only major global economy to reduce emissions. In fact, 100 percent of all electricity growth in the United States since 2003 has come from clean energy. Electricity from natural gas actually shrunk a little in 2017. At this point, it is a foregone conclusion that almost all net new electricity generation in the United States will come from zero-emissions fuel sources. United Nations Secretary General Ban Ki-moon insists, "Zero emissions from new electricity needs to happen by the second half of the century." We can now see that's possible by 2050.

There's certainly the capacity for tremendous expansion in renewables. Globally, solar has been deployed at a rate of $200 billion a year for the past few years, scaling faster than cellphones, and it's still the most underutilized proven technology to stave off the effects of climate change. How underutilized? Consider that Australia deployed solar on two million homes in less than five years, from 2009 to 2013. That's one solar system for every twelve people! In the United States, solar only reached two million solar homes in 2018. To match Australia, we would need to build almost 25 million new solar home systems in the next 5 years, a $400 billion investment.

This might sound like a daunting number, but is it? Not really, since most of the innovation in the clean energy space has come not from new technology but from the financing sector. In 2005, the company I founded, SunEdison, put together the first solar-only fund, along with Goldman Sachs, kicking off the global no-money-down solar boom. More recently, companies have been going directly to pension funds and the public markets to gain access to even cheaper capital. We have achieved the goal of attracting institutional investors who want better returns than they can achieve from buying corporate bonds. It has worked so well that today, anyone can invest in renewable energy projects by buying stock that pays a secure dividend. Institutional investors are now so sold on the stable returns of renewable energy that they have set aside large amounts of money for new projects and are finding that there are not enough projects to satisfy their appetite. *When it comes to renewable energy, there is more money chasing projects than there are projects to finance.*

Scientists say that to stave off the worst impacts of climate change, we need to reduce our emissions by 80 percent. Solar, wind, nuclear, biomass, geothermal, energy efficiency, clean transportation fuels, local food, water-saving technologies, and many others are now cost effective for communities across the United States.

People Power

What recent elections have shown us is that even intellectuals can be upended when people come together around a common cause. Today, we are fighting inertia. What we need are passionate people with good local connections. To reach our $10 trillion goal, we need only 10,000 local groups of people convincing their communities to buy at least $10 million worth of money-saving climate change solutions every year. That's easier than it sounds: A $10 million project is the equivalent of helping a school district put solar on their rooftop, geothermal heat pumps in their basement, and LED lighting in their classrooms. It can also represent ten electric buses.

Today almost every climate change solution above can be sold using pay-as-you-save financial models. By now, most folks have heard of no-money-down solar, but the same financing mechanism can be used for energy-efficiency projects, electric vehicles, and energy-saving heat pumps. Customers pay absolutely nothing upfront but start saving money from day one. Every month, the customer pays a portion of that savings back to the investor, until the investor is repaid with interest. This is an important tool for cash-strapped municipalities that don't want to raise taxes to pay for these upgrades.

If citizens in communities across the country talk to their mayors, neighbors, friends, local business owners, and anyone else who'll listen about how to save money with climate-change solution technologies, it's not only possible but inevitable.

Changing the Narrative

In general, finding passionate people who are committed to solving climate change has been relatively straightforward, but the narrative has been established by environmentalists rather than marketing experts. Clean energy in the United States has been defined by earnest environmentalists who, to their credit, embraced it wholeheartedly; however, to our collective detriment, they spun an ideological, naïve story divorced from the reality of the energy economy transformation actually taking shape around us.

The result is that clean energy is mistakenly seen as a passive and precious solution for a future society, like a delicate sunflower waving in the face of a muscular coal miner or a pristine field of green and sky of blue set against a dirt mound penetrated by a fracking rig. It feels more utopian than aspirational, like more luxury than necessity.

In short, it doesn't feel American.

American is a brand that respected around the world. American means can-do, right-now, yes ma'am. Luckily, the actual transformation of the energy economy is as American as the Hoover Dam or the interstate highways and even more earth-shaking. If only the discussion among politicians, media, business leaders, and the American public reflected that reality!

Unfortunately, the clean energy conversation is profoundly and unnecessarily polarizing. Like climate change itself, it's become part of a larger culture war that fits neatly into the media's predictable tendency of false equivalence, pitting workers against activists, businessmen against academics, and common sense against idealism. As a result, according to recent surveys, public sentiment about the urgency of action to prevent climate change is split along political party lines, either declaring, "Let's do something!" or, "Meh."

The energy might be clean, but the work and the jobs are as rooted in dirt, sweat, and back-breaking labor as any American endeavor, and even more lasting.

We need to change the conversation to align with the deep emotional and aspirational narratives that speak to the American public. Clean energy can feel as all-American, cutting-edge, rugged, reliable, resilient, and tough as fracking. The same American ideals of independence, freedom, self-sufficiency, and opportunity can bring together green advocates and Tea Party stalwarts, labor and entrepreneurs, main street and Wall Street.

Independence is the heart of the American identity. Clean energy is independence turned into electrons: the application of cunning, sweat, and ingenuity to harness the restless power of the American landscape.

The American energy economy is changing, and changing rapidly. Clean energy and energy efficiency are where growth is happening. We can move millions of people from coal mining, low-tech manufacturing, and even oil and gas into good-paying jobs that don't negatively impact the health of people and the planet. In fact, we already have.

By rebranding clean energy, we can empower all Americans to work together for a stronger future. It's time to get down and dirty!

Climate Crisis "Keep/Stop/Start" Actions

Below for your consideration, are several author-suggested thought, behavior, and action items. Action nourish inspiration! For a further explanation of the Keep/Stop/Start organizational tool, please see How to Read This Book.

KEEP:

1) *Keep in mind* the next time you are looking to buy a car that, with a 250-mile range, all-electric vehicles are now a serious option to consider.

2) *Keep recycling* your garbage and composting your food waste.

STOP:

1) *Stop thinking of solar as a wimpy*, hippy-dippy power source. It is now state-of-the-art, high-tech, space-age technology, that works, as investors are realizing. According to the Department of Energy, the U.S. clean energy sector employed 3.38 million workers in 2016. That is 10 percent more than the 2.99 million employed by the fossil fuel industry. With your activism, we can build on that success!

2) *Stop overly expensive*, onerous permitting processes in local government that impede sustainable energy projects. Get involved in local politics by calling your representative, as a group of activists did in Santa Cruz, California.

3) *Stop governmental regulations that impede sustainable energy growth*, such as one that was implemented in (believe it or not!) the Sunshine State. The state of Florida passed laws to inhibit the growth of rooftop solar-array installations, unlike the state of Georgia, which has embraced solar energy.

START:

1) *Start considering solar* energy for your house. To match Australia, the U.S. needs to install twenty-five million new solar home systems in the next five years. Be one of them!

2) *Start or push your local school*, governmental, church, or commercial buildings to install solar equipment, geothermal heat pumps, and LED lighting and help them save on their energy bills for years to come (while enhancing environmental quality).

3) *Contact your employer's financial*, pension, or IRA managers and urge them to seriously consider including sustainable energy investments in their portfolios.

4) *When considering your financial plan*, think about investing in energy efficiency for your favorite nonprofits. These are solid investments, a gift that keeps on giving.

5) *If available in your area, change your power company* to one that purchases sustainable energy, or look into starting a CCA (community-choice aggregation) joint-powers authority that increases local control over energy procurement and allows for more renewable and efficient resources to be deployed in your region.

6) *Start sharing the reality* that sustainable energy is the present and the future. Part of increasing momentum is knowledge about its ability to create jobs and be cost-effective.

~ ~ ~

Jigar Shah, M.B.A. is a clean energy entrepreneur who creates market-driven solutions and eliminates market barriers to address climate change. Shah has recognized this as creating climate wealth and maintains that this occurs when mainstream investors team up with entrepreneurs, corporations, mainstream capital, and governments at scale to solve the big problems of our time while generating compelling financial returns, not concessionary returns. Shah is the co-founder and president of Generate Capital and author of *Creating Climate Wealth: Unlocking the Impact Economy, 2013 Icosa Publishing*. Shah serves as a board member of the *Carbon War Room*, a global organization he previously served as CEO. Shah has also become an outspoken advocate to end all energy subsidies, including those for renewable energy, to "create a level playing field." Jigar lives in San Francisco, California. Read more at: www.climateabandoned.com.

Chapter 19

Climate Crisis & Corporations

by

Jill Cody. M.P.A.

"If our governments actually intend to crack down on carbon emissions and make polluters pay, then they will first need to declare independence from the fossil fuel industry. The energy giants are motivated by one thing: profits. As long as those profits mean the extraction and burning of fossil fuels, they will never be part of the solution." – **Naomi Klein,** author of This Changes Everything: Capitalism vs. The Climate

"Corporations are people, my friend!" Mitt Romney made this statement at a campaign stop during his 2012 presidential run. A corporation, a business construct, cannot possess humility, compassion, remorse, or empathy that is possessed by a natural, normal human being. If a corporation were, in fact, a "person" then it would be a psychopath.

What is beyond annoying is that there was no legislation passed by Congress that made corporations into people. It is imperative to understand that in 1886, corporations achieved personhood based only on a Supreme Court clerk's notation.

In his book, Unequal Protection, Thom Hartmann wrote, "...corporate personhood was never formally enacted by any branch of the U.S. government." This enormously important issue that set the stage for today was written as a headnote by a court reporter, J. C. Bancroft Davis. Headnotes are not law, and they are not a decision. They are notes written by a commentator, in this case, a court reporter, or book publisher. Yet, today, our corporate legal structure is based only on this headnote. Astonishing. Also, from Unequal Protection:

> ...there has been no Supreme Court ruling that explicitly explains why a corporation-with its ability to continue operating forever, it's being merely a legal agreement that can't be put in jail and doesn't need fresh water to drink or clean air to breathe-should be granted the same constitutional rights our Founders fought for, died for and granted to the very mortal human beings who are citizens of the United States to protect them against the perils of imprisonment and suppression they had experienced under a despot king.

Citizens United

For a little background, the conservative tax-exempt, nonprofit Citizens United states that they are "dedicated to restoring our government to citizen control." In reality, though, they appear to be dedicated to serving up our government to corporate control. The organization produces television commercials and web advertisements and has created twenty-five documentaries, all with a conservative, anti-government corporate spin. Since it requires the power and funding of the federal government to combat the climate crisis with the urgency it now needs, an uncompromising anti-government ideology is a significant mindset that put us on the endangered species list.

Citizens United took issue with the Bipartisan Campaign Reform Act of 2002, also known as the McCain-Feingold Act. The nonprofit had made a movie about Hillary Clinton and planned to advertise and release it before the 2008 presidential election. The Federal Elections Commission (FEC) determined that it was a ninety-minute campaign ad and was in violation of the Bipartisan Campaign Reform Act of 2002. A lower court agreed and upheld the FEC ruling, but Citizens United appealed the lower court's decision to the Supreme Court.

For real, human, flesh-and-blood beings, things went frightfully wrong in the Supreme Court. Taking a giant leap, the court built on a 1976 decision known as Buckley v. Valeo that awarded free speech rights to money and a 1978 decision known as First National Bank of Boston v. Bellotti, which granted free speech rights to corporations. The end result was that, since corporations had already achieved personhood, corporations now had no limits when it came to campaign spending.

Justice John Paul Stevens stated, in his dissenting opinion:

> At bottom, the Court's opinion is, thus, a rejection of the common sense of the American people, who have recognized a need to prevent corporations from undermining self-government since the founding, and who have fought against the distinctive corrupting potential of corporate electioneering since the days of Theodore Roosevelt. It is a strange time to repudiate that common sense. While American democracy is imperfect, few outside the majority of this Court would have thought its flaws included a dearth of corporate money in politics

Nina Totenberg, an American legal affairs correspondent for National Public Radio, in an article titled, "When Did Companies Become People? Excavating the Legal Evolution," explained that corporations have no political speech and, in fact, that a corporate corruption scandal in 1907 caused Congress to pass a law banning corporate participation in federal elections, a law maintained for seventy years. The first so-called crack came in 1978, with a 5-4 Supreme Court decision (First National Bank of Boston v. Bellotti), stating that corporations have the right to spend money in state initiatives. Then, Citizens United, in another 5-4 decision gave corporations free speech rights and full rights to spend money however they want on candidate elections, at all levels of government. Specifically, many point to the inclusion of a single sentence by Justice Anthony Kennedy as the most impactful: "We now conclude that independent expenditures, including those made by corporations, do not give rise to corruption or the appearance of corruption." Though ten years later we can now see how incomprehensibly false and illogical this statement is, its inclusion in the Citizens United ruling made it the law of the land. Totenberg accurately assessed, "It thrilled many in the business community, horrified campaign reformers, and provoked considerable mockery in the comedian classes." This thrill was felt worldwide by psychopath corporations.

Just how thrilling was this for corporations? Look no further than the graph below illustrating the outside spending during election years in the last three decades.

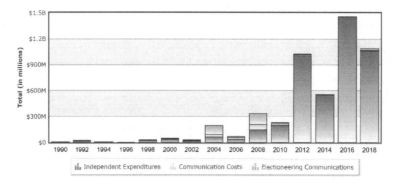

Outside Spending by Cycle thru December 29th of election year, Excluding Party Committees

Source: Center for Responsive Politics, found at https://www.opensecrets.org/outsidespending/fes_summ.php

Independent expenditures have more than quadrupled in both presidential election years and midterm election years just in the past decade. So much so, the 2018 midterm election saw more outside spending than the 2012 presidential election—something that was previously unfathomable prior to the Citizens United ruling. Parallel to this increase in spending has been the rise of super-PACs—political action committees that can raise unlimited sums of money from corporations, unions, associations and individuals, then spend unlimited sums to overtly advocate for or against political candidates. Given the unlikelihood of the conservative-leaning Supreme Court to overturn Citizens United anytime soon, we can only expect the influence of dark money and outside influence from corporations to continue to infect our politics for the foreseeable future—not ideal for the fight to mitigate the climate crisis.

Two Psychopath Oil Company Examples

Today, human civilization is on the endangered species list due to the psychopathic behavior of oil companies for the last sixty years. Dr. Kate Marvel, associate research scientist of the NASA Goddard Institute for Space Studies, said there is "no cliff, but there's sure a slope." Being on the slope puts us on the list and trajectory for a dismal future for civilization. If the executives and stockholders had taken the responsible path since they were first informed to what lay ahead, what their product was doing to the atmosphere, none of us would be on the "slope." None of us would even have known such words as climate change, global warming, and extinction.

What happened sixty years ago? Physicist Edward Teller, on November 4, 1959, was invited to speak on energy patterns of the future at the oil industry's 100th birthday celebration at Columbia University hosted by the American Petroleum Institute (API). The 300 attending oil executives and industry officials expected a benign lecture, but instead, they received a warning (https://www.theguardian.com/environment/climate-consensus-97-per-cent/2018/jan/01/on-its-hundredth-birthday-in-1959-edward-teller-warned-the-oil-industry-about-global-warming). An excerpt of Dr. Teller's address says it all:

> Whenever you burn conventional fuel, you create carbon dioxide. [....] The carbon dioxide is invisible, it is transparent, you can't smell it, it is not dangerous to health, so why should one worry about it?

> Carbon dioxide has a strange property. It transmits visible light, but it absorbs the infrared radiation which is emitted from the earth. Its presence in the atmosphere causes a greenhouse effect [....] It has been calculated that a temperature rise corresponding to a 10 percent increase in carbon dioxide will be sufficient to melt the icecap and submerge New York. All the coastal cities would be covered, and since a considerable percentage of the human race lives in coastal regions, I think that this chemical contamination is more serious than most people tend to believe.

How did these companies respond to Teller's warning? Apparently, their answer was what we are experiencing today, intensified hurricanes and wildfires, famine creating droughts, planetary temperature increase, sea level rise (fish swimming on the flooded streets of Miami!), animal extinctions, insect invasions, and melting ice caps, to name a few. However, knowing a bit more about what API and the oil industry did in the subsequent decades is eye-popping.

Stunningly, API "silently" commissioned a report that was completed in 1968 by the Stanford Research Institute and it stated that "significant temperature changes" would happen by the year 2000 and that there was "no doubt" the damage caused by this temperature change can be potentially severe. It also stated, "What is lacking, however, is [...] work toward systems in which CO2 emissions would be brought under control."

Shell Oil is a prime example of a corporation behaving like a psychopath. In 1991, twenty-seven years ago, Shell created a 28-minute educational film for schools and universities titled "Climate of Concern." It was quite alarming as it explained that the burning of fossil fuels will warm the planet and that we will experience extreme weather events, floods, famine, and refugees. Believe it or not, the film stated, "If the weather machine were to be wound up to such new levels of energy, no country would remain unaffected. Global warming is not yet certain, but many think that to wait for final proof would be irresponsible. Action now is seen as the only safe insurance." Say, what now? Action was the only safe insurance?

Yet, Shell executives who clearly knew the catastrophic planetary reaction their company's product would bring forth onto civilization, invested billions of dollars in Arctic exploration and tar sand operations. If this wasn't evil enough, they spent millions to fund lobbyists to fight any climate legislation that attempted to combat the very problem they had known about for decades (https://www.theguardian.com/environment/2017/feb/28/shell-knew-oil-giants-1991-film-warned-climate-change-danger).

Even before, Edward Teller's speech, Humble Oil (now Exxon Mobil) in 1957 published a paper regarding carbon in the atmosphere and in the ocean. The company's scientists understood the link between carbon and temperature rise by saying in the paper, "Although appreciable amounts of carbon dioxide have undoubtedly been added from soils by tilling of land, apparently a much greater amount has resulted from the combustion of fossil fuels."

Then, again, 20 years later Exxon's Products Research Division distributed an internal paper titled, "The Greenhouse Effect." This internal paper stated that "Present thinking holds that man has a time window of five to ten years before the need for hard decisions regarding changes in energy strategies might become critical."

Years later, Exxon's director of Theoretical and Mathematical Sciences Laboratory, Roger Cohen, wrote a memo stating,

> The consensus is that a doubling of atmospheric CO2 from its pre-industrial revolution value would result in an average global temperature rise of $(3.0 \pm 1.5)°C$ [equal to $5.4 \pm 1.7°F$]...There is unanimous agreement in the scientific community that a temperature increase of this magnitude would bring about significant changes in the earth's climate, including rainfall distribution and alterations in the biosphere.

What is unfrigginbelievable is that after Roger Cohen wrote those words, he later became a leading climate denier for an Exxon funded front group. A prime example of how one sells their soul.

In the next two decades, ExxonMobil goes all out in creating doubt. Below are abbreviated excerpts from a Greenpeace report, "Exxon's Climate Denial History: A Timeline":

- (1997) Exxon CEO Lee Raymond tells the 15th World Petroleum Congress in Beijing that the world's climate isn't changing, and that even if it were, fossil fuels would play no part.

- (2000) The ExxonMobil published an ad titled, "Unsettled Science."

- (2000) The CEO, Lee Raymond presents a study to his shareholders saying that fossil fuels may not be causing global warming.

- (2001) The company donates $100,000 to George W. Bush's inaugural fund.

- (2001) ExxonMobil sends President Bush a letter requesting that the administration oust Robert Watson, chair of the Intergovernmental Panel on Climate Change because, during his chairmanship, the IPCC released several reports linking climate change to human activity.

- (2001) President Bush announces the withdrawal of the Kyoto Protocol.

- (2004) ExxonMobil creates an ad titled, "Directions for climate research." It presents the position for "uncertainties that limit our current ability to know the extent to which humans are affecting climate and to predict future changes caused by both human and natural forces."

- (2006) Rex Tillerson becomes CEO of Exxon.

- (2006) Under pressure from the Royal Society, Britain's preeminent scientific organization, Exxon states it will stop funding climate-denying groups. Funding is cut to some climate-denying groups. Funding to others continues.

- (2013) The Exxon-funded American Legislative Exchange Council (ALEC) wrote the "Environmental Literacy Improvement Act" that, if passed, would have mandated the weaknesses of global warming theory.

- (2015) Exxon issued and investigated by multiple entities such as the US Department of Justice, the FBI, the US Virgin Islands, New York Attorney General, the Security and Exchange Commission, and Senate Democrats, for violating the Racketeer Influenced and Corrupt Organizations Act (RICO) as well as laws on consumer protection, truth in advertising, public health, and shareholder protection.

- (2016) The Exxon board confronts 14 proposed resolutions from shareholders, 10 of which are climate-related. All climate change resolutions are voted down.

- (2016) Rex Tillerson appointed Secretary of State

Exxon then capped off these two decades of sewing doubt by announcing a lobbying initiative to advance the case for a carbon tax (2018). If corporations truly were people, Exxon would not only be a psychopath, they'd be a schizophrenic.

Minuscule, near-zero, statistically non-significant impact

"When the preferences of economic elites and the stands of organized interest groups are controlled, the preferences of the average American appear to have only a minuscule, near-zero, statistically non-significant impact upon public policy." These words were written in a scholarly article titled, "Testing Theories of American Politics: Elites, Interest Groups, and Average Citizens," completed in a partnership between Princeton and Northwestern Universities. It was an article that genuinely rocked its readers. The report also stated:

> Multivariate analysis indicates that economic elites and organized groups representing business interests have substantial independent impacts on U.S. government policy, while average citizens and mass-based interest groups have little or no independent influence.

The report authors, Martin Gilens and Benjamin I. Page, studied 1,771 bills passed by Congress. Notably, they completed their analysis before the Citizen's United Supreme Court decision. One may easily assume the situation has only worsened since that decision opened corporate funding floodgates. As former President Jimmy Carter once said, "The Supreme Court seems to be eager to see rich people become more powerful and to see corporations become more powerful."

And powerful they are with the fossil fuel corporations and the economic elite, such as the Koch brothers, being on top. The Guardian undertook an investigation and discovered that in 2013, the coal, oil and gas industries benefited from subsidies of $550bn, four times those given to renewable energy. Senator Bernie Sanders put it this way, "At a time when scientists tell us we need to reduce carbon pollution to prevent catastrophic climate change, it is absurd to provide massive taxpayer subsidies that pad fossil-fuel companies' already enormous profits."

Using the Subsidy Tracker https://www.goodjobsfirst.org/subsidy-tracker), one can quickly see that the taxpayer subsidies are diverse and numerous. In one example, the state of Texas gave ExxonMobil nearly half a billion dollars in just one megadeal! Then, looking at what the federal government and other states dish out under a variety of programs, ExxonMobil has been handed billions, and billions of our taxpayer money and ExxonMobil is just one example.

What Gilens and Pages' study sadly and shockingly highlighted for us is that what corporations and the morbidly-rich who own them, always get what they want and the individuals such as ourselves and mass-based organizations, such as Greenpeace, the Sierra Club or the National Resources Defense Council (NRDC) have little or no independent influence. It's like sitting at a poker table with someone who has a billion dollars in chips, and you have only $100 in chips. How long do you think you'd last in the game?

Yes, there are corporations with products such as solar and wind power who are on the angel side of the equation, and they are growing businesses. The US Department of Energy released a report, the 2017 US Energy and Employment Report, stated that clean energy jobs outpaced fossil fuel ones by five to one! If you then added part-time related jobs, it is 14 times more than fossil fuel employment.

It is evident that sustainable energy corporations should be receiving the multi-billion- dollar subsidies now, but the morbidly-rich Koch brothers and their ilk are heavily invested in the industry and don't want any of us to get off fossil fuels. So, with a head-spinning whiplash, we are right back to the people's impact on Congress having only a minuscule, near-zero, statistically non-significant impact upon public policy.

What is the answer? The answer is easy. Get corporate and billionaire money out of politics. However, achieving it is next to impossible. Move to Amend is working towards a Constitutional amendment. However, no state under Republican control would ever pass it. Changing enough states to Democratically held legislatures would take too long. As the most recent National Climate Assessment (NCA) report warned, we only have a few years remaining to get a grip on this challenge. The NCA was a congressionally mandated report by scientists from 13 federal agencies providing an overview of climate threats based on economic sector and by region (https://carbon2018.globalchange.gov). It stated that the United States is in an increasingly bleak and dire situation. It further noted that the "era of climate consequences for the U.S. is well underway, and only actions taken in the next few years can be effective in addressing the scope and severity of the problem." Thus, there is not enough time to change state legislatures to democratic control.

The strength we have as citizens is to elect people who won't vote for subsidies, at the state or federal level, to the fossil fuel industries AND who don't depend on billionaire money to win elections.

Climate Crisis "Keep/Stop/Start" Actions

Below for your consideration, are several author-suggested thought, behavior, and action items. Action nourishes inspiration! For a further explanation of the Keep/Stop/Start organizational tool, please see How to Read This Book.

KEEP:

1) *Keep educating yourself.* "We can fight climate change one person at a time." This video shares what you can do: https://www.youtube.com/watch?v=B_WSURyhXhM. The Sanders Institute, https://www.sandersinstitute.com, under their Issues and Climate tabs has several short and informative videos.

2) *Keep fighting for wealth and income equality.* The richest 1% of America now own more wealth than the bottom 90% (https://www.nytimes.com/2014/07/24/opinion/nicholas-kristof-idiots-guide-to-inequality-piketty-capital.html). In the last four decades (adjusted for inflation), American workers have seen an 11.2% increase in earnings, meanwhile American CEO's have seen a 937% increase in earnings (https://www.cnbc.com/2018/01/22/heres-how-much-ceo-pay-has-increased-compared-to-yours-over-the-years.html). Until Jeff Bezos decided to raise Amazon's minimum wage to $15 last year, he was making the average Amazon employee's annual salary every 9 seconds (http://time.com/money/5262923/amazon-employee-median-salary-jeff-bezos/). The income inequality in America is sickening, and as we have seen, corporations clearly do not have our best interest in solving the climate crisis. Though it is theoretically possible to solve the climate crisis without wealth equality, it is not possible to develop a truly sustainable society in the face of such drastic income inequality.

STOP:

1) *Stop providing taxpayer subsidies to fossil fuel* companies such as Shell and ExxonMobil. Currently, the American taxpayer hands over 21-billion to fossil fuel multinational corporations. Call every one of your elected representatives now and demand that fossil fuel subsidies be moved to sustainable energy companies and request that they also support "The Green New Deal" (https://en.wikipedia.org/wiki/Green_New_Deal).

2) *Stop falling victim to greenwashing.* Not only do many companies use "green initiatives" to market to you on an individual basis, they also do so on a larger scale. Namely, many corporations that you know and love for their public support of climate action also fund climate-deniers behind closed doors. Do your research and know where your money is going.

START:

1) *Start contacting your retirement programs* and financial planners and ask if your investments contain fossil fuel stocks. If they do, then request that the retirement program no longer does so and that they buy shares that do not include fossil fuel holdings. Like Desmond Tutu, Archbishop Emeritus explains, *"People of conscience need to break their ties with corporations financing the injustice of climate change."*

2) *Start checking the Subsidy Tracker Tool* to learn how much of your hard-earned money is being funneled to corporations, https://www.goodjobsfirst.org/subsidy-tracker.

3) *Start calling your elected representatives.* As one Shell spokesperson let out of the bag to The Guardian: "Shell supports and endorses incentive programmes provided by state and local authorities that improve the business climate for capital investment, economic expansion, and job growth. Shell would not have access to these incentive programmes without the support and approval from the representative state and local jurisdictions." Of course, Shell would support and endorse incentive programs. Why incentivize an industry that is killing us? So, fight those "incentive programs" dished out by your elected representatives. Common Cause will help you find your representative: https://www.commoncause.org/find-your-representative/

4) *Start supporting non-profits fighting* climate denying corporations. Check out Corporate Accountability at: https://www.corporateaccountability.org/climate/. Their goal is to "wage strategic campaigns that compel transnational corporations and the governments that do their bidding to stop destroying our health, human rights, democracy, and the planet." Their website contains a corporate Hall of Shame. Check it out and see who has been nominated. We need to make the Corporate Accountability organization more prominent and stronger with our financial support.

~ ~ ~

Jill Cody, M.P.A is an authorized presenter (2007 4th class) of Al Gore's "An Inconvenient Truth" slideshow. She is also the author of the award-wining book *America Abandoned: The Secret Velvet Coup That Cost Us Our Democracy* and is producer and host of KSQD 90.7FM "Be Bold America!" radio program. Jill conceived of the *Climate Abandoned: We're on the Endangered Species List* anthology when realizing friends and acquaintances did not fully comprehend the breath, dynamics, and complexity of the climate crisis. Jill is a distinguished alumni of San Jose State University's College of Applied Sciences and Arts and currently serves on the university's Emeritus and Retired Faculty Association Executive Board.

Chapter 20

Climate Crisis & the Future of Democracy

by

Solange Márquez Espinoza, Ph.D.

"Every time you are tempted to react in the same old way, ask if you want to be a prisoner of the past or a pioneer of the future?" **– Deepak Chopra**

Democracies worldwide are in danger, though not necessarily for the reasons you might think. Russia, Cambridge Analytica, and Facebook are three of the most common targets for finger-pointing when it comes to the recent detriment of Western democracy, but we cannot deny another reason to worry. That concern, which affects all of us in every corner of the globe, is climate change.

Increasing poverty levels, more frequent catastrophic natural disasters, more severe droughts and water shortages, uncertainty in food security, and many other consequences of a warming planet are primary drivers of armed uprisings, despair, and democratic decay. It is not a question of whether or not this will happen; it is only a question as to the extent to which democracies will face meltdowns on our ever-warming planet. There are important questions to be answered here: What challenges will democratic institutions face moving forward? And what does the future hold if democracies continue to fail or, worse, fall to authoritarianism?

Democracy and What Is Expected of It

Democracies are malleable and dynamic, in that they evolve based on political and cultural demands. The mechanisms of representation, a delegation of power, and control of that power are prominent features of democracy; however, over the years, democracy has encountered various dangers that have forced it to transform some of its structures and ideological foundations.

Today, in addition to climate change, democracy faces one of its most significant risks: populism. The emergence of populist movements around the world continues to increase, without any evidence of slowing. The past promises of democracy, those guarantees of a handful of rights, seem useless when the bread is scarce at the table and having a job is the dream of a few.

However, populist feints have an even more sinister origin. The democratic archetype is collapsing. Piece by piece, democracy is losing credibility and strength. Every broken promise, every dream turned into a nightmare, and every step back awakens and feeds the populist beast. Whether on the right or on the left, the autocracy ends up being the same, and it gives the same devastating results for freedom and equality.

The promises of a better future, bright employment statistics, economic growth, and higher income have always been present, but they represent major challenges for world democracies. Order and security seem to fail at the expense of freedom, but order based on the promise of a better future, one that was always distant but assumed attainable, seemed to decrease the discourse, thanks to well-crafted political rhetoric.

Today, though, many of those lofty promises are even harder to achieve due to the terrible damage caused by catastrophic climatic events. Hurricanes destroy even the most advanced cities and torrential rains drag homes, cars, and people from their roots. Floods displace millions, leaving them without homes and food. In addition to all this chaos, there are dire consequences for human health, even worse for those who have few resources to rely on.

The climate crisis is a global phenomenon. The consequences not limited to a single geographical area; clearly, they are spreading rapidly across the entire planet. Global climatic disasters will, sooner or later, have an impact on the global political climate.

The festering growth of social inequality and economic disparity is currently the most critical concern of modern societies, but future growth projections of these issues cannot be fully understood without examining the context in which we are affected. Specifically, those who will be most affected by the increasing frequency and intensity of climatic events are those who can least afford it.

The relationship between democracy and climate change involves two essential aspects. Each is as crucial as the other, though the final impact will be divergent. Here, we will analyze and offer solutions for both.

The Growing Demand for Energy

"Jobs, jobs, and more jobs!" promise order and progress to the small autocrats-turned-world leaders, those who came into power by manipulating or taking advantage of the rules of the system itself. They memorize and master all the strategies of the big game of democracy, so they can exploit the system, hoard power, and come out on top. It is something like playing Monopoly you're your investment banker uncle, the one who always wins. Of course, politicians are careful; they are keenly aware that society is tired of campaign trail promises when what voters really need is food, a roof over their heads, schools for their children, and jobs.

For there to be factories, companies, and industry to lighting streets and houses at night, to producing food ... energy is needed and delivering energy comes at a very high cost, both financially, and its impact on our natural environment. Energy must be produced and stored in large quantities, because that is what cities demand. Energy infrastructure must be readily available at all times to develop and provide even minimum services. Until recently, producing those massive amounts of energy could only be achieved through the burning of fossil fuels.

Therefore, it seems that democracy and the fight against climate change are antagonistic elements. After all, to reduce the use of fossil fuels, it is necessary to stop producing energy that allows maintaining the essential and most basic requirements of a population in constant growth, right?

According to projections from the International Energy Agency, the demand for energy in the world will increase by around 30 percent by 2040. In terms of size and consumption, this increase is the equivalent of adding another China to the global table. In this sense, the U.S. Department of Energy calculates that the worldwide production of energy will have to increase by 57 percent in the next 20 to 25 years to keep up with demand.

Where will this energy come from? If the supply is guaranteed today based on our historical reliance on traditional fossil fuels, the circle seems to close in favor of oil, thereby increasing emissions. However, at this point, we must ask ourselves about the role renewable energy plays in this equation. The quantitative leap, the one that will turn all our scientific and financial projections upside down, is precisely there.

The World Energy Outlook 2017 clearly highlights four elements that determine a change in the world scenario in terms of energy, starting with the rapid growth of clean energy that has led to an unprecedented fall in prices. Today, prices of solar energy are 70 percent lower than they were 7 years ago. That is the equivalent of a top-of-the-line MacBook Pro dropping in price from $1,500 in 2011 to $450 in 2018! The report proposes a scenario called Sustainable Development, a scenario that includes the energy-related objectives contained in the United Nations 2030 Agenda for Sustainable Development but does not include those related to the fight against poverty, hunger, or inequality.

The point of equilibrium, of achieving sustainability but not necessarily reversing climate change, invariably happens due to a decrease in global development, but there is a noteworthy paradox involved. The premise of global energy stagnation contained in the report foreseen for 2040 can be determinant in terms of the development of the countries (measured in terms of GDP) and the satisfaction of their basic demands and satisfaction with the model in general, mainly economical but ultimately with democracy itself. In other words, to reduce or at least stagnate global energy demands and curb greenhouse gas emissions, we must paradoxically decrease development in the face of an exponentially growing world population. This is where renewable energy technology comes into play. Furthermore, it is predicted that renewable energy costs will reach parity with non-renewable options.

In this sense, we would also have to respond if, according to the increasing data of renewable energy, these would be sufficient to respond to the demand for electricity. If the stagnation occurs equally, what effects will this have on the recent investments on clean energy? As most of us learned in those college microeconomic classes, the greater the supply, the lower the demand. This rings true for energy as well, for the greater the supply, the lower the prices of fossil fuels.

It is also worth noting that the impact on the increase in the use of renewables also has a geostrategic component. China continues to face the highest energy demands, yet they also invest the most in renewable energy. Specifically, China has pledged to put forth over $360 billion to renewable energy development by 2020. For the sake of comparison, that is the equivalent of the combined GDPs of Oman, Luxembourg, Costa Rica, Panama, Uruguay, Bulgaria, and North Korea—and that's just in the next two years.

Our final concern has to do with the relationship between increasing frequency and intensity of extreme weather events. This is an interesting relationship to examine because the two are closely related to two issues we will outline in the next section: the scarcity of resources and infrastructure damage. These are all fundamental elements for measuring the perception of citizenship, concerning the capacity of democracies in the world to respond to their most basic needs.

The Breakdown of Democracies and Climate Change

From rising sea levels to floods and hurricanes to typhoons and droughts, the impact of the frequency and intensity of natural disasters has become the norm in today's world. Climate Change even changes migratory patterns all over the world. A 2009 article in *The Economist* developed a name for these migrants: climate refugees.

Every second, a person in the world is forced to move due to natural disasters, according to calculations by the United Nations Agency specializing in refugees. That's roughly 90,000 people a day. The agency also calculates that, since 2008, more than 22 million have become climate migrants, a number expected to increase radically as weather events increase. To put that in perspective, we will essentially need to find new homes for numbers equal to the entire population of Florida every decade.

Migration has begun to take root, as there is a desperate need to escape the worst consequences of climate change. Many must leave small island countries due to increasing sea levels, and the massive and unexpected exodus to Europe has grabbed plenty of global attention.

The 97 percent of new displacements in 2016 was triggered by weather and meteorological disasters such as storms, floods, wildfires, and severe winter conditions. In 2016, more than 50 percent of the trips (12.9 million) were caused by storms. Droughts and desertification have also become a reason for migration, creating greater poverty, lack of resources, and pressure on host communities.

Several authors have cited the instability of democratic regimes as a problem in modern society. That said, if we widen our scope to include dimensions outside history class, we recognize a different context. Previously, threats to democracy came from angles that, viewed in the light of time, seemed to have nothing to do with current threats, but we all know that history does tend to repeat itself. We may not be able to make a clear distinction about how modern-day problems will affect future global democracies, but we have already seen that a lack of credibility in democracy and the abandonment of the democratic paradigm as the best deal for modern society is a recipe for instability, crisis, and failure.

Syria

To paraphrase former U.S. Vice President Al Gore, one of the leading causes of the civil war in Syria has been the worst drought in 900 years, which forced the migration of more than 1.5 million people and the death of more of 250,000. The issue is not minor and should not go unnoticed, despite the silence of mainstream media outlets.

In 2015, a study published by NASA and the University of Arizona found that the drought that plagues the Mediterranean region, Syria in particular, has been the worst in at least the last 900 years and definitely the worst in the previous 500. The drought began in 1998 and has since destroyed 75 percent of the farms, resulting in the death of 85 percent of cattle, all in the few years between 2006 and 2011. This collapse was the primary cause of the forced migration of farmers to cities, generating greater pressure, scarcity, and poverty and diminishing the quality of life even years before the armed conflict even began.

If we analogize the already existing discomfort against the Syrian political regime to be the bullet in the gun, we can say climate change-induced drought was the trigger that set off the civil war. That war has almost completely destroyed the country, wiping out almost all its infrastructure and generating lasting damage that will take several decades to recover from.

The Syrians wanted democracy. When the movements of the Arab Spring began in various countries, Syria was left out. Changes seemed unviable in a country that had maintained stability for many years, in spite of its lack of democracy. An understanding of the complex Syria situation can come only through careful examination of its history, the multiple cultures that have influenced it, and the ethnic and religious diversity that was the key to unleashing the spring in a country with a not-so-longstanding dictatorship.

In the 1960s, Syria was under the ideological aegis of Soviet socialism, which transformed it into a military, secular, socialist regime, an ideological aspect that would still be maintained by the coup perpetrated in 1970 by Hafez al-Assad, father of current President Bashar al-Assad.

Political control became the key to maintaining power. Torture and repression of any attempt to advance a change of hands was the harsh response of a regime that was by no means democratic and did not hesitate to massacre those who could oppose it.

From other perspectives, however, Syria flourished. The country was culturally rich, tourist-friendly, and open to the rights of women who could walk alone, study, and work without the limitations usually imposed on them in nations where religion and the State were one. Remarkably, in the years before the war, female students even outnumbered their male classmates in university enrollment for various careers.

In 2014, the U.S. Department of Defense began emphasizing the importance of considering the effects of climate change when assessing national security risks. In particular, it has referred to the interaction of these effects with other factors such as unstable politics and social tensions that could "accelerate conflicts and instability."

Of course, the case of Syria is unique, as it is not necessarily a failure of democracy. Nevertheless, we can talk about stability and development. The process of a democratic opening can only take place under conditions of peace. A violent fall of a regime can, in many cases, lead to the seating of regimes that are even more authoritarian than their predecessors or even give way to stronger dictatorships, with lesser intentions to begin liberalization processes.

Syria's chance to become a democracy ended when violence and civil war began in 2011, and that year became ground zero for examining the modern history of Syria. Now, seven years later, the violence is just starting to diminish, and opportunity is knocking for democracy once again.

In a 2015 report, the G7 countries (Canada, France, Germany, Italy, Japan, the UK, and the U.S.) warned that political and social instability is perhaps the most dangerous risk for countries impacted by climate change. It is all part of a vicious cycle that perpetuates poverty and inequality and eliminates the possibility of democracy. Seven years ago, who could have imagined that the peaceful uprising against the regime of Bashar al Assad would morph into the brutal, bloody civil war that has practically ended the country and closed the chances of democracy?

Brexit

The war in Syria not only represented a reversal in the country, but it also brought about enormous challenges for consolidated democracies such as those in Europe, where the massive migration of refugees has caused political conflict and economic and social instability and has seemingly awakened feelings of nationalism and xenophobia that were previously dormant for many decades.

Regarding European nationalism, it would be remiss not to mention the situation in the U.K. In the summer of 2016, Britons voted, albeit by a narrow margin, to leave the European Union to seek more economical and diplomatic independence, a movement known commonly now as Brexit. While Brexit was not a direct effect of climate change, it can indeed be said that it can be seen as collateral damage.

Throughout Europe, the refugee crisis exacerbated the already precarious situation of an economy hanging by a thread. The pressure was even higher for countries that lacked resources and those that received the most massive influx of people, due to their geographical position. These included Italy, Turkey, and Greece for example.

The story has been told and heard by thousands: It is the sad tale of refugees fleeing violence, scarcity, and poverty, only to find themselves in countries where extremism, jingoism, and prejudice now prevail. Europe, and now the United States, have increasingly seen their political speeches filled with hatred. Neo-fascists and neo-populists show their heads as new agents of the destruction of democracy. This example is only a reflection of the world in which we live, a world in which intolerance, hatred, and revenge have taken the place of the values of democracy that were installed years ago in our collective imagination: tolerance, respect, and solidarity.

The Arrival of Populism or Chaos

Considering the catastrophes in various countries around the world, it seems the ideological struggle of today is less right-versus-left and more nationalism-versus-globalism. After Brexit and Trump's triumph, it became clear that there is a growing anti-globalization sentiment developing in modern societies.

The speeches of various populist candidates have been inflamed by xenophobic rhetoric, blaming migrants for crime, economic failure, and the lack of jobs. Alternative facts, as they are called, allude to nonexistent crises, and invented problems are contrived, created with manipulated figures. Many of us question how such a dishonest strategy could be successful, let alone appealing, but as we have seen Donald Trump and his team accomplish in the United States, it is almost too easy.

For extremist nationalist movements, the pathological lie has served as a very effective positioning strategy. Now, it is more important than ever for us to understand why these trends have proven increasingly successful. We must realize that we are entering an age of altruism. The countries and governments of the world no longer govern for only themselves; they know that what they do or fail to do impacts others in the world. The responsibility for the future of the world is and must continue to be shared, and keeping down the path of nationalism will only create a tremendous setback. Of course, there is much to improve upon, but knowing that xenophobia and isolation are not the correct answers, we must work to find a feasible solution.

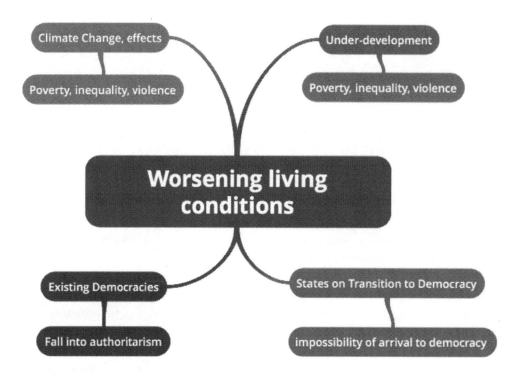

What the Climate Change Impact Means for Democracy

Social democracy presumes the inescapable characteristic of a State with the capacity to guarantee that all its citizens can count on equal rights that are not limited to individuality alone. Social rights imply an equal opportunity for their own growth and realization, and this, in turn, implies an acting State instead of a disregarded State.

A suitable, healthy environment for human beings, without a doubt, requires social excellence not in just one country but on a global scale. The effects of climate change, as we have seen, greatly hinder the possibilities of democratic states to meet social demands.

Moreover, these effects put at risk the possibilities of democracy to settle in transition zones; they put health, education, and even people's lives at risk, creating higher rates of inequality, increasing social differences, deepening conflicts, and increasing violence.

Therefore, we must recognize that a clean environment and maintaining modest increases in global average temperature are, in this sense, an indispensable requirement to sustain democracy in the world.

More droughts, greater floods, the disappearance of a large number of island states due to rising sea level, acidification of the oceans, loss of crops and greater food shortage, increasing frequency and intensity of climate events, and the proliferation of tropical diseases impact the health of millions of people.

Democracy, in strictly theoretical terms, is understood from its minimum point of view, in which representation and voting are viable options for citizen participation, though they typically leave out most of the concerns of the citizen. Citizens are interested not necessarily in *who* will be the next president as much as in *how* that president will contribute to solving their most pressing needs or concerns. Unfortunately, this is sometimes lost among the investigations, scandals, rumors, and other personal conflicts we obsess over every election season, but it is undoubtedly at the core.

It is tempting to blame democracy at a time when there seems to be no choice but to resort to autocracy in search of solutions. It is also tempting to point out that solutions to climate change find a better response in authoritarian regimes, where the necessary changes can be imposed from the top, particularly when they are not popular among society. In both cases, democracy is affected, and the question of its suitability facing a global crisis like the one we face now is inevitable. Its permanence as a political system is on a tightrope. Nevertheless, it is evident that the authoritarian option has historically offered far worse results. All of that said, democracy as we know it must undergo structural modifications to survive.

A democracy devoid of substance can only lead to further deterioration of the current situation. Democracy must be built in the same sense with which we build sustainability, or the battle will be lost. We need an even more global democracy, a more integrated one because the consequences of climate change are global. We cannot afford more discrimination or too many borders.

To provide an example of a country's capacity for innovation, new technology, and public policy, we can look to Sweden. That country is an excellent example of how, even with high taxes on coal and an extremely progressive agenda regarding climate change, companies can still compete with the rest of the world, chiefly thanks to a boost in the development of renewable energy. More than half of their domestic energy production comes from renewables, and Sweden was among the first countries to ratify the Kyoto and Copenhagen Agreements.

Their annual expenditures on carbon-neutral technology research and development typically exceed 3 percent of GDP. All of this and the fact that they still remain one of the EU's steadily growing economies demonstrates that environmental sustainability and economic prosperity can indeed coexist.

The consolidated European democracies, particularly in northern Europe, have shown that maintaining democracy and fighting climate change are possible. Democracy, as understood in international terms, in the style of the older European democracies provides a democracy that cares for others and for the future.

Climate Crisis "Keep/Stop/Start" Actions

Below for your consideration, are several author-suggested action items. For a further explanation of the Keep/Stop/Start organizational tool, please see How to Read This Book.

KEEP:

1) *Keep maintaining policies that promote* the generation of renewable energy. These, in addition to mitigating climate change, have proven to be a source of creation of better paid and *technified* jobs, increasing the quality of life. We are already well on our way to transitioning toward a clean, carbon-neutral, renewable energy infrastructure grid, but we are not there yet. Furthermore, this transition will not happen overnight. Until it does, we need to continue to pressure federal, state, and local governments, non-governmental organizations, and private enterprises to divest from fossil fuels and invest in the inevitable clean energy future.

2) *Keep educating yourself!* An active, informed republic is one of the fundamental pillars of democracy that is too often forgotten about. Though it doesn't take much to form an opinion, for a democratic system to have meaningful debates about developing solutions to common issues, we must have an informed public who engage with credible, intelligent, honest arguments based on fact and reason and are willing to share opinions and share open discussion on important matters that affect us all. We will never agree

on everything, but we can never have an honest debate about the role of policy and government with an uninformed or unmotivated public.

STOP:

Stop taxpayers financially subsidizing the coal industry, because they adversely affect the advancement of policies that fight against climate change. Don't support anyone who doesn't promote sustainable solutions in politics, socially, and, above all, the economy. Plus, the taxpayer money that goes to subsidize the fossil fuel industry can be redirected to sustainable solutions.

START:

Start discussing more openly the need to deepen what democracy must achieve. Lose the fear of leaving the discussion about votes, elections, and political parties. Talk about current individual and collective needs, particularly regarding the protection of the environment. Contemporary politics is too much red-versus-blue, and it should all be green! Avoid circular, futile, conversation. Political wins today aren't worth anything if they are merely repealed or replaced the next time the other party lands in office. Instead of focusing on micro-issues, expand conversations to discuss what is best for the overall system and best for the future. Compromise, foresight, and flexibility are forgotten dimensions of democracy, but they are essential to its sustainability; start re-inserting these factors into your political discussions.

~ ~ ~

Solange Márquez Espinoza, Ph.D. is a political consultant. Since 1996, she has given lectures about political participation, democracy and climate change in different countries such as Albania, Brazil, Spain, Germany, Thailand, the UK, and the U.S. She holds a Ph.D. in law and is a professor at the National Autonomous University of Mexico (UNAM). Her research focuses on democracy, political participation, and Parliamentarian issues. Read more at: www.climateabandoned.com.

Chapter 21

Climate Crisis & National Security: Alan Kurdi, Our Blinking Red Light

by

Bob Hallahan, M. A., Cmdr. (Ret.), U.S.N.R.

"Tenet told us that in his world, 'the system was blinking red.' By late July, Tenet said it could not 'get any worse.' Not everyone was convinced. Some asked whether all these threats might just be deception."

– 9/11 Commission Report" on warning signs seen by the CIA, reported by Director George Tenet in the months before 9/11, p. 259

Part I: Syria's Not as Far Away as We Thought

The Boy

Everybody stopped scrolling when the dramatic photo of the dead 2-year old Syrian boy, Alan Kurdi, rolled across newsfeeds in September of 2015. In that horrifying image, the child was face down on the beach, his obviously lifeless body just at the edge of the surf lapping against him. He just lay there, precariously straddling two worlds: the water in which he died and the land he came from, where he belonged. He was dressed in a bright red t-shirt, blue shorts, and smart shoes, like any other boy you might know, except this boy appeared abandoned.

Death rarely comes to 2-year-olds. In fact, any child who makes it past the first year of life stands a good chance of living a great deal longer. Rarer still is for us to see an actual picture of a dead boy. Even media outlets that lean toward sensationalism rarely publish such photos, either because they just don't receive them or because they wish to give the family space and time in which to grieve.

Alan Kurdi's image didn't follow those norms though; rather, it splashed across Twitter, then went viral on Facebook and the news wires because he was anonymous at the time. The world didn't know whose permission to seek. Everyone just saw an abandoned boy and took ownership and grieved for him. We later learned that little Alan was a Syrian refugee, but when the picture was taken he was just a handsome boy who deserved our love, like every other boy we've known.

In the aftermath of Alan Kurdi's death, there was a significant outpouring of compassion and increased awareness about people fleeing the Syrian civil war. Governments responded quickly: Germany, Britain, and Canada promised to open their doors to increased refugee flows. The Balkan states created a humanitarian corridor, and U.S. President Obama agreed to accept 10,000 Syrian refugees. Some of this resolve proved short-lived, but it's worth telling, the story of how the toddler Alan Kurdi penetrated world indifference after his death in a way he was unable to during his short life.

This chapter of the climate crisis tells the story of how global-warming pollution changed Syria, as well as what this might mean for America. Of course, it would be an overstatement to claim climate change *caused* the horrors in Syria, but evidence does show that climate change presented challenges to the Syrian government, that the challenges were surmountable even though the government failed to surmount them, and that tremendous suffering was the result. I believe this is a pattern we will see over and over again, if we permit it to happen.

The Refugees

Nobody wants to be a refugee. Few will abandon their homes and possessions casually or voluntarily. People make these hard choices because they must, because they feel they have no alternative. Alan Kurdi drowned during a night transit attempt in a fifteen-foot boat unsuited even for the relatively short, three-mile crossing his family attempted, from Turkey to the Greek island of Kos. He was from the Syrian village of Kobani. When brutal fighting erupted there, the Kurdi family fled, hoping to find solace first in Bodrum, Turkey.

Turkey is a NATO country and relatively safe place out of the war zone, but at the time it did not allow refugees to work. For the Kurdi family, this restriction meant it would be a temporary haven, a transient stop. Hoping to eventually unite with his sister in Canada, Alan's farther paid human traffickers €4,000 for passage on a tiny vessel. When the time came, the traffickers loaded the boat and ordered them on their way. Alan's father tried to steer the boat, but when it capsized in the darkness and waves, five people lost their lives: Alan, his 5-year-old brother, and the boys' mother were among the fatalities. Two Syrian men were later convicted for human trafficking in connection with the deaths.

The Kurdis' story is not all that unusual. Sadly, their journey was much like those taken by many thousands of others seeking the safety of Europe by crossing in unsuitable or overloaded boats, primarily to small Greek islands like Kos and Lesbos. Thousands of these people died; hundreds of thousands more did make it successfully to Europe.

The Drought and a New, More Hostile Climate

Syria is considered water scarce, with a generally arid climate. Even so, the now-troubled nation located on the eastern shore of the Mediterranean Sea has historically received enough rainfall that its people established a farming and herding culture, particularly in the northern and eastern areas. It is the center of the so-called Fertile Crescent, a cultural area historically hospitable to farming, which extends from the Persian Gulf, along the Tigris and Euphrates floodplains, then across the eastern Mediterranean to the Nile delta.

In the 1950s, the population of Syria was approximately four million, but this grew rapidly over the years, peaking in 2010 at over twenty-one million. Bread is a staple in the Middle East, and food and water security were managed closely for decades by the ruling Assad family, who prioritized wheat growth through various policies. As a result, Syria was able to export wheat to its neighbors, a rarity among Middle-Eastern nations and a point of some pride among Syrians, along with being a significant revenue source to the government.

Peter Gleick, a water researcher, noted that from 1900 to 2005, Syria experienced six significant droughts, five of which lasted one year and one that lingered for two. Then, in the autumn of 2006, rain did not come as expected, signaling the start of a multi-year drought so severe that it would permanently change the nation. By 2008, just two years in, the drought had cut Syria's total water resources in half, compared to what they had in 2002. At least 75 percent of farmers lost all their crops, and in the northeast, 85% of the livestock perished. By July 2008, the Syrian Minister of Agriculture was quoted in diplomatic cables as saying to the U.N. that the economic and social fallout from the drought was "beyond our capacity as a country to deal with."

As this disaster dragged on, hundreds of thousands of families eventually abandoned their land. In 2010, *The New York Times* published an interview of Syrian farmer Ahmed Abdullah, who lived outside Raqqah. "I had 400 acres of wheat," he said, "and now it's all desert. We were forced to flee. Now we are at less than zero — no money, no job, no hope." Another farmer and sheep herder, Khalaf Ayed Tajim, relayed in the same piece, "My uncle's well used to be 70 meters deep. Now it's 130 meters, and now the water became salty, so we closed it down."

When half his herd died and his fields dried up, Khalaf was forced to leave his native village and moved into a concrete bunker with his seventeen children, two wives, and his mother. The drought was described in stark terms by Gary Nabhan, a renowned ethnobotanist, as the "worst long-term drought and most severe set of crop failures since agricultural civilizations began in the Fertile Crescent many millennia ago."

Climate activists have recognized for years that we need to be careful about blaming any particular weather event such as a flood or drought on climate change. Everybody understands that weather features natural variability and that climate is basically the average of the weather. It's been the purview of statistics to measure specific weather events and determine the probability that such an event would have happened *before* we put all those greenhouse gases into the atmosphere. Nowadays, however, the atmosphere holds so much more energy and humidity than it used to that it's no longer an overstatement to assert that virtually every water-related weather event has been intensified by climate change.

Multiple research groups have, in fact, strongly implicated human-caused climate change in the Syrian drought. The U.S. National Oceanic and Atmospheric Administration wrote compellingly about this in 2011, as did Colin Kelley and a group at the National Academy of Science, who found that Syria had seen significantly reduced moisture levels since the late 1970s and that manmade greenhouse gasses were responsible for about half of this. Higher Mediterranean Sea temperatures have led to weather patterns more conductive to droughts in Syria: fewer days of wind blowing humid air toward Syria and higher air temperatures simultaneously caused increased evaporation on the land.

The Internal Migration and Building Unrest

The ongoing environmental catastrophe led to displacements of a million and a half people inside Syria by 2011. They left rural areas that could no longer support them and moved to cities or makeshift camps outside those cities, such as Aleppo and Dara'a. At the time, Syria had already absorbed nearly a million refugees from Iraq, so work was scarce, and it is quite certain that the camps did not offer the most comfortable accommodation.

Francesco Femia and Caitlin Werrell, at the Center for Climate and Security, researched and wrote much about that period. They note that due to ongoing conflict, there has not been much reporting about the dynamics that actually sparked the war. Overcrowding, unemployment, poor infrastructure, and crime likely contributed to unrest in the displacement camps, but beginning in early 2011, this was initially expressed through mostly-peaceful protests inspired by the Arab Spring movement.

The Syrian government reacted shockingly, answering demonstrations with tremendous violence. This caused the protests to escalate into an uprising, and that uprising led to a full-fledged civil war. I necessarily skip over details of the war, but suffice to say that the governmental crackdown, which involved brutal tactics that violated all tenets of sound military strategy and the norms of wartime behavior, including the use of chemical weapons. Violence from both sides created dire humanitarian conditions. By 2016, Syrian deaths were estimated at over a half-million, internally displaced over six million, and net migrations out of Syria over five million.

Most of those who fled headed to Europe. In 2015 alone, an estimated 1.8 million immigrants arrived on European shores, many through the same route Alan Kurdi followed, from Turkey to the small islands of Greece. At the peak of the migration, thousands landed on Greek beaches every day, overwhelming local resources and stretching Europe to the limits of its generosity.

The Syrian Government's Culpability

Thus far, I've asserted that climate change reduced Syria's water resources, which led to fighting and the eventual death of Alan Kurdi. The Syrian civil war has roots too complex to describe in detail here, but suffice to say that stubborn religious, demographic, and ethnic tensions, including Syria's significant population growth and the influx of Iraqi refugees, were involved. Also, two significant world-wide food prices spikes, in 2007-08 and again in 2011, contributed to Syria's challenges. These were caused in part by difficult weather in agricultural regions such as Russia, Australia and California, and highlight again, the reliable connections between climate, weather, agriculture, local politics, and international security.

By all accounts, poor governance compounded the problems faced by Syrians during their drought. For decades, the Syrian government significantly mismanaged agricultural resources by encouraging wasteful practices such as flood irrigation, as opposed to drip or sprinkler; favoring water-intense crops such as wheat and cotton; and allowing overuse of underground aquifers and overgrazing of rangeland. All of these contributed to the nation's vulnerability to drought.

Part II:
Syria, Climate Change, and American Security

"Assessments conducted by the intelligence community indicate that climate change could have significant geopolitical impacts around the world, contributing to poverty, environmental degradation, and the further weakening of fragile governments. Climate change will contribute to food and water scarcity, will increase the spread of disease, and may spur or exacerbate mass migration." – **Quadrennial Defense Review,** 2010" U.S. Department of Defense, p.85

People who think deeply about national security, including the intelligence community and the military, have long been writing and talking about how climate change puts America at risk. Syria is a perfect example as to why.

The Syrian tragedy, to the extent it is related to the climate crisis, is really all about water. Laws of physics and chemistry dictate that as the planet warms, the water cycle intensifies, so water-related weather events in the future will be more extreme than those we have experienced in the past. Warmer air increases evaporation, and also holds more water vapor. More humid air produces more precipitation, much of which will come in larger downpour events, leading to more floods. Therefore, we will paradoxically suffer more drying of the land in some places and more flooding in others.

All this is likely to be tremendously disruptive to human civilization because we humans have finely adapted to the relatively stable climate we've experienced since the end of the last ice age. Syria is an example of how a single devastating drought can stress a relatively stable government and society to near collapse; in their case, it led to a civil war and massive population outflow. This, in turn, created a very expensive mess in neighboring countries, with cascading effects throughout the world.

The wave of immigration to Europe during the Syrian civil war alarmed European voters, contributing to a resurgence of right-wing power there, as demonstrated in the 2016 UK Brexit vote, French support for Marine Le Pen's 2017 presidential campaign, and the September, 2017 election to German Parliament of right wing AfD party members. It also terrified many Americans, and that fear was clearly a factor that helped propel Donald Trump into the presidency. However, even the Trump administration's hostility to immigration has not insulated us from all consequences; in July of 2017, Trump pledged $140 million to help Lebanon with the million refugees who have arrived there, many from Syria.

Another water-related challenge to American national security will come in the form of melting glaciers. Consider India's Ganges and Brahmaputra Rivers; these alone support more people than the entire population of North America. The sacred waterways originate in Himalayan glacier fields that are melting faster than anywhere on Earth. While glaciers shrink, river outflow increases beyond historical norms, but as the glaciers disappear, river flows will drop very quickly and also become much more variable between seasons. About 625 million people rely on these rivers for daily needs. Will they neatly and effortlessly adjust, or might the losing of reliable water cause major cultural upheaval, as it did in Syria? How this will affect our national security is not clear, but doubting that it will is wishful thinking at best and naiveté at worst.

Global warming-related sea level rise is already underway, adversely affecting the United States and all other coastal nations in the world. This is climate related both because meltwater from doomed glaciers is gushing into the oceans, and because our warmer atmosphere heats the seawater, causing it to expand. American national security will be impacted in many ways. Most urgently, it threatens our major East Coast military complex, which includes low-lying bases in Norfolk, Hampton Roads, Langley, and Little Creek. This hurricane-prone area is home to the largest naval base in the world, and already experiences periodic sunny-day flooding due to the combination of subsiding land and higher seas.

While the U.S. Navy considers how to deal with flooding of some of our major bases, low-lying island nations such as Kiribati and the Maldives face threats far grimmer, as rising seas will literally render their land mass uninhabitable. Coral islands like these will lose their reliable freshwater supply long before the sea completely submerges them. In this way, climate change is incredibly unjust. For Americans, climate change is a severe inconvenience to which we will adapt, albeit at great expense, but for small oceanic nations, it will literally remove their land from the face of the Earth.

A New Kind of Threat

"The stresses that climate change will put on our national security will be different than any we've dealt with in the past. For one thing, unlike the challenges that we are used to dealing with, these will come upon us extremely slowly, but come they will, and they will be grinding and inexorable. But maybe more challenging is that they will affect every nation, and all simultaneously." – **Vice Admiral Richard Truly,** former astronaut and NASA Director, Military Advisory Board, Center for Naval Analysis

Climate change is a different kind of threat from the national security challenges faced by previous generations. It's very easy to marshal public reaction against a foreign attacker. Consider the American response after Pearl Harbor: Jolted awake, the entire nation mobilized to defeat Nazism and imperialism, then went about a new national project to restructure much of the world to prevent a repeat. During the Cold War, the Soviet Union perfectly played the villain, inspiring decades of American resolve. After 9/11, when terrorism instantly became a national obsession, America sailed around the world to wage war against a landlocked country for giving refuge to those who attacked us. In sharp relief, climate change – **even though it threatens to submerge our coastal cities,** bring our economy to its knees, sow chaos worldwide, and flood our borders with refugees – has elicited only delayed and meager response thus far.

If the American public were one day informed that a foreign power was caught raising the sea level with the goal of submerging our coastal cities and pouring chemicals into the oceans to kill off our fish stock, this nation would react with unstoppable outrage. Instead, our greatest enemy is our own behavior, our own foolishly wishful belief that we really don't have to sacrifice for the common good.

A certain, undeniable characteristic trait of human beings works against our enthusiasm to fight climate change: loyalty to one's own. Psychologist Jonathan Haidt described it perfectly in a TED talk. Strong kin loyalty was helpful thousands of years ago, when we were a young species living on the plains, facing threats from hostile neighbors with unknown intent. Outside threats banded us together to protect our families and our tribes. Nevertheless, while the trait is less relevant in today's more connected world, tribalism remains with us. We see it in our loyalty to sports teams and in our patriotism, loyalty to country. Tribalism is helpful in creating societies, but it undoubtedly hinders our embrace of solutions to the climate crisis. Why? Because carbon pollution is trans-national. Because we find it hard to see most victims of climate change as our own people, we're less motivated to protect them.

Risk Management

"We never have 100 percent certainty. We never have it. If you wait until you have 100 percent certainty, something bad is going to happen on the battlefield. That's something we know. You have to act with incomplete information. You have to act based on the trend line."

– Gen. Gordon Sullivan, Former Chief of Staff, U.S. Army, Center for Naval Analysis, Military Advisory Board 2007, urging action on climate change

Pushing more Americans beyond the false debate about climate change is crucial to solving it. One way to do this is to frame the issue in terms of risk management; this will allow us to consider solutions without having to agree on the problem. Risk management acknowledges that science is ongoing, and there can still be a lot of debate on the leading edge of it, even if the debate about broad principles is long over.

Military personnel and national security professionals are trained risk managers. They understand that risk is a product of the likelihood of any event occurring and the consequence level if it does occur. An event that carries a medium chance of actually happening but is inconsequential is a low-risk event. On the other hand, because the warming of our planet will bring extreme consequences, even if the actual probability of our behavior causing of it were only medium, the net risk would still be extremely high; therefore, considerable expenses are justified to avoid the risk.

Author Greg Craven concludes that "Are the scientists correct?" is the wrong question. Trained risk managers instead ask, "What should we be doing now, just in case the scientists are correct?"

Part III: Parallels to the Past...and Are We Going to Keep Letting This Happen?!

America's been caught underprepared before by risks it should have better managed. Thus far, our response to the climate crisis has been wholly inadequate; in many ways, it is quite similar to America's tepid response during the lead up to 9/11. "The 9/11 Commission Report," the U.S. government's official study of the events surrounding the attack, goes into great detail about the various activities of Osama bin Laden and al Qaeda in the years prior to 2001. The inescapable conclusion is that the attack, while not precisely forecast, should have been much more difficult for al Qaeda to carry out.

The 9/11 Commission laid out a long series of events, outlining bin Laden's rise and how he and his allies advertised their intentions increasingly clearly throughout the 1990s. The report notes bin Laden's initial reaction against the United States in response to American military presence in Saudi Arabia before the 1991 Gulf war. It describes his increasing hostility, including a 1996 declaration of war against Americans, which by 1998, he sharpened by issuing a religious proclamation asserting that it was the individual duty of all Muslims to kill all Americans anywhere on Earth, whenever possible.

The report continues to describe how al Qaeda put their words into action through deadly attacks, describing the 1998 dramatic, simultaneous bombings of two American embassies in Africa and the 2000 *USS Cole* bombing in the Yemeni port of Aden. Remarkably, despite these events, terrorism before 9/11 did not capture the imagination of the public:

"As best we can determine, neither in 2000 nor in the first eight months of 2001 did any polling organization in the United States think the subject of terrorism sufficiently on the minds of the public to warrant asking a question about it in a major national survey. Bin Laden, al Qaeda, or even terrorism was not an important topic in the 2000 presidential campaign. Congress and the media called little attention to it." – **The 9/11 Commission Report,** p. 341

The report next examined American government responses to al Qaeda, with the implied but damning conclusion that they were inadequate. It even titled one chapter "Adaptation- -and Nonadaptation- -in the Law Enforcement Community." To be fair, many in the U.S. government of the 1990s took the terrorist threat very seriously, including FBI Director Louis Freeh, who shifted emphasis even though he was unable to muster a concomitant shift in resources. "The 9/11 Commission Report" noted that by 2000, U.S. national security leadership had concluded that if they wanted to "roll back" al Qaeda, they would have to markedly increase their efforts.

We could undoubtedly reach similar conclusions about our government's inadequate response during the lead-up to Pearl Harbor; we had indications of Japanese intent and scattered warnings about the attack itself but still failed to translate these into an effective defense posture.

Since these same dynamics seem to be playing out in the climate crisis, we are confronted with the question: How do we fix this!? Those of us who see and acknowledge with alarm the infuriating combination of steadily rising carbon dioxide levels, government inaction, and public apathy must oppose them all in whichever ethical ways we can, everywhere we can, with as much energy and creativity as we can muster!

Institutional inertia was at the root of inadequate government response to 9/11. Leadership came too slowly to a proper understanding of new threats. America's inaction on climate is more problematic than simple institutional inertia, though, because – unlike terrorism – climate science faces the added complication of an active, well-funded denial movement.

What we see here is ultimately part of a bigger problem: Powerful and self-interested corporations have successfully substituted their agenda for the pursuit of the common good. They've coopted national priorities like parasitic cowbirds who lay eggs in other birds' nests, then push the host parents' eggs out. Through lavish lobbying budgets, sophisticated advertising, and misinformation campaigns, they've successfully convinced large portions of the American public and political and business leadership to value quarterly profits above what's good for the country.

The Trump administration's continued embrace of polluting fossil fuels is just one example of this. There are a host of others: America's healthcare system, which mandates the purchase of insurance from corporations; Medicare's legal prohibition from negotiating with drug companies; the setting of student loan rates above market values; the privatization of formerly public services, such as road construction and prisons; and the nationwide movement to roll back business taxes and public-interest regulations.

We can fix all of this, but only if we start to once again put the common good in its rightful place ahead of narrow corporate interests. Doing so will require reforming the way we finance campaigns, including overturning the Citizens United decision that has allowed unlimited corporate money to flow into elections and strengthen its hold over elected leaders.

It infuriates me to watch people drape themselves in the trappings of American patriotism, only to drag their feet when it comes to actually solving the country's major problems. Those who dismiss climate science or resist solutions are abandoning our national security and our responsibility to pass down a secure, functioning nation and world to our children.

Some will criticize me for starting this chapter with a discussion of Alan Kurdi. "Demagoguery!" they will howl, claiming that I "politicize tragedy." Alan Kurdi's death was a tragedy, of course, but beyond that, I suspect Kurdi himself wouldn't think much of such criticism. Rather, I think he would appreciate the fact that the image of his small body would at least advance the fight against the forces that contributed to his death. Those who are quick to take offense that a dead child's image could be possibly used for political ends would more appropriately direct their outrage at the unfairness his death embodies. As I write this, Earth's atmosphere sits near 410 parts per million CO_2, much of which originated in America, the richest nation that has ever existed in the history of the world and the only one whose national leadership has abandoned the Paris Agreement.

Climate Crisis "Keep/Stop/Start" Actions

Below for your consideration, are several author-suggested action items. For a further explanation of the Keep/Stop/Start organizational tool, please see How to Read This Book.

KEEP:

Keep taking children outside. I'll never forget something my friend Matt told me in college: "People protect what they value, and values are determined in childhood." Taking kids outside by going to parks, camping, rafting, and swimming will connect them to nature and the small creatures they will see in it. Kids brought up this way develop a lifelong desire to protect nature and, therefore, to resist climate change.

STOP:

Stop buying everything! Our economy will be just fine, our national happiness will go way up, and our climate will thank us if we spend more time playing board games with each other, volunteering with civic groups, enjoying the outdoors, and discussing books and far less time shopping and stuck in traffic. Today's American children watch about 40,000 TV commercials per year, yet most cannot tell the difference between fantasy, reality, and that advertising. Help your children navigate this consumer minefield by showing them "Buy Me That Too: A Kid's Survival Guide to TV Advertising" on YouTube, and make sure you pay attention to these eye-opening documentaries as well.

START:

1) *Talk about climate change with everyone you meet.* Without being overbearing or one-dimensional, try to work the topics of climate change and ocean acidification into your conversations. This falls under "Start" because, as a society, we are not doing this already. In surveys that question which issues most concern Americans, climate change is repeatedly ranked at the low end of our priorities. This must change, and one way to accomplish that is by talking about it. If you need something to spark those conversations, refer to the stories in this chapter and book.

In a famous psychological study, "The Smoke-Filled Room," subjects are brought into an empty room and told to begin working on a questionnaire. Smoke is then vented in. Individuals who are in the room alone tend to get up and investigate, or else they leave the room. However, when subjects are in a room where conspirators have been coached to sit passively and ignore the smoke, most subjects do that instead! People take their behavior cues from others. Therefore, if you express alarm and opinions about climate change, your friends will tune in. Let's make it a national, daily conversation.

2) *Join a Group.* Join your local Rotary Club, Lions' Club, Grange, or some other service organization that meets regularly and aims to make improvements in the community. Local institutions are our first line of defense against civil disorder.

Alexis de Tocqueville noted Americans' propensity to join institutions for the sake of community improvement. This set us apart from the European society he was used to, which relied on monarchs and princes for what governance they could provide. Americans' willingness to pitch in and lend a hand, to take initiative to do things that need doing in an egalitarian way, made us unique and strong. Working to build America's institutions makes America more resilient and secure. Taking ownership of your country in this way is a habit we literally cannot have too much of.

3) *Travel overseas.* If you can, get a passport and visit at least one foreign country. Stay long enough to get to know the place and some of the people there. Learn some of the language and flip through a local newspaper. Make sure to take your children, and of course you should buy carbon offsets for your trip.

Travel to foreign countries undermines our us-versus-them mentality, obliterating the idea that others are different from us. It fosters goodwill between individuals, promotes worldwide cooperation, and helps us see the climate crisis as something that affects us all, something we must all play a role in solving. When we grasp that we do have a common enemy, perhaps we will all be willing to join the fight.

4) *Read history.* The first war military officers are taught about in depth at the U.S. Naval War College is the Peloponnesian War, which took place from 431 to 404 BCE. This seems crazy, right? Why do we study ancient battles that seemingly have no connection to today's world and weapons? Basically, because students know little about these events, they have few pre-formed opinions about them, so they are more open to the lessons that can be learned. Similarly, many Americans are so polarized today that we are unable or unwilling to recognize when our preferred political team violates a cherished principle. Undermine this by reading history.

5) *Support campaign finance reform.* Because political campaigns are expensive, candidates must raise huge sums to compete. The unfortunate result is that winning politicians seem to give more loyalty to the corporations, lobbyists, trade groups, and wealthy donors who paid for their campaigns than to the public who pays their modest salaries. One way to defeat this is to publicly finance campaigns. It would be expensive, but we must view it as buying back our elected officials' loyalty, and we could cut the expense by shortening campaigns. This can definitely be done! Nations such as Japan, Canada, and famously, France measure their national leadership campaigns in weeks. We should also shine light on political donations by requiring more disclosure of spending on advertising.

~ ~ ~

Bob Hallahan, M.A., Cmdr. (Ret.) U.S.N.R. is a retired U.S. Navy commander who became passionate about the climate during his military travels, in particular during deployments to the tropical Pacific and Indian Ocean islands. He completed five major deployments, including service in Iraq and Afghanistan. He studied natural science at the University of Notre Dame, then trained as a Navy carrier pilot on Whidbey Island, Washington. He holds a master's degree in national security and strategic studies from the U.S. Naval War College. Read more at: www.climateabandoned.com.

Chapter 22

Climate Crisis & European Views

by

Annamária Lehoczky, Ph.D.

"We do not need magic to transform the world; we carry all the power we need inside ourselves already: we have the power to imagine a better life." – J.K. Rowling

Have you ever heard of Svalbard? In case you haven't, you might be interested to know it is a tiny island group in the Arctic Ocean, halfway between Norway and the North Pole, a winter wonderland full of glaciers and polar bears and home to the northernmost permanent human settlement in the world. I had the fortunate opportunity to work on a scientific project there at winter's end in 2014, tasked with taking atmospheric measurements. The long working days without sun were tough, but that was not the strangest experience I faced; on several days, rather than the expected bitter frost and temperatures well below -10°C, the mercury climbed above freezing, often for several hours a day. Sometimes, rather than flurries and snow squalls, we were actually drenched with rain. In such a place, this uncharacteristic weather was as sad as it was frightening, and we discovered later that that particular February was the warmest on record, dating clear back to when records were first taken, in 1899.

Human-induced climate change is undeniable; a myriad of evidence and a tremendous amount of data corroborates it. For that reason, my heart bled all the more when I witnessed the disappearance of noteworthy databases and websites that contained precious climate data, the telltale facts and figures diligently collected from all over the world for several decades. The decision to disband those information resources was made by the U.S. government, thereby undermining the authority of scientists and science itself. Denying climate change is as much of a crime as any, obstruction of justice in the form of hiding and destroying evidence.

The interaction between science and policy has never been straightforward, but this relationship has been further complicated by current post-fact debates.[1] The use of knowledge and facts by governments is as old as human history. Innovations and new discoveries have oft been stimulated by governments, one of the earliest examples being the invention of writing to record taxes in ancient times. The ability of governments to harness the latest in science and technology to help them make optimal decisions is at the heart of modern civilization and progress. Therefore, why do so many people choose to live with their heads in the sand, pretending that science is a question of belief rather than an observation of the clear-cut, provable facts.

Is America Out of Touch?

Americans hold patterns of thought and values that distinguish them from Europeans and other nationalities around the world. In 2006, Andrew Kohut and Bruce Stokes penned *America Against the World*, a work about typical American attitudes and the U.S. international image. This book was written with the aim of better understanding of so-called American exceptionalism, which makes many Americans believe they do not need the rest of the world.[2] Kohut and Stokes maintain that Americans are more nationalistic than most western European countries and that they tend to detach themselves from international issues when there is no direct threat, such as terrorist attacks. This relative lack of interest in foreign countries and international affairs has

---

[1] EU for facts. Evidence for policy in a post-fact world. Annual conference of the Joint Research Centre, 26 September 2017, Brussels, Charlemagne.

[2] Kohut, A. & Stokes, B.: *America Against the World*. Times Books, 2006. Henry Holt and Company, New York, USA. Andrew Kohut is the director of the Pew Research Center for the People and the Press, the leading non-partisan polling organization in America. Bruce Stokes is the international economics columnist for the National Journal and a consultant to the Pew Global Attitudes Project.

resulted in a lack of awareness of the new phenomenon of global interconnectedness and interdependence. U.S. President Donald Trump's "America first" slogan builds on the strong nationalistic feelings of Americans and exacerbates such detachment from international affairs. Kohut and Stokes warn that in today's globalized world, this indifference to others can lead to severe problems.

Americans' greatest strength is their relentless optimism. Individualism, self-reliance, and tolerance are also defining attributes, and these qualities have contributed to making the country a powerful, influential world leader. Nevertheless, this optimism can be dangerous when it is coupled with such disregard for global interdependence. It fosters a tendency to postpone tackling large problems because there exists a false belief that technology, even Americans themselves, can fix everything. Such high optimism is especially risky if the nation chooses to ignore international agreements, as well as environmental and human rights that will affect both its own future and its leadership role. By rejecting international cooperation and disregarding the country's economic, energy, and environmental interdependence, America limits its ability to address global issues such as climate change, terrorism, and other geopolitical challenges.

"Climate Change Is a Hoax," Says U.S. President Donald Trump

On June 1, 2017, President Trump announced the withdrawal of the United States from the Paris Agreement, a global effort to combat climate change, despite condemnation from leaders worldwide. He also rolled back previous U.S. climate change commitments and defunded dozens of other programs meant to help us improve energy efficiency and cope with pollution cleanup. According to the *Business Standard*[3], the "Trumpestuous" climate that has taken over the U.S. "threatens to gamble the future of humanity for the sake of false nationalistic ideals that breed on ignorance and mistrust."

This is not the first time the U.S. has rejected a multilateral agreement to address global problems though. The U.S. rejection of the 1997 Kyoto Protocol, which aimed to regulate carbon dioxide emissions as a means of controlling global warming, already demonstrated transatlantic differences on global environmental issues. This my-way-or-the-highway approach and the pursuit of its own interests has taken the form of isolationist and unilateralist action many times in U.S. history[2], resulting in many incidents of poor international relations.

After the second commitment period of the Kyoto Protocol ends in 2020, the Paris Agreement, as a new global accord, will establish rules and guidelines to reduce global greenhouse gas emissions. The climate accord was signed in Paris by President Obama in 2015 and by all countries of the United Nations Framework Convention on Climate Change (UNFCCC) by November of 2017. However, the current administration rejects this agreement, similar to the rejection we saw in 2001 of the Kyoto Protocol under the Bush administration. It seems that U.S. unilateralism and its love/hate relationship with the United Nations[2] are likely to continue, especially regarding climate change.

"The U.S. will lose power," said Kofi Annan, former secretary-general of the United Nations, Nobel Peace Prize laureate, and Christiana Figueres, former executive secretary of the UNFCCC, who led the talks that created the Paris Climate Agreement in December 2015. "I would call it a vacuous political melodrama," Figueres commented on Trump's decision during a press call organized by the World Resources Institute.

Setting aside the government's decisions, what do Americans think about climate change? On that note, how do Europeans feel about the climate crisis? Is there any difference in these perspectives? How are our views shaped? Included are some inspiring examples of decisive action, as well as crucial practical resources that address the most pressing questions, such as: Can we do anything to tackle climate change on an individual level? What can we do in our communities to promote more climate-friendly decisions?

---

[3] US exit from Paris Accord: Trump's action calls for civil society to unite – *Business Standard* (1 Aug. 2017) http://www.business-standard.com/article/economy-poli-cy/us-exit-from-paris-accord-trump-s-action-calls-for-civil-society-to-unite-117060300168_1.html

Public Perception of Climate Change: United States vs. Europe

As the conclusions of Kohut and Stokes suggest[2], Americans don't see the world and the U.S. role in global affairs in the same way as non-Americans see it. This difference in perspective, especially between the U.S. and Europe, may stand as a recurrent obstacle to dealing with joint challenges around the world.

For the European public, climate change is a significant concern: 9 in 10 Europeans think it is a severe problem.[4] The European Union (EU) is comprised of countries with varying cultural, social, and economic backgrounds; therefore, there are also noticeable regional differences in climate change opinions among EU member states. In the three EU Nordic countries (Sweden, Denmark, and Finland), climate change is regarded as the single pressing challenge, but in others (such as Estonia, Latvia, and Portugal), it is seen as a less significant issue: In parts of eastern Europe, armed conflict and economic difficulties still concern more people.

At the same time, more than one-third of European citizens accept some form of personal responsibility for tackling climate change. They expect action, especially from national governments, business and industry, and the EU. Close to one-fifth of respondents accept the need for personal action. Also, there is a strong consensus that collective global action is required; the vast majority believe that fighting against climate change and its harmful impact can only be effective if all countries act together. Indeed, this idea is reflected in the core concept of the Paris Agreement, but that has been rejected by the Trump administration based on its "America first" foreign policy.

Turning our attention to the U.S., we see a different picture regarding climate change perceptions. Significantly fewer Americans (7 in 10) accept climate change as an existing problem, and only half say they are at least "somewhat worried" about it.[5] In addition, fewer than four in ten Americans believe it will cause them personal harm, while about seven in ten support regulating carbon pollution from coal-fired power plants and even more help regulating carbon dioxide as a pollutant more generally. Of course, the White House and Congress may do the opposite; President Donald Trump's proposed 2018 United States federal budget will, in fact, defund the Clean Power Plan and other environmental programs.

Concerning American views on scientific facts, even though slightly more than half claim that they understand that global climate change is mostly due to human activity, only about one in eight understand that nearly all climate scientists (more than 90 percent) agree that climate change is mainly caused by human activities. Additionally, one in eight Americans deny climate change altogether and believe the president's observation that it is a hoax. There are many reasons behind these misconceptions and this widespread lack of understanding.

Are We Frogs in a Pot?

One common root at the foundation of such an issue is how people perceive risk. The climate crisis is precisely the kind of threat that we, as human beings, are horrible at dealing with. It is a problem that has vast consequences in the long term, but little is visible in the short term, making it difficult to relate to everyday life. As humans, we have a natural instinct to react with flight or fight in the face of imminent danger, but we are not really motivated to act when it comes to slow-moving or somewhat abstract problems, even if dire consequences are already on the horizon.

This can be brought to the forefront of understanding by the case of the boiling frog. As the analogy goes, if a frog is placed suddenly into a pot of boiling water, the animal will naturally jump out. However, if that same frog is dunked in tepid water that is then slowly brought to a boil, the frog will not perceive the danger and will

---

[4] Eurobarometer Report – Public Opinion on Climate Change (2015). Survey requested by the European Commission, Directorate-General for Climate Action and coordinated by the Directorate-General for Communication. http://ec.europa.eu/COMMFrontOffice/publicopinion/index.cfm/Survey/getSurveyDetail/instruments/SPECIAL/surveyKy/2060

[5] Leiserowitz, A., Maibach, E., Roser-Renouf, C., Rosenthal, S., & Cutler, M. (2017). Climate change in the American mind: May 2017. Yale University and George Mason University. New Haven, CT: Yale Program on Climate Change Communication.

be foolishly and naïvely cooked to death. This metaphor aptly explains how creeping, gradual changes result in undesirable consequences because of our ignorance, inability, or unwillingness to react.

For example, we are naturally more worried about the day-to-day weather changes that have a direct impact on our lives than we are about something invisible and intangible, something we cannot yet see and struggle to perceive as a risk to our lives or the lives of our children and grandchildren. Consequently, most of those surveyed consider climate change a relatively distant threat, something that may harm future generations of people, particularly the impoverished or those in slowly developing countries; plants and animals; and the planet as a whole. Most respondents are less likely to think current U.S. citizens, their own children or grandchildren, or their own communities will be harmed by climatic changes. In general, Americans consider themselves the least likely to be harmed.

The Role of Political Views and Personal Experience

Views on climate change and perceived hazard depend on various factors: one's education, scientific literacy, political views, age, or experience of any damage or harm caused by climate/weather events. A study by the Pew Research Center pointed out that even though the majority of Americans are worried about climate change and some have already experienced some impact if a partisan watchword was included in a question, people answered through a different frame.[6] A nationally representative survey found that respondents tend to offer answers representative of an affiliated societal group, resulting in a sharp opinion contrast on climate change between groups supporting different political parties.[7] For example, 78 percent of Democrats and 54 percent of Independents believe that global warming should be a high or very high priority for the president and Congress, while only 24 percent of Republicans think so.[7]

What's more, political views are also very linked to people's trust in (climate) science. As John Abraham explained in *The Guardian*, "certain political and religious ideologies correspond to viewpoints on science: the more liberal your politics are, the more likely you are to accept the science and the solutions to combat climate change."[8] The survey data clearly shows that the idea that scientists nearly unanimously agree that human industrial activity is a root cause of global climatic changes is accepted or rejected by respondents depending mainly on their political beliefs; only 13 percent of self-identified conservative Republicans think scientists agree on climate change, as compared to 55 percent of liberal Democrats.[6] A stunning gap!

In Europe, public opinion on climate change is not sharply divided between left and right parties. Although European political parties on the left, including the Green parties, are generally supportive of measures to address the climate crisis, conservative European political parties maintain similar views, most notably in western and northern Europe. The shared sentiments between the political left and right on climate change further illustrate the divide in perception between the U.S. and Europe on climate change.[9] For instance, conservative German Prime Ministers Helmut Kohl and Angela Merkel differ with other parties in Germany in their opinions of how to meet emissions reduction targets, not whether or not to establish or fulfill them.

It is clear that besides political beliefs, the perception of climate-related risks is strongly influenced by firsthand experiences. Texas, for example, is one of the most vulnerable regions in the U.S. when it comes to climate change, and its residents will experience some of the worst repercussions in the U.S., due to rising sea

---

[6] The Politics of Climate: October 2016 Report by Pew Research Center http://www.pewinternet.org/2016/10/04/the-politics-of-climate/
[7] Leiserowitz, A., Maibach, E., Roser-Renouf, C., Rosenthal, S., & Cutler, M. (2017). Politics & Global Warming, May 2017. Yale University and George Mason University. New Haven, CT: Yale Program on Climate Change Communication.
[8] Reflections on the politics of climate change – *The Guardian* (2 June 2017) https://www.theguardian.com/environment/climate-consensus-97-per-cent/2017/jun/02/reflections-on-the-politics-of-climate-change
[9] Schreurs, M.A. & Tiberghien, Y.: Multi-Level Reinforcement: Explaining European Union Leadership in Climate Change Mitigation – *Global Environmental Politics* Volume 7, Issue 4 (November 2007), p.19-46.

levels, heat extremes, and droughts.[10] Nevertheless, residents of the state are split when it comes to how much they worry about it.

Based on the Yale Climate opinion maps, west and south Texas, as well as the Gulf Coast, exhibit more concern about climate change than the rest of the state; an average of 57 percent of Texans in these areas are at least somewhat worried.[11] *The New York Times* argues wisely that politics alone cannot explain this.[12] November 2016 election results clearly showed that south Texas favors Democrats, west Texas is decidedly more mixed, and the Gulf Coast is solid Trump territory. So, what is the shared element of these politically different parts of the state?

The most significant factor is that people have seen, with their own eyes, and felt, on their own skin, the severe consequences of shifting weather patterns, including remarkably higher temperatures, coastal hurricanes, and western droughts. On some occasions, dry spells were so intense and prolonged that some west Texas towns were forced to recycle wastewater for drinking, according to *The New York Times*.[13]

Everybody Talks About the Weather, but What About the Climate?

While many Americans claim to care about climate change, few take time to discuss it. In fact, only one in three Americans reports talking about global warming with family and friends "often" or "occasionally," and most "rarely" or "never" discuss it.[5] In addition, the lack of news coverage on climate change also contributes to misconceptions, as Americans take their cues about the importance of issues based on the volume of media coverage. The aforementioned Yale survey results[5] revealed that fewer than half of Americans hear about climate change in the media at least once a month, and only one in five listen to others talking about climate change on a monthly basis. These findings might make us wonder: why do people shy away from talking about climate change?

A possible reason for the silence on the issue is that it just doesn't come up in conversation, so we don't bring it up ourselves, as *Grist* suggests.[14] The "spiral of silence" is a term coined by German researcher Elizabeth Noelle-Neumann, who was attempting to understand why so many people remained silent under the Hitler regime. She found that when everyone is silent, no one dares to speak up, for fear of isolation and losing friends or social status. Surveys repeatedly suggest that Americans indeed care about the issue of climate change, but few are willing to speak openly about it.

Breaking the silence is only the first step though. The next must be action, and that requires a deeper understanding of the mutual benefits of reducing environmental pollutants such as greenhouse gases and strengthening communities so they can adapt to the unavoidable changes.

Let's Move the Needle on the Climate Deal!

While the "Trumpestuous" government busies themselves with ignoring climate science and hindering climate action, the vast majority of the world seeks resolution and has already begun to act. The exponential growth of the renewable energy industry is a clearly visible global trend. Outside of the U.S., people have woken up to the indisputable fact that fossil fuels are no longer an option. Falling prices on solar panels are just one example, indicating that the renewable energy revolution has started to unfold right in front of our eyes.

[10] Hsiang, S., Kopp, R., Jina, A., Rising, J., Delgado, M., Mohan, S., Rasmussen, D.J., Muir-Wood, R., Wilson, P., Oppenheimer, M., Larsen, K. & Houser, T.: Estimating economic damage from climate change in the United States – *Science* (30 June 2017) http://science.sciencemag.org/content/356/6345/1362
[11] Yale Climate Opinion Maps – U.S. 2016 by Yale Program on Climate Change Communication http://climatecommunication.yale.edu/visualizations-data/ycom-us-2016/
[12] How Americans Think About Climate Change, in Six Maps – *The New York Times* (21 March 2017) https://www.nytimes.com/interactive/2017/03/21/climate/how-americans-think-about-climate-change-in-six-maps.html?mcubz=0
[13] Water Flowing from Toilet to Tap May Be Hard to Swallow – *The New York Times* (8 May 2015) https://www.nytimes.com/2015/05/12/science/recycled-drinking-water-getting-past-the-yuck-factor.html
[14] Spiral of Silence – *Grist* (3 Oct 2016) http://grist.org/briefly/most-americans-care-about-climate-but-dont-talk-about-it-why/

Indeed, the numbers speak for themselves. In 2016, the installed capacity of renewable energy set a new record of 161 gigawatts; in 2015, investment levels reached $286 billion worldwide, more than 6 times the figures in 2004.[15] Over half of that was for projects in developing and emerging economies, as the Frankfurt School reported.[16] The fossil-free economy is already profitable, and it also creates jobs that contribute to a global green economy.[17] A 2017 report by the International Renewable Energy Agency and the IEA suggests that efforts to stop climate change could, in fact, boost our global economy by a whopping $19 trillion![18]

Although Trump's decision to pull the U.S. out of the Paris Agreement was a highly disappointing step in the eyes of all those who have been working relentlessly to see it succeed, many have carved out their own way to support the worthwhile goals. For example, the #WeAreStillIn campaign, which involves thousands of companies, mayors, governors, and universities from across the United States, has ensured that they will do their part to reduce carbon emissions to fulfill the promise of the global agreement.

French President Emmanuel Macron responded by recycling Trump's own slogan into "Make our planet great again," soon after the announcement that Trump was pulling the U.S. out. The often-tweeted phrase rapidly went viral, sprouting up all over social media, accompanied by Macron's invitation to American scientists and entrepreneurs disappointed by their government to work together in France to find climate solutions. This strategic move from the diplomatic novice Macron aims to "restore and amplify France's global voice at the heart of a stronger Europe."[19]

Tracking Individual Action on Climate

To know where we are and in which direction we are heading is essential to reflect on our beliefs and actions. Sometimes, we must be willing to approach both from a different angle. Taking a closer look at individual action, we see quite a promising picture. In Europe, per the Eurobarometer survey[4], nearly every second respondent (49 percent) said they have personally taken climate action. In the U.S., that number was similar (42 percent), with four in ten Americans claiming their family and friends make at least "a moderate amount of effort" to reduce global warming, according to the Yale survey[5]. It is likely that this number is actually much higher, as there are actions that do make a positive difference, even if they are not directly associated with tackling climate change. When prompted with a list of specific actions, almost all European respondents said they have undertaken at least one. What do these results suggest? These outcomes highlight that there is a gap between reported and actual behavior, and many are taking individual actions without even considering that they are having an impact on the climate.

There are various benefits that climate action can bring us. One in four Americans state providing a better life for their children and grandchildren is the most important reason to act on climate while protecting people's health and wellbeing or creating green jobs and a stronger economy were considered important reasons by only 3 percent of respondents. This illustrates that Americans perceive that their natural, technological, and human resources are abundant enough to help them solve any problem, while they are hardly aware of the benefits of climate actions such as improving the quality of their daily lives or strengthening the economy.

In general, European awareness of the importance of tackling climate change and their willingness to participate in bringing about change is quite strong[20]. It is even more fascinating when we consider young change makers. As an astonishing example of united action to clean up the atmosphere, when 9-year-old Felix Finkbeiner

---

[15] Figueres, C., Schellnhuber, H.J., Whiteman, G., Rockström, J., Hobley, A. & Rahmstorf, S.: Three years to safeguard our climate – *Nature* (28 June 2017)
[16] Frankfurt School–UNEP Centre/BNEF. Global Trends in Renewable Energy Investment 2016.
[17] IRENA. Renewable Energy and Jobs: Annual Review 2017 and www.clean200.org
[18] IEA/IRENA. Perspectives for the Energy Transition 2017
[19] 'Make our planet great again': Macron's response to Trump is praised – *The Guardian* (3 June 2017) https://www.theguardian.com/world/2017/jun/03/make-our-planet-great-again-macron-praised-for-response-to-trump
[20] Eurobarometer survey[4] requested by the European Commission conducted in May-June 2015

felt frustrated by adults just talking about climate change rather than doing anything about it, he was struck with a brilliant idea: Plant trees *immediately* to offset carbon dioxide emissions. We know trees are the best natural machines for storing carbon, as each absorbs approximately 10kg of carbon dioxide per year. In 2007, inspired by the actions of Wangari Maathai, young Felix decided that children should plant one million trees in every country on Earth. Towards this goal, the one-millionth tree was planted in Germany three years later. Soon, the initiative developed into a global movement called Plant for the Planet, which has already engaged 63,000 worldwide.[21]

The slogan of Plant for the Planet, "Stop talking. Start planting", promotes the idea that talking alone will make no difference and that now is the time for action. The movement aims to plant 1,000 billion trees worldwide by 2020, and they are currently on track to do so. If children can take action and move others in such an inspiring way, why do world leaders need so much time to see the point?

Final Remarks

Climate change is an exceptionally difficult issue for only one party to "own," as Robinson Meyer wrote in *The Atlantic*.[22] Reducing greenhouse gas emissions is the baseline for tackling climate change, but many other adjustments are also necessary regarding U.S. energy, taxes, foreign, transportation, and industrial policies. Stimulating change in these areas has always been a bipartisan affair, throughout U.S. history. How can the blue vs. red dynamic change on the topic of climate change? As has been concluded from climate polls,[23] and as David Roberts suggested in *Vox*,[24] if Republican political leaders will recognize the reality of climate change, bipartisan majorities on climate action will follow.

Climate action is not only about climate. It is about us enjoying better lives, without harming others or the environment. Climate action includes a wide range of activities, such as choosing energy-efficient products, selecting better electricity suppliers based on renewable energy profiles, using environmentally friendly transport alternatives, eating local and seasonal food, recycling, and cutting down on the packaging. Through these small actions, we will see healthier food on our tables, we will make a lower environmental impact, we will contribute to maintaining a healthy and functional environment, and we will create green jobs on the local and national levels. It's up to us, but the most important thing is to act now!

Climate Crisis "Keep/Stop/Start" Actions

Below for your consideration, are several author-suggested action items. For a further explanation of the Keep/Stop/Start organizational tool, please see How to Read This Book.

KEEP:

1) *Keep reading and watching international news*! When reading national news, try to glean relevant information from various sources. To stay up to date on climate and sustainability, check out these sites:

- CarbonBrief: This award-winning website covers climate science and climate and energy policy, specializing in clear, data-driven articles.

---

[21] Plant for the Planet: https://www.plant-for-the-planet.org/
[22] What Americans Really Think About Climate Change – *The Atlantic* (22 April 2017) https://www.theatlantic.com/science/archive/2017/04/climate-polling-burn-out/523881/
[23] Carmichael, J.T. & Brulle, R.J.: Elite cues, media coverage, and public concern: an integrated path analysis of public opinion on climate change, 2001–2013. *Environmental Politics* (5 Dec 2016)
[24] This one weird trick will not convince conservatives to fight climate change – *Vox* (29 Dec. 2016) https://www.vox.com/science-and-health/2016/12/28/14074214/climate-denialism-social

- Innovation Review by EIT Climate-KIC: The Innovation Review publishes inspiring stories, news, views, and insights into how Europe's leading innovation community is tackling climate change.

- GreenBiz: GreenBiz advances opportunities at the intersection of business, technology, and sustainability.

- Sustainia: The publications of Sustainia provide truly inspiring sustainable solutions for different sectors that can be used today.

- Climate Progress: Think progress created a climate information source, Climate Progress, that is a climate news site dedicated to providing readers with rigorous reporting and analysis.

2) *Keep being engaged* with a local community that empowers citizens on climate action. Spread the word and share the resources mentioned above. A more sustainable lifestyle is achievable through small changes that will also save you money. To be inspired on sustainable solutions you can apply today, check out these practical resources:

- Action Guide by Planet Vision: This handy guide empowers you with simple, everyday choices to help build a healthier, more sustainable life.

- If You Want To website: Browse green services available near you and even add your own sustainable project.

STOP:

1) *Stop waiting for top-down instructions.* Think global and act local! Get in contact with your government representative! Lobbying your local, state, and federal representatives can ensure that climate change stays at the forefront of politicians' minds.

2) *"Stop* ignorance, we need openness and awareness: "Look up and around the world. America is the outcast and alone in ignoring the facts about the onslaught of the climate crisis. America is the only country that is not participating in the Paris Agreement. What do all those countries know that we don't know? Even war-torn Syria signed the Agreement. America also didn't sign the Kyoto Protocol."

START:

Start conversations about climate change and the solutions in your family, at work, and in your local community. Be the one who breaks the silence! Remember that you don't have to know all the facts and figures. Just start the conversation. If you want to learn more on the topic, check out these trustworthy resources:

- The Climate Council website offers a wealth of readily accessible information on climate change.

- The Yale Program on Climate Change Communication will inspire practical action.

- The Yale Climate Connections is a comprehensive source of knowledge on climate change, including exciting and entertaining videos and podcasts.

- The Skeptical Science website is dedicated to explaining climate change and rebutting climate myths. Their information will help you answer questions.

- The Intergovernmental Panel on Climate Change (IPCC) reports are recommended if you want to take a more in depth look into the science. The IPCC is the leading international body for the assessment of climate change and its impacts. Thousands of scientists contribute to their reports, and the sixth is currently in preparation.

- Climate Voices is a network that allows climate scientists to meet with communities to talk about the local effects of a changing climate and possible ways to address it. Invite one to give a talk in your area!

- More than Scientists is a collaboration between scientists, advocacy organizations, and the public. It offers personal, inspiring videos from scientists to explain climate change, what it means for the future of our families, and how we can take and motivate action.

- The Drawdown Project lists 100 solutions to reverse global warming.

~ ~ ~

Annamária Lehoczky, Ph.D. is a climate scientist and a passionate environmental journalist, holding a Ph.D. in Climate Change. She covers environmental-social issues, reporting about climate politics, sustainable solutions, and climate innovations. In 2014, she traveled to the arctic island of Svalbard to study the impacts of climate change on glaciers. Her report won first place in a travel writing competition by the Campus Hungary Programme. Read more at: www.climateabandoned.com.

Chapter 23
Climate Crisis & Children

by
Luis Camargo, M.F.A.

"But man is a part of nature, and his war against nature is inevitably a war against himself." – **Rachel Carson**

As early humans, we roamed the valleys and mountains in search of food and shelter, walking alongside the changes of seasons and migrations, just like every other species that inhabited our shared planet. Our needs were undeniably dependent on Earth and its systems, and we were intrinsically connected to the earth and all living beings that coexisted here. We lived, felt, and knew we were interconnected, interdependent on everything that surrounded us. As newborns came to the world, their development and learning were deeply connected to their natural environment. Knowing how nature worked and the relations between species and seasons was a critical factor in survival. Sadly, the passing of time has changed this reality.

Throughout history, especially after the Industrial Revolution, humans have gathered in cities. In 2014, 54 percent of the population lived in cities, in rather high concentrations: 84 percent in North America, 80 percent in Latin America and the Caribbean, and 73 percent in Europe. Africa and Asia have remained mostly rural, with only 40 percent of the population urbanized. Nevertheless, it is estimated that by 2050, over 66 percent of the population will reside in cities.[1]

Cities are human-created environments, and they can quickly become all-encompassing, relatively self-sufficient. Large American ones like Los Angeles and New York, as well as cities such as Ciudad de México or Bogotá D.C., become artificial environments where children are born, develop, and grow. While some people are interested in exploring the surrounding natural landscapes, few have the opportunity, much less the means, to venture out into the wilderness. Growing in these manmade environments steadily reduces opportunities to build relationships with nature, and rare is our glimpses of the sky and clouds. Only the lucky ones can still enjoy the spectacular sunset or sunrise.

While walking the urban landscape, children learn to recognize street names, buildings, and landmarks such as stores, restaurants, and crossroads. Sporadically planted trees, many of which are exotic (brought in from other parts of the world), remind us of our long-forgotten past, of those paths our ancestors trekked through the thick forests. As children grow, they bond with their immediate environment, and these bonds create the emotions associated with the place we call home. Access to direct experiences with nature lessons with each generation, indirect experiences are reduced, and we are dwindling it down to the point where the most common way is virtual and vicarious at best, through interfaces such as television sets or video games. Internet-connected devices have become the only means through which children see magical places and incredible beings. While access to information and images of species and ecosystems have never been so available it will never replace the real thing. We have forgotten the smells, the glimmering changes in the sunlight throughout the day, and the feeling of soft, earthy ground under our feet. As children lose that precious bond with nature, their empathic capacity toward life beyond humanity, as well as their capacity to understand the basic principles of interconnection and interdependence, weaken tremendously.

---

[1] United Nations, World Urbanization Prospects: The 2014 Revision.

Cities do give us a sense of control. We can provide for ourselves: food in supermarkets, clothing in stores, and all our necessities brought to us via our technology and inventions. It seems that all we have results from our capacity to create stuff. Notions of a human-centered (anthropocentric) world are reinforced as we become godlike, as we bask in the glory of our power to both create and destroy. Children who grow up under this influence come to believe they have absolute reign over nature and natural systems, weakening their capacity to understand the deep connections that exist. They may feel they can do little to make a difference in the natural world and are out of touch with, one they no longer understand.

Beyond localized examples such as those associated with pollution, until recently, the results of our way of life have not been evident. Now, for the first time, climate change as a systemic reaction reveals the deep connections between our way of life and natural systems that long for dynamic balance.

Abandoning our planet will rob us of our sense of place; in fact, this is currently seldom associated with Planet Earth. Instead, we see ourselves as belonging only to our cities, neighborhoods, or countries, all of which are human concepts. We no longer belong to the forests, valleys or savannas. Our landscapes have changed due to political unrest and war for control over oil. There has been a massive transformation of ecosystems for economic production, and these changes have forced changes in nature, devolving formerly hospitable areas of the world into bare, inhospitable land. Broader senses of place are also affected by the increasing migration into cities and amplified by forced displacement.

Abandoning Earth hurts, and solastalgia, a term coined by Glenn Albrecht, defines this profound sense of loss as "the pain or sickness caused by the loss of, or inability to derive solace from, the present state of one's home environment. Solastalgia exists when there is recognition that the beloved place in which one resides is under assault."

Education

Education allows us to learn and appropriate a worldview defined by the way we relate to the world. Our educational system has remained relatively unchanged for over 100 years. It was initially established as a factory model, of sorts, with its roots in methods for the transfer of religious and, later, state doctrine. This was further strengthened to generate all that was needed for social and industrial bureaucracies to work. Standardized learning, discipline for work, and task specialization became the main focus of the educational system, as these were fundamental in the new industrialized era. During the last several decades, competitiveness increased, so schools found it necessary to produce highly competitive, allegedly prepared students. Time spent inside the classroom has increased, as has time spent doing homework or after-school tutoring. Why? Because students must comply with ever-increasing testing standards. At the same time, unstructured play and creative learning have been demoted, replaced by highly directed, strictly structured learning.

As if this is not threatening enough, the perception of education has also changed. Parents and many schools feel that time well-spent means structured focused learning for reading and math, for students to solve complex problems earlier in life. On average, American kids spend a little over six and a half hours in school daily, plus an additional one to two hours for homework. In other words, boys and girls are spending eight hours a day studying inside.

If that sounds crazy, it only gets worse. The onset of technology has added a strong, almost unbeatable demand on children's little free time. Screen time (TV, internet, video games, social networking, and devices) has rocketed in the last decades. It is estimated that children 8 to 12 years old spend an average of 6 hours staring at screens per day, and this does not even take into account their schoolwork or homework. For teens (ages 13 to 18), this increases to 9 hours.[2]

---

[2] Common Sense Media.

Simple math indicates that up to fourteen hours a day is spent on school and screen time, leaving our children little time to share and build relationships with family or nature. Studies have concluded that most children are spending less time playing outdoors than what is considered the minimum acceptable time for prison inmates in maximum-security detention centers.

How does this add to climate change? As children grow and learn, out of touch and context with the natural world, the key concepts of interdependence and interrelations between living systems are lost. More so, the capacity to feel a connection to Earth and the workings of natural systems to sustain life is lost. Lastly, the lack of interaction with living systems other than humans undermines the recognition of participation in the cause-effect chain.

"In the end, we will conserve only what we love, we will love only what we understand, and we will understand only what we are taught." – **Baba Dioum**

Caring about and protecting anything or anyone requires a deep emotional connection. The stronger the bonds, the more building blocks of relationships are created, especially during our earlier, formative years. We learn to relate to others, value family, and establish bonds with nature in early childhood. These relationships define the way we interact and the decisions we make and actions we take. When children are asked about climate change, their responses are frightening: Most have limited understanding and are entirely unaware of the possible dire consequences we will face if things do not change. Their actions are less guided by academic knowledge than by the relationship they have created with their environment, the value they attribute to different items in it, and their capacity to feel their connectedness to it. Unfortunately, nature is just no longer a big part of their day-to-day environment. They cannot solve problems in an environment they are not familiar with, so education must focus on teaching them about this environment and exposing them to it more often, rather than confining them to their classrooms and computer screens.

Consumption

Children and youth are growing in a brand-rich environment where nature has faded. Brand names, superheroes, and virtual (not quite) reality abound. In several studies, the evidence points to the horrible fact that the same children who have lost their capacity to recognize and name real, live species in their close surroundings can easily identify brand logos and the names of hundreds of Pokémon. Nature has become a stranger, their own back yards a foreign place, one they no longer recognize or understand and certainly don't care to enjoy or conserve.

Even though it is possible to care for nature without knowing the scientific names of the things within it, the less we know and understand about our natural surroundings and the way it works, the more difficult it is to participate intentionally in sustainable, restorative behavior. In essence, nature has been belittled to just another prop on the stage of life, lacking the primary quality that differentiates it, which is life and its connection to us.

The hidden connection between the necessary resources and actual origins of things are forgotten under layers and layers of packaging and brand names. The Earth gets little credit for being our base provider, so our capacity to connect with and recognize it as such is limited. Adding to this, places on Earth where resources are scoured, such as the tar sands or large, open-pit mining projects, as well as most garbage disposal sites, are hidden from plain view and, thus, our perception. The connections between Earth and the stuff we use are almost invisible. We no longer acknowledge where things really come from, and we don't care where they end up, nor do we realize the impact all that stuff makes during its use, the consumer-led impact that affects us all.

Where does food come from? How does it grow? Food and water are, perhaps, our most essential resources, things we need every day. Without them, life would not be possible. Still, modern children have little understanding of the significant way Earth provides so much of our livelihood. Very few have the chance to see a plant grow from a tiny seed to fruit or maturity, to harvest it and see how that living element, started in a bit of soil, transforms into food. Water still pours from the sky, reminding us of the often-forgotten cycles that move Earth's elements about, but if not for this, we might be entirely convinced that all water comes from plastic bottles and faucets, with no involvement on Earth's part whatsoever.

Despair and Disassociation

Doomsday is coming, or at least it sometimes feels this way. Children are becoming increasingly afraid of nature and its forecasted power of destruction. In 2017, storm after storm pummeled the world, in epic proportions, destroying towns and infrastructures and flooding houses and streets. Fires raged through the mountains, devouring entire regions and polluting them with unbreathable smoke for weeks at a time. News media and social networking sites regularly report on these catastrophes, hitting us with an onslaught of terrified faces and heart-wrenching statistics about the precarious situation of the impacted people and places.

Nature's cycles do sometimes involve extremes, but for the last several centuries, these cycles have managed to maintain at least some balance and limits. As the system is pushed out of balance, extreme events become more common and more extreme, and young children see only the aggressive, savage side of nature. Unfortunately, the current narrative usually blames nature, as these events are reported in our headlines as "natural disasters brought on by the rage of Earth." Many children, especially younger ones, do not fully comprehend the science or the cause-effect circles involved; with wide eyes, they can only stare at a world they've not been exposed to, one they fear because they see only destruction, suffering, and imminent threats that scare them.

Climate despair is currently an important issue in children and youth. Earth has great power, and we are helpless against the force of nature: rising sea level, mega-storms, drought, fire, and avalanches, just to name a few. These events affect ecosystems and species, but they also have a great impact on human communities, bringing about suffering and hardship, even in those big cities we've built. As children grow up under these conditions, fear, and uncertainty of the future become an increasing issue. Two main responses are observed: negation and despair. A third, affirmative and productive action, is seen far less often. Narratives from parents, media, and education that interpret these events with a negative spin have a profound impact on children's understanding and reactions.

For various reasons, some people do not believe in the science behind climate change. Perhaps their beliefs do not consider the capacity for man to change planetary conditions and outcomes, because those beliefs are based on faith or fate rather than the causes of our actions. Sadly, this doesn't result in action. Children who grow up with this mindset do not believe human-caused climate change exists. This position of negation renders the individual indifferent, and such an individual will not engage in actions or attitudes that will genuinely address the problem.

As individuals recognize the threat and reality of climate change, the overwhelming feeling of uselessness, the fear associated with the destructive power of climate events, and the bleak future that is foreseen as temperatures rise, deep despair and depression ensue: "Doomsday is coming!" Another result can be indifference: "We're all going to die anyway, so what's the difference? Why bother?" Both of these feelings render youth inactive, and life becomes nothing more than surviving right here, right now, thinking only about the current moment rather than the future. They focus on protecting themselves and their families from the day-to-day threats. In a study done with Mololo children in the Philippines, the perception of despair and risk was higher in poor communities. Those children expressed helplessness and exasperation and felt as if they were unable to intervene when faced with climate threats. In fact, most studies point out that more impoverished communities will suffer the most from the effects of climate change and will have more difficulty adapting to it and surviving.

A reduced number of young individuals see the opportunity to change the direction we are currently heading, to be the change as activists and innovators. These will spearhead the movement, with the voices and actions of change.

Coming Back to Earth

In order to build the capacities of children and youth to face climate reality, harsh as they are, it is crucial that we make shifts and intentionally redirect education and learning toward recuperating our sense of belonging to Earth and to our expanded communities (local to global, interdependent, inclusive to all life). Many of the issues that cause breaks in these connections are generated by our current way of life. To address them, we have to break the cycles close at hand and create bridges to overcome the structural elements that are beyond our control.

As we abandon Earth and climate, we stop feeling a sense of belonging. To reverse this, we must first regain a sense of place and belonging, a feeling of connectedness and awe regarding nature and life systems. As we have explored in this chapter, children build their relationships during their growing years, influenced by where they live, whom they live with, how they live, and what they do.

Most of the modern population occupies cities and urban centers; as such, the majority of humans on Planet Earth right now are influenced by modern life. Even as we embark in city life, it is fundamental to start changing the environments our children are growing up in. Greening our houses, neighborhoods, schools, and cities is just part of it. Greener cities are more resilient to climate events and are generally healthier places to live. We are limited in what we can ultimately achieve, mainly because the level of change we must generate is continuously on a larger and larger scale. However, we can focus on our immediate surroundings and grow our actions from there.

There are aspects we cannot change in the short run, but we can look for alternatives. Where are the greenest city parks close to you? Find them and visit them! Start enjoying outdoor, natural recreation areas with your family, and introduce your children to nature. Frequent private nature reserves, nature centers, and state and national parks as often as you can. Just go enjoy the day or stay longer by camping out.

Greening parks and schools are just the beginning. We must create nature playgrounds and naturalized spaces that re-create small corners of wilderness inside the city. Exploration and play are critical components in reconnecting children to Earth. Recuperating spaces and times for unstructured outdoor play adds significant value to children's development, even beyond reconnecting with nature. It has been found to improve social skills, such as the capacity to relate and empathy as well as increases in creativity and reduction in childhood obesity.

Education can significantly benefit from incorporating mechanisms for children to reconnect with the Earth and build a stronger sense of place outside the classroom. A child's learning becomes more meaningful when it is associated with the real world. Learning environments can and should shift from classrooms to forests; students can get up from their desks and actively participate, constructing their own experiential knowledge.

Children and youth need to disconnect to reconnect. While much can be learned about the natural world online, it is crucial to disconnect in order to give context to that knowledge. Recovering the joy of exploring the outdoors will help create the bonds needed to understand the value and magical qualities of all life on the planet. It will allow our children to feel that they are indeed part of nature. In *The Nature Principle: Reconnecting with Life in a Virtual Age*, Richard Louv explores many of the additional benefits of contact with nature to human wellbeing and creativity.

As children increase their contact with nature, nature literacy is recovered. Waking the naturalist within us all goes a long way toward stimulating exploration, observation, inquiry, and artistic representation of nature. During this process, our devices can serve as tools to collect information, to identify plants and animals, and to be platforms to tell stories. In this way, we disconnect from the devices and gadgets that hold us hostage, those inventions that dominate our world experience interphase. Once we experience the real, three-dimensional world for ourselves, those devices can resume their places as tools that allow us to communicate and share those experiences.

The Children and Nature Movement (USA) and The Wild Network (UK) are dedicated to generate awareness and share research about the impacts associated with children's abandonment of nature. They also create useful resources and tools to reconnect to nature.

Developing Nature Empathy

"We must fall back in love with Mother Earth." – Zen Master Thich Nhat Hanh

Children are born as semi-blank slates, and they almost immediately begin developing through their interactions and experiences. In their infancy, they experience smells, colors, and shapes, and they interact mainly with close family members. As curious toddlers, they want to explore everything around them; they want to touch, feel, and taste whatever they can get their little hands on. As this process happens, everyday recurring experiences related to the feeling of safety start creating essential relationships of belonging and their definition of home. Making sure children have access to time outdoors will incorporate the elements of nature into the initial processes of building relations.

As children age, they develop empathy, which can be defined as the capacity of an individual to project within its own experience the feelings, thoughts, and experiences of others. It is commonly described as being able to put yourself in another's shoes, to understand their reality from their perspective. Empathy is considered one of the most critical twenty-first-century skills since it unites and forms a foundation upon which many life and learning skills are built and strengthened. Empathy allows us to be better leaders, more effectively utilize teamwork, and enjoy the benefits of stronger social bonds and interactions.

The positive impact of empathy has been broadly studied and documented. Nevertheless, very little has been researched when it comes to dispositional empathy toward nature. Dr. Kim-Pong Tam, who examined the subject in 2013, expressed that connectedness to nature is a contributing factor that facilitates the capacity for empathy with nature, but it is not the same. In his 1996 Beyond Ecophobia, David Sobel says, "If we want children to flourish, to become truly empowered, then let us allow them to love the Earth before we ask them to save it." As an environmental and outdoor educator, I have seen the impact of fostering nature empathy, especially during the earlier years of a child's life, when he or she is more open. At this stage, little ones feel innately attracted to bugs, critters, and all of nature's wonders.

In the process of recovering climate from abandonment, recovering the connections children have with Earth should be a priority of all educators, parents, and society in general. Also, if empathy for nature is developed, we will produce citizens who have the sensibility and capacity to go beyond the limited goals of sustainable development and into regenerative symbiotic development and the healing of a changing planet. Climate change requires immediate action from children, youth, and adults alike. More importantly, it needs visionaries and changers who are able and willing to create and establish alternative ways of life that foster a healthy, thriving planet, one on which all species can flourish and live in balance together.

Immediate Action: Hope

A new type of youth activism is on the rise, and the power of youth is fundamental in our forward progress. While it has been said and sung, "I believe children are our future," I believe this is a case of the wrong syntax. In fact, children and youth are actually our present, but they will hopefully help mold a better future.

One of the most significant gatherings of Earth's People (Nations) was seen in 2016 at Standing Rock Camp, in resistance to the construction of the Dakota Access Pipeline meant to pass through sacred Native American territory. They desired to protect the water supply, and they offered prayers to Earth. This massive response was initiated by a group of youth leaders who believed firmly in doing something. Among them, Jasilyn Charger and White Eyes led the One Mind Youth Movement and set up a prayer camp. Within a few weeks, this mission grew to the Standing Rock #NoDAPL worldwide movement that saturated the news. Even though the encampment was ultimately dissolved by the government and the pipeline was pushed through, on June 14, 2017, a federal judge ruled that the permits allowing the pipeline to cross the Missouri River violated the law. The federal judge "it did not adequately consider the impacts of an oil spill on fishing rights, hunting rights, or environmental justice, or the degree to which the pipeline's effects are likely to be highly controversial."[3]

In Juliana vs. the United States, twenty-one young plaintiffs sued the U.S. government for failing to act on climate change, an action supported by our Children's Trust. In describing the case, one of the teenage plaintiffs, Youth Director of Earth Guardians, Xiuhtezcatl Tonatiuh Martinez, stated, "The federal government has known for decades that CO2 pollution from burning fossil fuels was causing global warming and dangerous climate change. It also knew that continuing to burn fossil fuels would destabilize our climate system, significantly harming my generation and generations to come. Despite knowing these dangers, defendants did nothing to prevent this harm. In fact, my government increased the concentration of CO2 in the atmosphere to levels it knew were unsafe."[4] An Oregon district court established that the "United States is subject to a public trust duty to protect the atmosphere and that this remedy is not displaced by the Clean Air Act"[5] Despite actions by the president and government to generate a dismissal of the case, the trial date is still set for early 2018.

Like many other youth activists of the new generation, Xiuhtezcatl Tonatiuh started speaking on Earth's behalf when he was very young, just 6 years old. In one of his early speeches at a national, global warming event, he said, "Every choice we make is for or against our future." Since then, this brilliant young man has led other youth environmental activists under the banner of Earth Guardians. Kids from around the globe are joining in the fight for Earth and their future. In Portugal, after fires ravaged the country, fueled by extreme summer heatwaves, a group of 5- to 14-year-olds prepared to sue 47 European Union governments over climate change and the adverse effects on their livelihood. At the time of this writing, the Global Legal Action Network (GLAN) supported the case and began using crowdfunding as a means to jumpstart it. In Colombia, 25 youth between 7 and 25 years of age with support from DeJusticia a social justice collaborative established legal action towards the government for threatening a future healthy environment as their fundamental right.

Another way youth are voicing their opinion and proposing solutions is through their individual actions and life choices. Sustainable farming and small, sustainable, cooperative villages are gaining strength as youth look for alternate ways to build community structures that respect Earth and foster peaceful relationships. Inspiring examples of young urban farmers and rural agricultural cooperatives like Rural Hub, created in Colombia, or the Ecovillages NextGen initiative, bring a resurgence of the Global Ecovillage community model.

In his study, "Life Trajectories of Youth Committing to Climate Activism," Scott Fisher found that an essential factor in getting youth engaged is illuminating the link between climate change and social justice. It is really not

---

[3] Case 1:16-cv-01534-JEB Document 239 Filed 06/14/17
[4] Press Release, August 12, 2015.
[5] Case: 17-71692, Filed 09/05/17

only about the planet; fundamentally, it is about all life forms and systems, reducing suffering, and fostering a just, balanced future.

Last Words

For decades, we have abandoned our climate and our Earth, creating an era called The Anthropocene[6] where human dominance and influence over natural systems has resulted in climate change. Now is the moment to return to and embrace the capacity of children and youth to lead changes in the way we relate to each other and to the planet. These changes are necessary if we want to revert and re-envision the way we, as humans, inhabit a magical, nature-rich Mother Earth. Listen and feel. Think of development beyond sustainability. Seek regenerative systems that celebrate life and its connections; these, in turn, thrive on symbiotic relationships between living beings. As Glenn Albrecht says, "The next era in human history should be The Symbiocene."

Yesterday was already too late. We need to shift now!

Climate Crisis "Keep/Stop/Start" Actions

Below for your consideration, are several author-suggested thought, behavior, and action items. Action nourish inspiration! For a further explanation of the Keep/Stop/Start organizational tool, please see How to Read This Book.

KEEP:

1) Enjoy the outdoors with your kids!

2) Be intentional in your choices and adapt them as you learn more about their impacts. All choices are a way to voice a position and generate change.

STOP:

Stop your consumption habits. Buy local, organic, fair trade, and sustainable, with minimized packaging.

START:

For Families:

1) Go for morning walks together. Take notice of all the colors, shapes, and plant and animal life in your neighborhood.

2) Find and map natural areas close to you. Know the programs and services they offer and visit them often.

3) Set up a house garden of herbs and vegetables that kids can grow and harvest.

4) Establish connection-free moments, device-free zones. Enjoy a family dinner or an afternoon soaking up the sun in the park, as if it's 1969.

---

[6] Crutzen and Stoermer, 2000

For Educators:

1) Start teaching classes outdoors. Encourage students to look at the clouds, feel the wind, and discover critters.

2) Incorporate elements of nature into your class structure.

3) Create a nature literacy program.

4) Link outdoor experiences with exploration, reflection/discussion, and critical engagement.

5) When teaching about climate change, find a midpoint that conveys our critical situation but also offers proactive strategies to generate change day by day.

6) Use digital storytelling based on outdoor settings and experiences.

For Youth:

1) Explore and learn about your surroundings. Use local naturalists and apps to help you (iNaturalist, Project Noah, eBird, LeafSnap, etc.).

2) Realize that the future will be your present, then embrace climate change action today.

3) Get involved! Join a youth action group.

4) Talk with friends and community about climate change and your concerns. Find collective actions that involve everyone in creating better quality of life and experience.

~ ~ ~

Luis Camargo, M.F.A. is the Founder and Director of OpEPA (Organization for Environmental Education and Protection), an award-winning Colombian organization focused on reconnecting children and youth to nature so they can act in a more environmentally responsible manner. Luis is also a speaker and writer on nature-based education, biodiversity, and climate change. He has been recognized as an Ashoka Fellow, Young Global Leader (2008), and Climate Reality Leader. Luis works with students, teachers, and ecotourism guides to find mechanisms that strengthen a sense of place and help to break climate despair cycles by building empathy for nature and reconstructing relationships that acknowledge interdependence and interrelations within the planet. Luis lives in Bogotá, Colombia. Read more at: www.climateabandoned.com

Chapter 24

Climate Crisis & Education

by

Maria Santiago Valentin, M.A.

(To the climate activists in New Jersey and to the people of Puerto Rico)

"The ignorance of younger generations about climate change, is a weapon that will be used for the benefit of the same sectors that destroy the environment."

In 1517, students were taught the theory of *geocentrism*, the idea that Earth lies at the center of the solar system and all celestial bodies revolve around us. In 1817, many academics learned that geological features of Earth are permanent, and the idea that the continents could drift was considered stupid and ludicrous. Now, in 2017, some teachers are instructing students that climate change is nonexistent and, even if it is, this is indeed not due to any human activity. What do these three paradigms have in common? They are all perpetuations of blatantly false information, baseless, and they both impede and obstruct the development of real scientific knowledge and, thereby, the betterment of our species.

In public schools, climate change education has become an issue of controversy and conflict. That, unfortunately, hurts its very mission and, in a way, hinders the education our children receive. Our public schools are particularly vulnerable, and climate change deniers are blocking climate change education as a tool to exercise control over future generations.

We the people pay the taxes that fund education. As such, we are financing a type of climate change education that goes against our own social and environmental interests. Due to certain sectors of society that we choose to call the sources of oppression, we are living times of uncertainty. There is a rampant, undeniable attack underway on the paradigms, institutions, and structures of democracy, governmental openness, and honest discourse.

When I studied Earth science in grade school in the seventies and eighties, my teachers focused on natural disasters and pollution. These, as well as the ozone hole, were the primary issues addressed in class. In the 1990s, I became aware of the benefits of recycling and our duty as concerned global citizens to protect the planet. Every Earth Day, we celebrated the occasion by planting a tree on the campus where I used to study. I don't think it is too far of a reach to believe that many of you have similar memories.

When we look into the 1990s K-12 science curriculum of Earth systems education, we find that the science curriculum presented global climate change as part of Earth science. The curriculum developed by teachers and scientists created seven principles:

1) Earth is unique, a planet of rare beauty and great value.

2) Human activities, collective and individual, conscious and inadvertent, affect Earth's systems.

3) The development of scientific thinking and technology increases our ability to understand and utilize Earth and space.

4) The Earth system is composed of interacting subsystems of water, rock, ice, air, and life.

5) The Earth is more than four billion years old, and its subsystems are continually evolving.

6) The Earth is a small subsystem of a solar system within the vast and ancient universe.

7) There are many people with careers and interests that involve the study of Earth's origin, processes, and evolution.[1]

Of the seven principles, let's take a look at the second, which addresses the manmade impact on our environment. My questions and confusion come from the evidence: If we already have in place a clear principle that places responsibility on our actions, acknowledged twenty-two years ago, why and how have we drifted from that wise premise crafted by teachers and scientists?

Clearly, over two decades ago, we saw a link between Earth's system and biodiversity with human activities. Ten years later, in the 2000s, science teachers preferred to teach about other topics, such as acid rain, the ozone hole, and air pollution. In other words, climate change was not in the top twenty topics discussed in Earth sciences classes.

Climate change education, sadly, is, at best, a subtopic in Earth and environmental science education. In some places, the need for a specific climate change curriculum has been challenged altogether. The criticism of climate change education is not new. In fact, the first documentation I found, which should not be exclusive of criticism in previous decades, was published in 1992, almost a quarter-century ago. The main points of that piece suggested that environmental education:

- Is often based on emotionalism, myths, and misinformation.

- Is often issue-driven rather than information driven.

- Typically fails to teach children about basic economics or decision-making processes, relying instead on mindless slogans.

- Often fails to take advantage of lessons from nature, and instead preaches socially or politically correct lessons.

- Is unabashedly devoted to activism and politics rather than to knowledge and understanding.

- Teaches an anti-anthropocentric philosophy that man [sic] is an intrusion on the earth and, at times, an evil.[2]

Déjà vu, anyone? Although written over two decades ago, these paradigms are strangely similar to the statements being shared, posted, tweeted, and spat out by our current administration, including EPA Administrator Scott Pruitt.

Fortner argues that teachers of science agreed that environmental education is fundamental for the teaching of climate change; however, in the fifth and sixth principles, we can clearly see the foundation for the rhetoric we are hearing from modern climate change deniers. What's my point here? Simple: This misguided position did not appear out of nowhere in the late 2000s, as many purports There is nothing new under the warming sun, folks.

Climate change must be taught from scientific, environmental, and social perspectives. The controversy, however, is not scientific in nature; that is spawned by politicians and corporations, as the teaching of climate change significantly impacts their profits. Yes, denying access to climate education makes it a social justice issue.

[1] Mayer, V.J. & Fortner, R.W., Science is a study of Earth. Columbus, (Ohio: Earth Systems Education Program, The Ohio State University, 1995)
[2] Fortner, *Climate Change in School: Where Does It Fit and How Ready Are We?* (Ohio: School of Natural Resources, The Ohio State University 2001)

The good news is that many states, legislators, and educators are trying to bring accurate knowledge to our youth, attempting to shine a light in an otherwise dark tunnel. In fact, at the time of this writing, twenty red, purple, and blue states have adopted the Next Generation Science Standards (NGSS). Believe it or not, the NGSS were not developed by the federal government; instead, they were created by twenty-six states in 2013, with private funding, and they were endorsed by the National Science Teachers Association. Nevertheless, resistance remains in other states to adopting the climate change standards embedded in NGSS curriculum.

The resistant states wish to eliminate the teaching of standards that stand in opposition to their denial. The NGSS that address weather, biodiversity, Earth systems, and human activity include:

ESS3-MS-5: Ask questions to clarify evidence of the factors that have caused the rise in global temperatures over the past century.

ESS3.C: Human Impacts on Earth Systems

LS4.D: Biodiversity and Humans

ESS2.D: Weather and Climate.

Of the five standards, two specifically link climate change impacts to human activity. One of these is ESS3-MS-5:

> Human activities (such as the release of greenhouse gases from the burning of fossil fuel combustion) are major factors in the current rise in Earth's mean surface temperature. Other natural activities (such as volcanic activity) are also contributors to changing global temperatures. Reducing the level of climate change and reducing human vulnerability to whatever climate changes do occur depend on the understanding of climate science, engineering capabilities, and other kinds of knowledge, such as understanding of human behavior and on applying that knowledge wisely in decisions and activities.)

ESS2-D discusses global temperatures in regard to the amount of greenhouse gases we generate, another indication that our activities do make an impact:

> Current models predict that, although future regional climate changes will be complex and varied, average global temperatures will continue to rise. The outcomes predicted by global climate models strongly depend on the amounts of human-generated greenhouse gases added to the atmosphere each year and by the ways in which these gases are absorbed by the ocean and biosphere.

The polarization began when some states decided to not adopt the climate change portion of the Next Generation Science Standards or, even worse, not adopt the NGSS at all. Others are still pushing the debate between opposing viewpoints: intelligent design or natural cycles versus human activity as the culprit in impacting climate change.

Another problematic situation is that legislators (who are not educators) propose that we should present students with both the disadvantages and the advantages of climate change. Those who have no clue about pedagogy and lack even basic scientific knowledge of climate change suggest that we should send our children a mixed message. This is completely outrageous and unacceptable. Climate change taught from such a dual perspective is an example of *cognitive dissonance*, the stress a person suffers when they hold two or more conflicting values and act in a way that contradicts those values. The fact of the matter is that the consequences of climate change caused by human activity offer no benefits to any living being on the planet.

Going back to the good news, I want to share some optimism. Twenty states have adopted the NGSS and the climate change standards: New Jersey (my State of residence), Vermont, New Hampshire, Connecticut, Rhode Island, Delaware, Washington DC, Michigan, Ohio, Kentucky, Illinois, Iowa, Florida, Arkansas, Kansas,

Washington, Oregon, California, Nevada, and Hawaii. Collectively, these represent 40 percent of our nation's public school districts. However, the impact of the NGSS is not felt equally since the states differ in size and population. Legislation at the state and local level, as well as anti-climate change instructional decisions made by teachers in the classroom, represent a grave threat to its accessibility to students.

States that aggressively deny the human impact on climate change are modifying the integrity of the NGSS. Curriculum and standards are developed to foster social change, to increase students' awareness of and empathy about their surroundings, to achieve a higher level of consciousness, and to make connections from the concept to the application of the concept in real situations.

This approach can be called the Three C's of Climate Change Education: change, consciousness, and connections. The climate change curriculum should foster social change. Educators must understand that, in order to change daily practices to reduce our carbon footprint, we need to mobilize the powerless in our communities, raise the level of student awareness (consciousness) amount what can be done at the individual and collective level as concerned citizens, and connect the climate change curriculum and standards with what is happening in our underserved communities. Also, educators have to make interdisciplinary connections with other curricular areas, including technology, engineering, marketing, and others. In these areas, the concept of climate justice can be applied pragmatically by students.

This model mimics the world language curriculum frames for their standards. Climate change is a social justice issue, and the perspective of societal change is critical. Younger generations should read about decisions made by legislators, government entities, and private corporations, especially when they lack the social justice component. If the human, social issue is not dealt with, naïve detachment from climate change will continue. Here is where empathy and social and emotional learning play an important role. The process by which our students learn and apply new knowledge, simultaneously discovering how to manage the impact of that new knowledge on their emotions; feel empathy for others; and develop a sense of responsibility to create positive goals is called social and emotional learning.

The communities most impacted by climate change are not the primary drivers of this global change. Ignoring climate change causes setbacks and suffering in the communities least capable of fighting back, those comprised typically of residents of lower socioeconomic status, usually with an annual income of less than $15,000. Unfortunately, this is how risk is distributed mostly in the United States. In fact, one could view all this pushback against teaching climate change in schools as a concrete example of environmental racism.

> Whether intentional or not, the impact is elitist when the effects of environmental policy and reforms systematically distribute the costs of climate change and environmental pollution onto the less privileged or are inconsistently enforced across communities.[3]

Choices and resources for communities of disadvantage to opt for a greener lifestyle are limited. This is a clear-cut indication of the powerlessness and poverty inherent in our allegedly democratic institutions.

We need to connect change to consciousness so we can inspire future generations to mobilize and act. With an emotional and social learning approach to climate education, both current and future generations will develop the necessary empathy to care for the communities that have long been ignored, due to lack of empathy from politicians and corporations.

The climate change curriculum should explore and foster projects that allow and encourage students to improve the quality of life in their local communities, as well as on a global scale. In addition, stories and narratives from friends and family who have been impacted by storms, forest fires, floods, or environment-related health

---

[3] Xiao, Why Climate Change Is an Issue of Social Justice for Vulnerable People 2016. https://everydayfeminism.com/2016/10/climate-change-for-vulnerable-ppl/ (accessed on July 25, 2017)

concerns like asthma are fundamental in helping our youth make the connections from the concept to pragmatics and develop compassion. The real-life experiences of others help students connect the knowledge with their emotions and feelings about climate change. As useful as these terrible tales are for fostering empathy, even more horrible and unfortunate is the fact that we are beginning to see more and more relatives and friends who are suffering the impacts of climate change. Whether it is a cousin in Houston rebuilding after a flood spurred by Hurricane Harvey, a former co-worker in Washington State fleeing her home to escape the extensive wildfires, or an uncle in California attempting to sustain his family farm through the severe droughts of recent years, these traumatic stories are becoming more common. If it keeps up this way, it won't be too much longer before we're all affected in some way.

When my older daughter was completing her studies for her bachelor's degree during her college years, she shared with me the reaction one of her political science classmates had when a debate broke out to determine whether climate change is a hoax or not. When one of her peers supported the hoax theory, another responded that such a statement completely opposed reality. He was offended by the hoax theory because his family lost their home during Hurricane Sandy in 2012. It was as awkward for my daughter as it was eye-opening.

Empathy had brought social change through action before in our country, as Reverend Barber mentioned when the Jim Crow Laws were declared unconstitutional in 1964. Empathy is also what mobilized Jill Cody to compile this book, a collection of intellect and heartfelt input from multiple authors of many different backgrounds, to discuss the complicated issues of climate change. Morality is part of American DNA.[4] We, as adults, need to advocate to preserve and maintain climate change education.

We must be open to organically change with new knowledge and new behaviors that either foster hope or resistance from us, the climate change education advocates. The last C needs to happen in multiple layers and levels for students to embrace climate change. Climate change must be connected to self, science, other disciplines, local communities, global communities, water, health, food, social justice, environmental justice and racism, legislation, clean energy, renewables, and green jobs, just to name a few.

Did you know that from 1979 to 2003, excessive heat exposure contributed to more than 8,000 premature deaths in the United States?[5] As depressing as that is, it is essential for our children to be aware of the facts. It also exposes the true connection between climate change and human health, and it makes the concept tangible. Unbelievably, in May of 2017, the Environmental Protection Agency (EPA) removed their Climate Change Kids site. Clearly, the pushback from the current legislation won't stop, but we must continue to resist these attacks on our education, our children, and our more deprived communities.

In the Classroom

The adopted climate legislation, policies, and resolutions impact the instruction of anthropogenic climate change in the classroom. A 2016 study from the National Center for Science Education revealed that one in three teachers teach climate denial, arguing that many scientists claim that climate change is not directly caused by humans. Surely, these teachers have not read the 2013 Intergovernmental Panel on Climate Change (IPCC) Report, which stated that 95 percent of those in the scientific community agree that human activities have accelerated climate change.

On the other hand, the same study from Nuticelli, in 2005, revealed that three out of five teachers were unaware that the papers and research done by deniers harbor some fundamental flaws. These include: lack of a control group, inappropriate statistics, incomplete systemic methodology, and a "clear lack of plausible physics."

[4] Reverend Barber, These Times Require a New Language and a New Fusion Coalition 2016. https://thinkprogress.com/rev-barber-these-times-require-a-new-language-and-a-new-fusion-coalition-c741b9eb1b47/ (accessed on July 25, 2017)
[5] EPA Encouraging Behavior Change in the Community: A Tale of Two Programs 2015 , https://archive.epa.gov/epa/statelocalclimate/encouraging-behavior-change-community-tale-two-programs.html (accessed on July, 25, 2017).

This is unsettling because it indicates that teachers are disseminating information to our children that are inaccurate and incorrect. How would you feel if your child came home from school one day and proudly said, "Mommy, today we learned that the sky is purple, two plus two is six, and the Declaration of Independence was signed by the Backstreet Boys!" Maybe you don't think of lack of climate change education or, worse, false teaching on the topic as that drastic, but I'm sure you get the point.

Our teachers should intertwine the impact of climate change in economics, urban neighborhoods, and communities of color. Why? Because those are the communities most impacted, yet they have the least access to resources to recover quickly from major events like storms, floods, droughts, fire, heatwaves, and the health conditions like asthma and post-traumatic stress disorder (PTSD) that can result from these natural disasters. The communities that are already challenged economically face drastic climatic consequences and are simply not equipped to stave them off. For them, proper education may be their greatest hope.

Besides teaching the facts, schools districts have the moral obligation to include opportunities for students to think about the issue and create solutions for climate change mitigation and adaptation, along with the scientific evidence that proves it. Children and youth can learn new behaviors that will influence their decision-making processes about issues concerning the wellbeing of our planet when they become adults. For this reason, students must be taught that climate change is a hot topic not only for scientists but for nearly every other industry in the world. For example, coastal developers must anticipate sea level increase when designing architecture and engineering in those areas, if it is even safe to build there at all. Insurance companies must carefully examine the risk of insuring properties that will face flooding or complete destruction, and the risk some communities face determines their decisions.

It is important to note that legislation, policies, and resolutions are a way to intimidate the teachers and the administration in deciding what and how to teach about climate change. If a teacher works in a district that resists the climate change curriculum, it is very likely that he or she will avoid the topic for the sake of job security. Not only that, but many fear receiving parent complaints. If the administration does not support the teaching of climate change, it is very likely that the topic will not be covered at all. This is where we, as parents, must step in. If the local school board, state legislation, and administration in our districts do not help educators delivering instruction about climate change, we need to strongly advocate for the curriculum and support the individuals in charge of that instruction. We pay the taxes that fund public schools, so the loudest voice in any school district should be the voice of the justifiably concerned parent. We can make the difference with our activism. Yes, we can!

Climate Change Denial Legislation at the State level

Any good news aside, the current situation in climate change education in the other 60 percent of our nation is horrific at best. In 2010, South Dakota adopted HCR-1009, which states that the current curriculum is politically biased, and it describes climate change as a theory, not a fact. Yes, you read that right: This was in 2010, not 1910! However, maybe this is less surprising when you realize how many states have a history of promoting the teaching of *creationism*, the religious belief that all life is created with divine intervention, and this belief is used to influence legislation.

In Idaho in February of 2017, the House Education Committee voted, eleven to three, to remove curriculum standards from the NGSS (ESS3-MS-5, ESS2:D, LS4.D, ESS3.C) reference human interaction with the environment as the cause of climate change. The legislation approved the NGSS without reference or language related to climate change. Strangely enough, and worthy of note, it was never presented to the State Board of Education before that vote was conducted.

Idaho's cousins—Mississippi, Louisiana, Tennessee, and Indiana—have created legislation that gives teachers academic freedom to teach multiples angles regarding climate change, though the definition of that freedom is somewhat relative. Alabama, Iowa, South Dakota, Oklahoma, and Texas have also created legislation favoring academic freedom, but those bills died before a final vote. Some concerned citizens and educators say this legislation opens the door for climate change denial instruction.

Florida is the icing on the proverbial cake. In 2017, the Sunshine State approved two controversial pieces of legislation. One grants teachers the freedom to talk about religious beliefs. The other, which is openly supported by the Florida Citizens Alliance, allows parents to reject classroom materials selected by educators. Finally! Now parents can rightfully reject the dangers of their children learning taboo subjects like math or history. This legislation has received opposition from a variety of organizations, including the Florida School Board Association and the National Science Teachers Association.

In New Jersey, Climate Parents circulated a petition in 2014 asking the state to adopt the NGSS. On July 9, 2014, the New Jersey State Board of Education adopted and implemented the NGSS in New Jersey public schools. Democracy really does work when it is handled properly!

The diligence and activism of climate change education advocacy groups has much to do with the adoption of the standards. Deniers reject and ignore the fact that our planet is 4.5 billion years old and that, while climate has changed naturally over the years, we are now experiencing a crisis because of the significant and quick changes that have occurred over the last fifty years, changes that are clearly not natural occurrences.

Climate Change Denial Policy at the Local level

Believe it or not, one in ten teachers is a denier. In 2010, a group of Colorado parents gathered signatures and delivered a petition requesting that the school district teach the other side of climate change, protesting after their children complained about not being able to vocalize their religious views in class. Their efforts met the opposition of other parents and teachers, and that request was dismissed.

In 2011, The Unified School District in Los Alamitos, California adopted a policy requiring teachers to use sources of both sides when teaching controversial issues. The policy was revised, and currently, teachers are required to present multiple sides of an issue. One side teaches that the release of carbon dioxide has increased greenhouse gases in the atmosphere, while the other claims that the fluctuations of the temperature of the Earth have always happened.

Several articles imply that teacher preparation exposure to climate change is not enough. If the teacher does not feel confident teaching the topic, the teacher will avoid it altogether. The NGSS are not enough at this point either. Content knowledge training, continued education, and professional development is necessary. Our teachers must share current facts, not those gathered in 1990, which mimic climate crisis denial.

As of Today

Current events such as President's Trump withdrawal of the United States from the Paris Climate Agreement, have doubled the attacks on climate change education. Trump suggests that he left the Paris Accord because it put the USA at a disadvantage, but nothing could be further from the truth.

That misinformed, foolish decision by the current administration is a catastrophic backpedal from important promises we made to the world. It was reckless, irresponsible, and harmful to all global citizens. It was also shameful and embarrassing since our country is also principally responsible for the pollution and toxins that are impacting other communities who have not harmed the environment.

As a result of this new mindset in a new era, some teachers have received a textbook from the Heartland Institute, *Why Scientists Disagree About Global Warming*. The title of the book may as well be *Why Scientists*

Aren't 100 Percent Sure Smoking Causes Cancer. The Heartland Institute consists of deniers, and the group president encourages teachers to share only their viewpoint on climate change. He also openly accuses environmental activists of pushing their agenda with the NGSS and climate change education.

Two critical organizations in the field of national education, the National Science Teachers Association and the National Center for Science Education, encourage teachers to recycle these books, perhaps as winter kindling. Whatever is done with the books, the best course of action would be to never, ever teach anything from them.

It is very suspicious that the current administration refuses to accept scientific fact. It could be assumed that they wish to continue building an economy with practices that put global environments, our populations, and climate at risk. It is very convenient to keep current and future generations misinformed for profitable and marketable reasons.

At this pivotal moment in time, when we can still do something about it, climate change education is not an issue for debate. There should be no question about whether or not it should be taught, and it is unacceptable that we even have to argue about this critical knowledge for our children.

Benefits of Climate Change Education and Careers

It is no-brainer to realize that it is imperative to teach climate change and add it to other content areas in science, engineering, technology, architecture, welding, plumbing, or math. Our children need to apply mitigation and adaptation concepts in projects when sea level is expected to rise, weather conditions are expected to worsen, and where the potential for flooding, drought, heatwaves, etc. is continually increasing.

Teaching climate science and climate change helps students developed higher thinking skills, such as interpreting data, drawing conclusions, developing empathy and compassion, connecting to the global communities, and applying problem-solving skills. With the appropriate instruction of climate change, the infrastructures future generations will create and build will reflect the climate crisis adaptation and mitigation principles applied in their fields. Also, informing students of careers that require climate science and climate change education is critical; this knowledge will bring about a variety of new jobs related to green energy and renewables.

More and more, communities are moving into living greener, and careers in transportation, real estate, energy, agriculture, and engineering are in demand. A recent perusal of indeed.com, a popular job site, revealed some of the following listings: climate defender; climate change associate for a business for social responsibility (New York or Paris), climate and energy program associate, senior greenhouse/gas climate change specialist, climate change and health science fellow, policy analyst, climate fellow in green America, and many more.

Climate change education is critical, and if our goal as educators is to equip children for a brighter future and successful careers, perhaps no other topic is more critical today!

Climate Crisis "Keep/Stop/Start" Actions

Below for your consideration, are several author-suggested thought, behavior, and action items. Action nourish inspiration! For a further explanation of the Keep/Stop/Start organizational tool, please see How to Read This Book.

KEEP:

1) *Keep signing petitions* that request support of Next Generation Science Standards and climate change education.

2) *Keep being active and vocal!* Write to legislators and local newspapers, advocating for climate change education.

3) *Keep discussing climate change issues* with your children and ask them about the information they are learning in class, if any.

STOP:

1) *Stop engaging in arguments with deniers* of climate change education, those who refuse to accept scientific evidence.

2) *Stop isolating yourself* from Boards of Educations at the local, state, and national level. Be involved when they discuss climate change education.

START:

1) *Start contacting advocacy groups* in your municipality or state, such as Climatemama, Mrs. Harriet Shugarman, and Sierra Club. At the national level, try Climate Parents.

2) *Start reading science education* and climate change education advocacy blogs. Follow these groups on Facebook and Twitter to stay current with legislation changes, events, rallies, hearings, etc. A few you might visit include:

   - https://tiki.oneworld.net/global_warming/climate_home.html

   - http://www.eschooltoday.com/climate-change/causes-of-climate-change.html

   - http://easyscienceforkids.com/all-about-climate-change/

3) *Start encouraging your children* to engage in community programs that connect climate change with social justice. Some include:

   - iMatter (http://www.imatteryouth.org)

   - Cool the Earth

   - Green Kids

   - Inconvenient Youth

   - Climatechangeeducation.org

   - Greenambassadors.org

- Boston Latin School Youth Climate Action Network

- Roots and Shoots

- Action for Nature

- Alliance for Climate Education

- ECO2school

4) *Start advocating for climate change education* that is delivered with the principles of social and emotional learning. This is fundamental and critical to change the way future generations will view others and our planet.

5) *Start thinking about running for office* in your local community as a Board of Education member to advocate for climate change curriculum.

6) *Start working with the Climate Reality Project* (CRP) Leadership Corps in your state and local community. Offer to give middle and high school CRP presentations with age- appropriate language and example. More importantly, help students make connections through social and emotional learning, to make their learning meaningful and lasting.

~ ~ ~

Maria Santiago-Valentín, M.A. is a doctoral student in Education (Reading, Literacy, Assessment), a learning disabilities consultant/behavior technician, currently one of the bilingual columnists and writers of the digital magazines *My Trending Stories* and *Negocios Hispanos USA*. She has served as an educator for over twenty-two years. In 2010, Maria served on the Board of Trustees of the Barack Obama Green Charter HS in Plainfield, New Jersey. Maria became a Climate Reality Leader in 2013 and a Climate Reality Mentor in 2017. Maria organized and co-lead the 2017 and 2018 NJ People's Climate Rally. She was in the Steering Committee of the 2018 NJ March for Science. Maria was appointed in September 2017 as Treasurer of the New Jersey Chapter of the National Association of Hispanic Journalists. Maria lives in Clark, New Jersey where she is an alternate member of the Clark Environmental Commission. Read more at: www.climateabandoned.com.

Chapter 25

Climate Crisis & Millennials

by

Kevin Quinn, M.L.S.

"Society grows great when old men plant trees whose shade they know they shall never sit in."
– Ancient Greek Proverb

We get it. Seriously. We get it. Climate change is an impending, dooming, looming threat that, if nothing changes, will affect us in some way or another, whether we like it or not. Maybe it's because we grew up with apocalyptic themed cinema like *Mad Max, Armageddon, and Wall-E,* imagining what that futuristic wasteland could look like. Maybe it's because we're the generation to receive the latest and greatest STEM education (barring Gen-Z soon entering the workforce and blowing us out of the water).[1] Maybe it's because of our meme-obsessed, ignorance-repudiating internet culture that fosters a healthy knowledgebase about the world around us. Regardless of why, there is an overwhelming majority of millennial Americans who believe climate change *is* real and, therefore, something must be done about it. In a seemingly dissonant political atmosphere in which it's difficult to get a roomful of people to agree on any one thing—or even to get them to agree to be in a room together in the first place—climate change may be the unifying issue that spans the aisle.

Why, then, does it seem we are backpedaling? We've pledged to pull out of the Paris Climate Accord, attempted to defund nearly a third of the Environmental Protection Agency federal budget, and repealed environmental regulations on a massive scale. This doesn't seem to gel well with most Americans, let alone millennials. Although a major underlying reason is that our generation is still merely on the cusp of taking charge of government, I argue that there are many undiscussed nuances of our culture and society that ultimately drive the rift between attitudes of millennials and other generations. I hope to explore these throughout this chapter.

Before we begin, I must make something clear: this is not a chapter about bashing baby boomers or any other generation (though it is tempting...). Furthermore, you will not read phony, unwarranted praise for my own generation. I will not claim we are the best and the brightest group of young people to ever grace Planet Earth. My aim here, rather, is to take a hard, honest look at the relationship between the millennial generation and the climate crisis.

First, let's set the groundwork and objectively look at what millennials believe about climate change. This is easier said than done as I write this from behind my computer, nestled in the liberal, green bubble that is Denver, Colorado, but we'll look not only at what my generation believes but also how it compares to other generations and how it breaks down across the political spectrum. Next, we will transition into the why: Why do we believe these things? What are the drivers behind our opinions? Is it economic consequences? Extreme weather events? Impending global instability? Or is it simply just greater feelings of stewardship? Lastly, we'll discuss how to move forward. We'll take a long look at the grim future that will occur if we continue business as usual and, contrastingly, we'll learn what tangible changes and benefits can be reaped if we take on an aggressive mitigation attitude.

The bottom line is this: I understand that I'm not going to inspire every one of you to drop everything, link arms with Al Gore, and head off into the Arctic to save the polar bears. Rather, my goal is to illustrate the dynamic

[1] The Journal. (2017). *Survey: Generation Z's Best Students Seek Careers in STEM and Healthcare.* Retrieved from https://thejournal.com/articles/2017/07/12/survey-generation-zs-best-students-seek-careers-in-stem-and-healthcare.aspx

relationship between my generation and the climate crisis and demonstrate that although we millennials have a healthy understanding of climate change, it's still not enough, and we all need to act with much more urgency. Frankly, we're facing a five-alarm fire unfold before our eyes and we're still trying to put it out with a pint glass. Though it may sound cliché, we truly lie at a monumental decision point for the human race: we can either utilize the technology at our fingertips, garner the political will, and develop the innovative solutions to solve the climate crisis, or we can have front row seats to an ecological and humanitarian collapse like we have never seen. If I can just inspire you to rethink your own lifestyle, and consider what you can do to help mitigate this global catastrophe, then all these mocha lattes, avocado toasts, and late nights will have been worth it.

I must also add my disclaimer that I understand the difficulties of writing this chapter. Instead of linking a single subject to the climate crisis like many of my fellow chapter authors, I am going to attempt to analyze an entire generation's opinions toward climate change. This is especially difficult when you consider how much millennials hate being generalized and stereotyped.[2] To aide me in this endeavor, I will be leaning on a bit of playful jest and sarcasm—the intrinsic language of my generation. Overall, I urge you to read this chapter with a critical and serious lens in perusing my analysis of the climate crisis, and a forgiving and softhearted lens in in reading through my attempted humor. To my fellow millennials, keep in mind that I know I cannot generalize all of us. To everyone else, if something rubs you the wrong way, just keep in mind, millennials suck.[3]

Two percent: That's about the proportion of total body mass occupied by your brain. It's also the approximate percentage of the world's population who have naturally green eyes. Two percent also represents the total amount of time spent talking about climate change throughout the twenty-five hours of presidential debates over the course of the last five election seasons. Is this absence of coverage directly related to public belief? Luckily, no, but it certainly doesn't help.

Now, as with most attempts to gauge public opinion, a little more than five minutes of research will likely yield some contradictory results. However, if you spend hours, days, weeks, or months researching this subject like I have, you'll likely arrive at the same conclusion: The large majority of the millennial generation believes that climate change is real, humans are to blame, and something must be done about it. If you dive deeper and look at our proposed solutions for the climate crisis, things get pretty messy pretty quickly, so let's back up and start with what we do agree on.

Depending on which article, poll, or news source you Google or even when you read it, you'll learn that anywhere between 70 and 90 percent of Americans under the age of 35 believe climate change is occurring.[4,5,6,7] Not only that, but of this majority, the bulk believe it is manmade, anthropogenic. Furthermore, we need to make an effort to mitigate the damage. This could be why nearly a third of young voters say energy and environmental issues influence their vote in some way or another.[8] We understand that beyond the obvious environmental concerns, we will have to face economic, public health, and social consequences brought about by a warming planet. Burying our heads in the sand and postponing action on climate will only be sustainable for a few years. Older generations can afford to take this less urgent approach, but the younger will be left to deal with the consequences at some point or another. That, no matter where or when you Google it, is undeniable.

Maybe you're already saying, "Of course you're saying all your peers believe in climate change. You probably only talk to other environmentalists!" And I would tend to agree. All too often, it is easy to put ourselves in an echo chamber, to practice selective hearing. Many times, I fear I do exactly that; it brings about a false sense

[2] Influencive. (2016). *Here Is Why We Can't Generalize Millennials.* Retrieved from https://www.influencive.com/cant-generalize-millennials.

[3] New York Post (2016). *I'm a Millennial and My Generation Sucks.* Retrieved from https://nypost.com/2016/07/04/im-a-millennial-and-my-generation-sucks/.

[4] Pew Research Center. (2016). *The Politics of Climate.* Retrieved from http://www.pewinternet.org/2016/10/04/the-politics-of-climate/.

[5] University of Texas at Austin Energy Poll. (2016). *Millennials' Strong Views on Climate Change and Other Energy Issues Could Drive Presidential Election Results.* Web. Retrieved from https://news.utexas.edu/2016/10/27/millennials-views-on-climate-change-could-impact-election.

[6] YouGov. (2016). *Republicans split by age on climate change.* Retrieved from https://today.yougov.com/news/2016/02/03/republicans-split-age-climate-change/.

[7] Kuppa, S. (2018). *Do Millennials See Climate Change as More Than Just a Meme?* Washington, DC: Johns Hopkins University, Energy Policy and Climate Program.

[8] University of Texas (n5)

of assurance that everyone I encounter is just as politically left-leaning as I am and, therefore, believes climate change is an imminent threat that we must act on with urgency. In this case, though, the notion actually rings true. Okay, okay, not everyone is as much of a radical lefty as I am, but data makes it clear that my generation's concerted opinion on climate change tends resonate throughout the political spectrum, regardless of ideology. To illustrate the magnitude of this, let's dissect it both politically and generationally.

Unsurprisingly, the majority of democratic voters, regardless of age, accept the fact that the planet is warming and that this untimely and dangerous heat wave is due to human activity. It gets really interesting when we look at conservatives. Although approximately 70 percent of republican voters over the age of 45 concede that the climate is changing, only a third in that camp admit that it is due to human activity.[9] Furthermore, there is still a small section of conservative voters over the age of 45 who purport that the climate isn't changing at all ("that deny fundamental and proven science"). I know this is probably not that shocking, but it does draw a stark contrast when compared to young conservatives. In fact, the amount of millennial republicans that believe in anthropogenic climate change is twice that of the baby boomer generation.[10] This might not seem that drastic, but when you consider that we live in a world in which political parties are more blindly allegiant to themselves than high school football teams, any rift like this should be noted.

There is much more to say about how my generation overwhelmingly believes in climate change. Truly, I could go on for days. And if you don't believe me, feel free to dive into the research yourself. However, to avoid the risk of putting you to sleep with more statistics and proof, let's shift our focus and look at why this belief is so strikingly prominent among young people. In my opinion and experience, I highlight three major points of reasoning: economic sensitivity, educational opportunity, and cultural influence. I know this is an extremely broad-brushed approach, but bear with me as we examine each rationale in more detail. . .

They say money makes the world go 'round. Some physicists might disagree and blame it on angular momentum or inertia, but here in America, money is definitely what spins our heads. The benefits of earning it and the consequences of owing it truly define many of the decisions made by millions of people every day. Although our economy has been crafted over hundreds of years by the financial decisions made by older generations, my generation is no stranger to its dynamics. We may not be old enough to have lived through the economic boom following World War II or the following recession of the 1970s, but we are all too acquainted with delicacies of the economy, as illustrated by the energy crisis of the early 2000s, the housing market crash of 2008, and the ludicrous inflation in college tuition in recent years. Furthermore, we understand the link between climate change and the economy because we are the ones who will have to deal with the economic repercussions brought on by a warming planet.

That said, I don't want to scare you off with all the gloom and doom. There are many immense benefits to be reaped from acting on climate as well. Innovative technology, redefining the energy grid, and endless employment opportunities are just a few examples of ways we can take advantage of the climate crisis. We'll get to all the happy stuff in a minute, but first, let's take a quick look at the tangible consequences we could face if we fail to mitigate climate change.

According to a study released by NextGen Climate and Demos, if we don't attempt to act on climate change, if we continue with business as usual, a 2015 college graduate who earns a median income will lose over $126,000 in lifetime income, $187,000 in wealth.[11] For comparison, losses from the Great Recession cost the median-earning, college-educated household $112,000.[12] Thus, the housing market crash and recession that followed

---

[9] Pew Research (n4)

[10] Pew Research Center. (2018). *Many Republican Millennials differ with older party members on climate change and energy issues.* Retrieved from http://www. pewresearch.org/fact-tank/2018/05/14/many-republican-millennials-differ-with-older-party-members-on-climate-change-and-energy-issues/

[11] NextGen Climate & Demos. (2016). *The Price Tag of Being Young: Climate Change and Millennials' Economic Future.* Retrieved from http://www.demos.org/sites/default/files/publications/NGC%20Demos%20The%20Price%20Tag%20of%20Being%20Young.pdf.

[12] Ibid.

in 2008 cost an entire household less than what failing to act on climate change will cost *a single person* of the same affluence. I don't know about you, but as a 2016 college graduate, that fact made me pee my pants a little.

To give you a bit of context, there are roughly 17 million student loan borrowers in the U.S. under the age of 30 that collectively owe approximately $376.3 billion.[13] Now consider that the average entry-level job out of college pays $50,390, and you'll quickly see how and why the millennial generation is so in-tune with our finances.[14] We already have it bad enough trying to compete in an oversaturated job market, attempting to pay back our tens or hundreds of thousands of student loan debt, and now we stand to lose over 100 grand more, just because somebody turned up the thermostat on the planet where we earned those degrees and diplomas. Great...

This $100,000 estimation stems from greater societal costs that will have to be accounted for as we deal with climatic consequences like sea level rise, drought, floods, wildfires, extreme weather events, food shortage, water scarcity, and many, many more. That same study predicts that as a whole, the millennial generation will lose nearly $8.8 trillion in lifetime income.[15] Let me put that in millennial terms you can better understand: Failing to act on climate change might cost my generation the equivalent of 4.4 trillion bottles of craft beer, 927 million lifetimes of Spotify Premium, 517 billion brunch outings, or 7 million roundtrip flights to Europe.[16] Come to think of it, that last one would actually save a lot on CO_2 emissions, but I digress. Scary, right? I'm not saying that everyone in my generation accurately knows the exact extent to which we will suffer immense economic consequences due to climate change, but when people start talking about furthering our already suffocating generational debt, we start paying attention. Now, as promised, on to the happy stuff.

As mentioned before, my generation is facing a highly competitive job market, regardless of college education or lack thereof. Who knew two bachelor degrees would still only qualify one for *unpaid* internships? But what if I told you there is a virtually untouched industry, currently worth over $1.35 trillion worldwide and still growing? Allow me to introduce you to the renewable energy sector. Welcome to the wonderful world full of solar, wind, geothermal, hydro, and many more clean and sustainable energy jobs. Employment opportunities range from wind turbine service technicians to solar panel manufacturers to sustainable architects. In fact, workers in the renewable energy industry outnumber those in the fossil fuel industry by a ratio of nearly five to one.[17] As more and more states join the U.S. Climate Alliance, a bipartisan group of states committed to upholding the objectives outlined in the federally rescinded Paris Climate Accord, the renewable energy sector clearly doesn't intend to slow down anytime soon. The transition to a clean economy is well on its way, and with it come many opportunities for innovation and employment for my generation.

Speaking of the growing renewable energy sector, we are well on our way to redefining the energy grid. No matter where you live, you've likely seen a few of those electric car-charging stations here and there. If you live in the southwestern U.S., you've probably seen some of those weird, white, robotic flower-looking things, those wind turbines, decorate the horizon on your morning commute. If you live in an even more progressive area, you might have even seen some of your neighbors installing solar panels on their roofs. These are all relatively new installments, but they are simultaneously saving the planet as well as our wallets. Of course, all of this assumes you don't live in Petroleum Center, Pennsylvania. (This is seriously what they named the town in the center of the late 19th century petroleum boom in Pennsylvania—creative eh? Look it up if you don't believe me.)

To demonstrate, let's deviate from our narrative for a moment and explore a brief example of how truly immense the economic benefits of renewable energy can be. To do this, let's take a trip back to my home state.

---

[13] CNBC. (2017). *Here's how much the average American in their 20s has student debt.* Retrieved from https://www.cnbc.com/2017/06/14/heres-how-much-the-average-american-in-their-20s-has-student-debt.html.

[14] CNN. (2018). *Starting salary for the class of 2018: $50,390.* Retrieved from https://money.cnn.com/2018/05/14/pf/college/class-of-2018-starting-salary/index.html

[15] NextGen (n12)

[16] Personal Estimates

[17] 2017 U.S. Energy and Employment Report

Georgetown, Texas is considered by many to be the most conservative city in the most conservative county in the most conservative state, and I'd have to agree, having visited my less-than-liberal grandparents there many times. Georgetown has also succeeded in becoming one of the first cities in the country to commit to 100 percent renewable electricity. Wait... What? How can a county that hasn't voted for a democratic president since Jimmy Carter suddenly love the idea of implementing clean energy and ditching fossil fuels? As you may have guessed, it all came down to economics. Georgetown Mayor Dale Ross is a certified public accountant who ran the numbers and discovered that his local economy would prosper by switching to 100 percent renewable energy. The decision to switch was not only more economically affordable as clean energy technology continues to improve, but the economic forecasts of renewables are much more stable and predictable than oil and gas. Surely if a city overwhelmingly full of Trump-supporters saw the benefits of renewable energy, we can confidently say we're well on our way to redefining the energy grid and facilitating a viable job market for the millennial generation.

Outside the various economic consequences and benefits prompted by the climate crisis, I believe my generation holds confidence in our attitude toward climate change because of our education. In addition to an overall improvement of our scientific knowledge base, I argue that this is closely tied to cultural influences and the technological innovations of recent years, which have led to our ease of access to information and an acute familiarity with the internet that we've learned to use to our advantage in understanding climate change.

Have you ever been at a bar, having a fun, trivial argument with some friends, only for the conversation to meet a quick end by someone proudly blurting some internet fact they pulled off their phone? Believe it or not, these incidents are a relatively new occurrence, at least for older generations who did not grow up ever connected to Wi-Fi. In this day and age, we have absurdly immediate access to endless information via the World Wide Web. I would argue that this has greatly improved the overall knowledge base and education of the human race. Just fifty years ago, we landed three men on the moon with technology that was less advanced than the iPhone in your pocket. Since then, both our ability to acquire information and our general knowledge base have escalated tenfold. Highly specialized information that was once only available to those with PhDs and higher education is now taught in middle and high schools nationwide. Not to discredit anyone with a science degree from the 1960s, but you've got to admit we've come a long way since then.

Fifty-something years ago, we were still struggling to comprehend the importance of seatbelts, the dangers of smoking, and how to make a calculator small enough to fit in your pocket. We can't really blame the older among us for failing to understand that we have suffocated ourselves with our own carbon emissions and poked massive holes in the ozone. Well, actually, we can definitely blame those formerly at the helm of Exxon and the Reagan Administration, but I'll resist pointing fingers at the average joe of the 1960's, -70's, and -80's.[18] These days, though, there is no excuse. We can easily access a peer-reviewed study full of factual evidence of climate change on our smartphones in less time than it takes to microwave a Pop-Tart. That said, we can just as easily access a junk-science Heartland Institute climate change denial article, so there must be something more to this.[19]

We millennials were born in the analog era and raised in the digital age. By growing up *with* the internet, we've grown up alongside the development of tactful online-misinformation campaigns, digital trickery, and clickbait sensationalism. This, I would argue, is why we possess a unique level of skepticism that affords us the ability to accurately differentiate truth and fiction better than our elder counterparts.[20] As it applies to climate change, this healthy skepticism has allowed us to come to terms with the harsh reality ahead of us, and has provided us with tangible ways that we can change our future for the better.

---

[18] New York Times. (2018). *Losing Earth: The Decade We Almost Stopped Climate Change*. Retrieved from https://www.nytimes.com/interactive/2018/08/01/magazine/climate-change-losing-earth.html

[19] Heartland Institute. (2018). *Why U.N. Climate Report Cannot Be Trusted*. Retrieved from https://www.heartland.org/news-opinion/news/why-un-climate-report-cannot-be-trusted

[20] Pew Research Center. (2018). *Younger Americans are better than older Americans at telling factual news statements from opinions*. Retrieved from http://www.pewresearch.org/fact-tank/2018/10/23/younger-americans-are-better-than-older-americans-at-telling-factual-news-statements-from-opinions/

Not only has this intimacy with the internet made us more accepting of the dooming climate crisis before us, the cultural influence of science has grown in prominence over the years and that has increased our trust in science. Neil deGrasse Tyson, Bill Nye, and Stephen Hawking have become household names, infecting our cultural sphere and influencing many in our generation to become fascinated with science or, at the very least, pay better attention in chemistry or biology classes. You'd be hard pressed to find someone in my generation who didn't feel at least a little excited every time they walked into their third-grade classroom and saw a TV at the front, with The Science Guy's face plastered across the screen.

Science and nature are cool again—I'll do you a favor and omit the awful global warming pun I thought of. Programs like *Planet Earth*, *Cosmos*, and *The Blue Planet* are all regularly chosen entertainment by those in my age demographic. What's more, films like Leonardo DiCaprio's *Before the Flood*, Jeff Orlowski's *Chasing Ice*, and Al Gore's *An Inconvenient Truth* not only entertain but, more importantly, educate. In a way, you might argue that we have been tricked into learning about science, nature, and climate, but if the outcome is a generation that not only trusts but understands the importance and necessity of science, there's nothing to complain about.

Hopefully, I've gotten my point across: our generation is very intertwined with the climate crisis. Whether it is due to our economic sensitivity, our ability to differentiate real news from fake news, or our familiarity with and trust in science, we really do get it. Now what though?

Whether you're in my generation, my parents', or my grandparents', there are steps we can all take to combat climate change. The both horrific and beautiful thing about the climate crisis is that change will affect it, for the worse or for the better.

To conclude, I'm going to make one more assumption: You probably picked up this book hoping to read some diverse perspectives on the climate crisis, from people much more qualified than me. Maybe you wanted to educate yourself about the various ways we are abandoning the climate and ruining the Earth. But I hope after reading my chapter, you have a bit more optimism moving forward. I know things don't look so good right now, but I truly believe my generation and the even younger people of today can make a change for the better. The information and technology are available. We just need to seize the opportunity and take action, even if that means we won't be around long enough to reap the benefits. We must be the generation who plants trees, even if we will never sit in that shade. Then hopefully, a hundred years from now, instead of reading *Climate Abandoned*, people can pick up a copy of *Climate Saved*—and that's where you *will* read warranted praise for my generation.

Climate Crisis "Keep/Stop/Start" Actions

Below for your consideration, are several author-suggested thought, behavior, and action items. Action nourishes inspiration! For a further explanation of the Keep/Stop/Start organizational tool, please see How to Read This Book.

KEEP:

1) *Keep being informed.* The fact that you are reading this book leads me to believe you are proactive in educating yourself about the issue of climate change and learning what we can do to fix it. Keep doing that. With the immense amount of information available today, you can remain vigilant in affording your attention to this issue and even press others to be informed as well. Read articles. Watch documentaries. Research. Talk to stakeholders. Attend lectures. Any and all of these will help you stay up to date on the science, politics, and policy of climate change and monitor the developments over time. Make sure to absorb at all information with a critical eye. As noted, there is a large subsection of fundamentally

incorrect information out there, propaganda used by corporations and politicians to sway voters into thinking climate change isn't a big deal. I'm not suggesting that you need to personally verify the validity every article you read, but understand that this is a politicized issue and that it is covered accordingly.

2) *Keep reducing, reusing, and recycling.* I'll also assume you've already heard the benefits of the three R's; however, we must emphasize the order in which they come. In all cases, we should reduce our consumption habits. This is easier said than done in our capitalistic, buy-now-waste-later society, but reducing the amount of things we consume in the first place is the best step toward reducing emissions, pollution, and waste. Next, we should aim to reuse things as much as possible before tossing them out. This includes anything from plastic water bottles to donating used clothes to a charity or thrift store. Lastly, once we reduce consumption and reused what we bought, we should recycle. Recycling is a great way to repurpose raw materials for re-consumption, but the fact of the matter is that our recycling infrastructure is not large enough to support our extensively wasteful habits. It is great to take care of the three R's, but by no means are these and end-all substitute for landfills.

STOP:

1) *Stop thinking someone else will solve this problem.* This may be the most important thing in this chapter, because it is the biggest dilemma in the entire issue of climate change. There is so much cognitive dissonance with the climate crisis that even those who understand how immense of a concern it is can easily put it out of sight and out of mind. "Well, I haven't seen any extreme weather events yet... I haven't dealt with any food or water shortages... This summer hasn't even been that hot." I hear these classic paradigms nearly on a weekly basis from climate change believers and skeptics alike. It is not called state or country warming; it is called global warming because it affects the entire planet. Regardless of which side of Earth you happen to occupy at the current moment, you will be affected in some way or another. Thus, it is our collective responsibility to solve the climate crisis. Furthermore, experts predict that we only have about 12 years to get our act together if we are going to limit warming to only 1.5°C. That's roughly 135 months; 586 weeks; 4100 days; just less than 100,000 hours to avoid global catastrophe. Saying we need to act urgently is putting it lightly.

2) *Stop wasting (water, food, energy, etc.).* Surely the other brilliant writers in this book have already touched on this, but it never hurts to reiterate. Although it is true that many of our resources are renewable, if we don't use them sparingly, they will be quickly depleted in the near future. Fertile agricultural land will be even harder to come by in the coming years as we deal with more drought and extreme weather. Water resources are already running low, and with a continually growing population, we must work harder to conserve what we have. As mentioned before, we are well on our way to redefining the energy grid and developing the renewable energy sector, but until we have a fully 100 percent carbon-neutral energy grid, we need to resist wasting energy and continuing to emit harmful greenhouse gases into the atmosphere. Reducing our consumption is perhaps the easiest indirect way we can combat climate change.

START:

1) *START being politically active.* Ok scratch what I said before, *this* might be the most important thing you read in this chapter. Fortunately, or unfortunately—depending on how you look at it—we live in a

democracy. Unlike a dictatorship or monarchy with top-down regulatory control, a democracy requires citizen participation to facilitate the proper checks and balances of government. Elected officials (should) answer to their bosses: the people that elected them. As such, it is part of our civic duty to engage in governmental matters. At a bare minimum, this means voting; **and yes, I'm speaking directly to you, millennials** (see chart to right). It is outright embarrassing to see how many people in my generation complain about politics on a daily basis yet have never bothered to cast a vote. If you get nothing else out of this chapter, please, for the love of god, just do me a favor and vote in the next election. Assuming you do that, in considering who you will vote for, take into account their stance on climate change. If you vote for those who are concerned about the climate crisis, you will send an unwavering message to our institutions that this is an important issue on which you insist they take action. Besides voting for candidates who pledge to act on climate, you can also contact your representatives by phone or letter or email. As mentioned, whether on the local, state, or federal level, elected officials are obligated to hear your concerns—trust me, I work for one. Furthermore, be active in persuading the public on climate policy. Write letters to the editor. Organize meetings. Make speeches. Make waves—before the waves of the rising sea levels overwhelm us all...

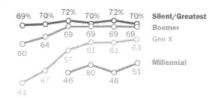

Voter turnout rates in presidential elections, by generation

% of eligible voters who say they voted

1996 2000 2004 2008 2012 2016

Note: Eligible voters are U.S. citizens ages 18 and older. Millennial refers to the population ages 20 to 35 as of 2016. Source: Pew Research Center tabulations of the 1996-2016 Current Population Survey November Supplement (IPUMS).

PEW RESEARCH CENTER

2) *Start gaining more self-awareness.* Don't take this the wrong way, but many of us, myself included, have a distorted view when it comes to how we personally affect the environment, and we seldom take the time to mitigate our own habits. As noted before, recycling and reusing are great, but don't let these actions satiate your desire to save the planet. A great way to become more self-aware is to calculate your carbon or ecological footprint. Do a quick Google search, and you'll find many online resources to aid in this endeavor. Quizzes will ask you things about your dietary, transportation, and energy consumption habits and calculate your estimated annual CO2 emissions. Many even show how many Earths we will need to support our population if everyone in the world shares our consumption habits. That was a real wake-up call for me.

3) *Start holding corporations accountable.* Aside from elected officials that fail to regulate them, the companies that fail to be regulated should be one of the biggest targets of climate activism. 100 companies are responsible for 71% of the global emissions since 1988.[21] As meaningful as it is to curb our own life-styles (as described above), we will need to severely alter the political and economic strategies of the corporations that monopolize our day to day lives if we are ever going to have a chance at solving the unnerving crisis before us. As much as I might like it to, American capitalism isn't going away anytime soon. Therefore, we must take on an attitude of conscious consumerism that demands change from the bottom up. Furthermore, we must do whatever we can to lobby the levers of power to act in the betterment of humanity—not their shareholders.

[21] The Guardian. (2017). *Just 100 companies responsible for 71% of global emissions, study says.* Retrieved from https://www.theguardian.com/sustainable-business/2017/jul/10/100-fossil-fuel-companies-investors-responsible-71-global-emissions-cdp-study-climate-change

~ ~ ~

Kevin Quinn. M.L.S. is a conservationist with a background in environmental science and policy. He holds bachelor's degrees in Environmental Studies and Ecology & Evolutionary Biology from the University of Colorado, Boulder and has recently earned a Master of Legal Studies in Environmental and Natural Resources Law and Policy from the University of Denver. Currently, he serves as Chief of Staff to Colorado State Representative Chris Hansen, where he supports a progressive legislative agenda that aims to promote clean energy, reduce carbon emissions, and protect the natural environment, among many other important issues. Read more at: www.climateabandoned.com

Chapter 26

Climate Crisis & Health: A Physician's View

by
Kasper Eplov, M.D.

"The evidence is overwhelming. Climate change endangers human health. Solutions exist, and we need to act decisively to change this trajectory." **– Margaret Chang,** physician, WHO Director General

"Many things have to change, but above all, it's we humans who need to change." **– Pope Francis**

"I don't give a **** if we agree on climate change," Arnold Schwarzenegger said in 2015, prior to the Paris Climate Summit. He went on to discuss a certain scenario: Imagine being required to stay inside one of two closed garages for one hour. In one, a car with a combustion engine is running on fossil fuels; the other houses an electric vehicle. Which garage would you choose? I think we all intuitively appreciate that being enclosed in the garage with the combustion engine would be hazardous.

It is a telling analogy, as the atmosphere surrounding our planet is really a confined space, just like the garage. When we burn fossil fuels, gases remain in the atmosphere, and the resulting particle pollution is inhaled, ultimately producing adverse health effects. Arnold pointed out that over six million people die each year due as a result of air pollution. Whether we are in denial about climate change or not, switching to green technology will obviously solve some air pollution issues.

Besides the fact that The Terminator can be quite intimidating in looks alone, there are many good reasons to pay attention to him. First of all, Arnold holds the title of Mr. Universe; in fact, he won that hard-earned title twice! More importantly, he is a progressive voice on environment and climate change, and he hails from an unlikely source, the Republican Party. As California governor, Mr. Schwarzenegger imposed some of the most progressive, far-reaching policies in the U.S., and these pushed California to the forefront of the sustainable energy revolution. The state has demonstrated that there is no contradiction between the transition to green energy and economic growth. Rather, it is quite the contrary.

From 400km above, at a speed of roughly 17,000 miles an hour, the house-sized International Space Station orbits Planet Earth. Witnessing the beauty and vulnerability of our little place in the universe is a profound experience, and many of the 556 people who have been to space describe a deep feeling of awe and compassion for our common home.[1] Some of these testimonies can be found on the Overview Institute website, which is dedicated to communicate the sentiments of astronauts who've journeyed to space.

From the space station, one would likely notice the thin layer of atmosphere surrounding Earth, the sea of air in which we all live our daily lives. That vantage point gives a clearer perspective, an appreciation of the fact that we all need a habitable environment with clean air, food, water, and survivable temperatures. Running a combustion engine inside that space station would certainly be unhealthy and dangerous, and the same holds true for our planet.

Yes, the bubble in which we live is much bigger than a garage or a space station. Nevertheless, we have the same needs as the astronauts on the station, and climate change is on a trajectory to disrupt the process, fundamentally making life on our planet harder.

---

[1] https://www.worldspaceflight.com/bios/stats.php

229

Climate Change Is a Medical Emergency

Esteemed medical journal *The Lancet* formed a commission on climate change, an academic, international cooperation between researchers. These brilliant teams have contributed to two reports so far, one in 2009 and one in 2015, and the commission plans to follow up on the progress since the Paris Accord and evaluate at five-year intervals moving forward. Co-chair Professor Hugh Montgomery stated in 2015, "Climate change is a medical emergency," and the report that year concluded that the risks associated with climate change were already unacceptably high then and that transition toward a fossil fuel-free society was needed.

Climate change affects human health in a number of ways, both directly and indirectly, and many determining factors will decide just how awful the side effects will be. Direct effects include direct exposure to extreme weather events such as storms, drought, floods, heatwaves, and wildfires. Clearly, being in the midst of a wildfire or tornado can be life threatening, an obvious health threat. As climate change worsens and extreme weather events continue and increase in ferocity and frequency, we are doomed to see an increasing number of casualties.

Another direct effect of climate change is its impact on the quality of our air and water supply. For example, we experience changes in land use, due to rising sea levels. Ecological changes also occur, such as distortions in the balance of marine life due to increase in sea temperature and ocean acidification. This severely disrupts the fishing industry, among many other issues. Flooding can lead to contamination of freshwater resources, as well as epidemics of waterborne infectious disease. Wildfires can lead to severe air pollution. And, sadly, all of this can lead to mental health problems for humans, including depression and anxiety.

For the people who populate Earth, much suffering can ensue. The severity of this suffering is often based on many determining factors. For example, infants and the elderly are more vulnerable to heat stress. Beyond age and gender, socioeconomic position also matters. Do you have resources to flee or evacuate if you must? Do you have air-conditioning? Persons with preexisting medical conditions are even more at risk. The availability and resilience of the healthcare system, especially its robustness or lack thereof, determines whether it can function properly under extreme circumstances, and we all know the U.S. healthcare system has been sick for quite some time. In all of these ways and more, climate change directly affects human health, from the quality and security of our food and water supply, to the mass spreading of infectious disease, to the safety of the very air we breathe.

Infectious Disease

What is an *infection*? Every day, we are exposed to a plethora of microorganisms. Some are harmful; others inhabit our bowels and skin and benefit us. In fact, you might be surprised to know that you have more bacteria on your skin and in your bowels than you have cells in your body! Your immune system is constantly at work to maintain balance and keep all these microorganisms in check. When we are presented with a new microorganism or virus, this is known as an infection. The immune system is forced battle that pathogen then store its ID so it can be recognized and attacked if it happens to return in the future. This is the mechanism behind vaccinations: A small amount of pathogen is presented to the immune system, thereby equipping it with a ready response if the infection happens to recur.

The transmission of infections can be direct or indirect. We've all caught a common cold, which is caused by a virus and spread through direct contact such as being sneezed on or touching door handles or other objects recently touched by an infected person. You can also ingest the pathogen by drinking contaminated water or eating bad food.

Some infectious diseases are spread by *vectors*, other animals, usually parasites, which transfer the infection. The most common vectors are blood-feeding *arthropods* (insects, ticks, or mites) that carry the disease from one host to the next, usually with amplification of the virus or bacteria in the vector. Others include flies, aquatic snails, fleas, and bugs. In this way, disease can be transferred to humans from animals and then from human to human, particularly when the vector bites one victim, then another.

Vector reproduction and parasite biting generally increase with increasing temperature. Due to climate change, living conditions for some vectors improve, facilitating their spread into new locations. Increased precipitation is also a factor. Malaria is one of the most common vector-borne diseases. It´s transmitted by different species of mosquitoes and is a major health problem in large parts of the world. Other climate-related diseases include Lyme borreliosis, tick-borne encephalitis, Chikungunya, West Nile, malaria, dengue fever, Schistosomiasis, Leishmaniasis, and Hantavirus.

As new populations with no prior immunization are exposed, the disease burden rises significantly. For example, Australia and New Zealand may become endemic to dengue fever. Several studies predict that many more people will be affected by malaria in the near future as a result of increased geographical spreading of the vector, particularly in Africa.[2,3,4] Likewise, and just as horrifying, if climate change continues to worsen, dengue fever may affect up to 50-60% of the global population in 2085, compared to the 35% of the population who are expected to suffer if the climate remains constant.[5] The result of the increased disease burden will be increased strain on health infrastructure in regions that already lack ample health services.

Mapping the connection between climate and disease is no easy task. Future prognoses can only be based on approximations, and exact outcomes are difficult to predict, if not impossible. The World Health Organization (WHO) and World Meteorological Organization (WMO) published an atlas on health and climate in 2012, mapping the relationship between health and weather patterns. Clearly, health and meteorological data can be linked to climate change; there is an obvious relationship between epidemics and weather patterns at the time when those epidemics occur.

The spreading of vectors is based on many factors. In addition to climate change, another driver behind the spread of disease are our modern global travel patterns. Zika is just one example of a virus that has spread geographically in recent years, rapidly affecting a large population. The spread of West Nile fever into the United States beginning in 1999 is another. Ticks carrying Lyme disease and encephalitis have been seen even farther north, in Canada and Sweden, due to warmer climates there. As conditions for the vectors improve, more and more cases can be mapped in more and more parts of the world that were not prone to these diseases before.

Diarrhea is another major problem, as it is the second leading cause of death of children under five, resulting in 525,000 deaths per year.[6] Cholera is a very severe form of diarrhea, and while it only occurs sporadically in the developed world, epidemics of it are common in areas with poor sanitation and unsafe water use. Globally, one in ten persons lacks access to safe water, and one in three has no access to a toilet.[7] Believe it or not, there are more mobile phones in the world than there are toilets![8] This can often result in the transmission of the waterborne diseases explained above.

In West Africa, a meningitis belt stretches from the west to the east. This region sees regular epidemics of the disease, such as an outbreak in 1996 that affected over 250,000 people and resulted in 25,000 deaths.[9] The epidemics are related to the amount of dry dust in the air, as the pathogen, meningococcus, is carried on these

---

[2] Lindsay SW, Martens P. 1998. Malaria in the African highlands: past, present and future. Bull World Health Organ 76(1):33–45.

[3] Leedale, J., Tompkins, A., Caminade, C., Jones, A., Nikulin, G., & Morse, A. (2016). Projecting malaria hazard from climate change in eastern Africa using large ensembles to estimate uncertainty. Geospatial Health, 11(1s). https://doi.org/10.4081/gh.2016.393

[4] Martens, W J. "Health Impacts of Climate Change and Ozone Depletion: An Ecoepidemiologic Modeling Approach." Environmental Health Perspectives 106.Suppl 1 (1998): 241–251. Print.

[5] Hales, S.L., Wet, N.D., Maindonald, J.H., & Woodward, A. (2002). Potential effect of population and climate changes on global distribution of dengue fever: an empirical model. Lancet, 360 9336, 830-4.

[6] http://www.who.int/news-room/fact-sheets/detail/diarrhoeal-disease

[7] WHO, UNICEF, Progress on sanitation and drinking-water: 2015 update and MDG assessment

[8] https://news.un.org/en/story/2013/03/435102-deputy-un-chief-calls-urgent-action-tackle-global-sanitation-crisis#.VGvCDTTF9EI

[9] https://www.cnn.com/2016/05/03/health/meningitis-elimination-vaccine/index.html

particles. For this reason, meningitis infections are expected to increase in the future, because drought brings the onset of more dust particles.

In addition to the diseases we already know about, there is always a chance that new infections can arise. For example, there is a risk of pathogens emerging as permafrost melts due to global warming. In the summer of 2016, there was an outbreak of anthrax in Siberia.[10] Out of the dozens of people who were hospitalized, one boy lost his life, and it is worth noting that 2,300 reindeer also died from the infection. That outbreak was the first in seventy-five years, and it is believed that the source was spores of bacteria from the melting permafrost layer. It may be an early warning that ancient bacteria or viruses can be released from the frozen layer, rendering us vulnerable to a whole buffet of infectious diseases.

Flooding and Storms

"Runoff and flooding resulting from increases in extreme precipitation, hurricane rainfall, and storm surge will increasingly contaminate our water sources." – **The U.S. Environmental Protection Agency**

Water is fundamental to life. Only 6 percent of water on Earth is fresh, and almost 70 percent of total water usage is for irrigation.[11] As temperatures increase, more irrigation is required, so this percentage will inevitably go up. The water level in the large aquifers of the world is diminishing, and a global water crisis is on its way, as temperature increases and population growth increase the demand for fresh water. At the same time, more drought, changed precipitation patterns, and disappearance of glaciers disrupt otherwise reliable water sources.

If you hang your clothes on a line, you know they dry more quickly on hotter days. This is because of the so-called steam pressure of the air; hotter air contains more humidity. Downpour occurs when warm air is cooled and, therefore, loses its humidity via rain. Due to increasing temperatures, air contains much more humidity, so downpours are more extreme when they do occur. The result is extreme precipitation, with months' or even what seems like years' worth of rainfall over the course of a few hours, resulting in flash floods or rain bombs. During recent decades, we have seen a significant increase in extreme precipitation, in direct correlation with rising overall temperature.

There is risk of drowning and other trauma during a flood, but apart from the devastation and destruction by the flooding itself, freshwater resources are often contaminated. Developed countries mostly enjoy reliable and safe distribution of fresh water, with sanitation systems in place to ensure that it is not contaminated. However, many people in large parts of the world are dependent on open water, and these areas are particular vulnerable to flooding and the adverse effects of it, especially in regions with unsuitable sanitation and where open defecation is common. Sewage water may also overflow and contaminate water resources with poisonous and infectious agents, putting cleanup crews especially at risk. In the aftermath of flooding, outbreaks of dysentery, E. Coli, cholera, Leptospirosis, and Cryptosporidiosis are common, just to name a few. Those who reside in buildings that were once flooded may also suffer aggravating respiratory disease such as pneumonia or asthma, due to mold and the indoor climate of a flood-damaged building.

Most heat from global warming is trapped by our oceans, and this causes the water to expand, leading to rising sea levels. This is simple physics, and the prognosis for rising sea levels can be calculated rather precisely. Added to this is the flow of water into the sea from melting glaciers. If all the ice on Greenland and Antarctica were to melt, the sea level would rise seventy meters, enough to flood all major cities on Earth, particularly since

[10] https://www.npr.org/sections/goatsandsoda/2016/08/03/488400947/anthrax-outbreak-in-russia-thought-to-be-result-of-thawing-permafrost
[11] Food and Agriculture Organization of the UN, Water at a Glance The relationship between water, agriculture, food security and poverty. http://www.fao.org/docrep/016/ap505e/ap505e.pdf

many are located in coastal regions. Naturally, this occurs over a prolonged period, but we are already seeing rising sea levels.

As almost half of Earth's population lives within sixty kilometers of one coast or another, dire consequences and mass immigration may be next on the agenda. Even modest sea level rise has a major impact, because it can lead to storms and flooding. When seawater invades coastal lands, freshwater is often contaminated, and farmland may be useless, impacting the resilience and livelihood of the local population. The WHO estimates that the number of people affected worldwide by flooding will more than double by 2030: 77 million a year as compared to 29 million today.

Seawater temperature is the main driver behind cyclones and hurricanes: The hotter the water gets, the more severe the storms are. All extreme weather events increase the risk of trauma and death and the destruction of infrastructure and society institutions, including health services, leaving them incapable of functioning when we need them most.

Hotter seawater is shown to increase algae growth, and these harmful blooms are associated with disease. Some harmful algae are neurotoxic, such as the pseudo nitzschia, which produces demonic acid, a potent neurotoxin that causes amnesiac shellfish poisoning in humans. Marine life is also exposed, and syndromes observed just off the California coast include epilepsy, hippocampal (the memory center of the brain) dystrophy, and impact on memory and learning. In 2015, fishing was temporarily prohibited in a large area spanning from Mexico to Alaska due to these toxic algae.[12] Also, cholera has long been associated with the seasonal algae and plankton blooms, and increasing temperatures exacerbate these blooms, resulting in increased risk of epidemics.[13]

Heatwaves and Drought

The climate crisis also affects human health when we are exposed to more frequent and more intense heat events. As temperatures increase, weather patterns change, due to increased moisture in the air and changed wind patterns. This leads to more frequent and more severe heatwaves, as well as prolonged drought. Since 2000, record-breaking temperatures have become the new norm, and almost every year is hotter than the last. This trend is alarming!

When it comes to human health, one of the most serious complications associated with rising temperatures is heatstroke. When exposed to extreme heat, the body can be unable to thermo-regulate and simply overheats. Heatstroke is a serious and life-threatening condition if emergency treatment is not available. Infants and the elderly are more at risk, as are those with preexisting medical conditions, especially cardiovascular disease and illnesses that affect plasma volume or ability to sweat. Outdoor workers are also at risk, as they are very exposed. The mentally ill, underprivileged, and impoverished can also be vulnerable because many lack access to air-conditioning during heatwaves.

Certain urban areas are in even more danger, as these urban heat islands tend to absorb more heat, due their proximity to and the use of certain building materials such as asphalt, concrete, and steel. In fact, these areas can be up to twelve degrees Celsius (53.6°F) hotter than a suburban location under similar conditions, so inhabitants of these areas suffer heatstroke more often.

During a heatwave, an increase in death tolls is often seen, particularly in older age demographics. Some studies show that for every 1°C (32°F) increase above the median climatic temperature, there is a 5-8 percent

---

[12] https://www.nationalgeographic.com/magazine/2016/09/warm-water-pacific-coast-algae-nino/
[13] Epstein PR. Algal blooms in the spread and persistence of cholera. Biosystems. 1993; 31(2-3):209-21.

increase in mortality of those 65 years of age or older.[14,15,16] As an example, during the 2003 heatwave in Europe, more than 70,000 died.[17] On hotter days, air quality is generally worse, because harmful particles build up in warmer air. emergency rooms can vouch for this, as they tend to treat respiratory problems far more as the mercury rises. This is a major public health issue, as the WHO estimates that asthma affects roughly 235 million people worldwide.[18]

Prolonged heat with decreased precipitation results in more drought. This means areas already negatively affected will face more of a burden, and millions may face consequential famine. Deserts grow and farmland shrinks, leading to malnutrition, which already cause 3.1 million child deaths annually.[19]

Another effect of drought is the spreading of many of the waterborne diseases we've already discussed. As clean water resources dry up, desperate people seek to quench their thirst with even contaminated water. This can lead to diarrhea, which is highly dangerous for those who may already be starving. Waterborne infections due to unsafe water sources and inadequate sanitation, as well as increased infection, are especially heinous in crowded areas where people have been displaced.

Drought is also a driver for panic, which often results in conflict and regional war. The war in Syria started following a long period of drought, and it´s been suggested that climate change has played a role in the tragic conflict. According to some models, large parts of the Middle East and Asia may become uninhabitable due to extreme heat even before the year 2100, and these are already war-torn regions.[20]

Mental Health

Victims of natural disaster are at increased risk for anxiety, depression, post-traumatic stress disorder (PTSD), and suicide. Somewhere between 25-50 percent of people exposed to extreme weather see declines in their mental health as a direct consequence.[21] Following Hurricane Katrina, nearly half of the survivors developed mood or anxiety disorders, one out of six was diagnosed with PTSD, and the suicide rate and suicide ideation more than doubled.

Mental health can be defined as the ability to cope with the daily stressors of life and function within the community with social behavioral and physical wellbeing. Mental illness affects an individual's thoughts, feelings, and behavior, as well as their ability to function normally. The immense stress from an extreme weather event can lead to mental difficulties for anyone, but those with preexisting mental problems are especially vulnerable. Mental illness may lead them to ignore obvious warning signs, struggle during evacuation, or fail to react appropriately, hydrate, or seek shade or nutrition.

The impact of climate change on mental health may be immediate or more gradual. For example, the slow-but-sure worsening of living conditions due to sea level rise can be a chronic stressor. Climate change leads to a myriad of triggers: more infections, worsening of asthma, and even changes in the ecosystem. Temperature rise and acidification of oceans render fishing impossible in many locations, threatening food security and the livelihood of affected populations.

---

[14] Anderson, Brooke G., and Michelle L. Bell. "Weather-Related Mortality: How Heat, Cold, and Heat Waves Affect Mortality in the United States." Epidemiology, vol. 20, no. 2, 2009, pp. 205–213. JSTOR, JSTOR, www.jstor.org/stable/20485691.

[15] Rupa Basu, Jonathan M. Samet; Relation between Elevated Ambient Temperature and Mortality: A Review of the Epidemiologic Evidence, Epidemiologic Reviews, Volume 24, Issue 2, 1 December 2002, Pages 190–202, https://doi.org/10.1093/epirev/mxf007

[16] Benmarhnia, Tarik & Deguen, séverine & S Kaufman, Jay & Smargiassi, Audrey. (2015). Vulnerability to Heat-related Mortality: A Systematic Review, Meta-analysis, and Meta-regression Analysis. Epidemiology (Cambridge, Mass.). 26. 10.1097/EDE.0000000000000375.

[17] Stott, P & Stone, DA & R Allen, M. (2005). Human Contribution to the European Heatwave of 2003. Nature. 432. 610-4. 10.1038/nature03089.

[18] http://www.who.int/respiratory/asthma/en/

[19] https://www.theguardian.com/global-development/2013/jun/06/malnutrition-3-million-deaths-children

[20] https://news.nationalgeographic.com/2017/08/south-asia-heat-waves-temperature-rise-global-warming-climate-change/

[21] Jyotsana Shukla, "Extreme Weather Events and Mental Health: Tackling the Psychosocial Challenge," ISRN Public Health, vol. 2013, Article ID 127365, 7 pages, 2013. https://doi.org/10.1155/2013/127365.

There may be indirect effects following displacement or loss of property, which can lead to loss of identity and sense of self. Loss of social identity and diminished social interaction can lead to alienation, and the increased psychological stress on the community may result in a range of adverse effects, such as more substance abuse and domestic violence. Large groups of disillusioned individuals may find themselves immersed in violent conflict, civil wars, or even extremism and even terrorism. Some may feel discouraged to participate in normal life activities such as starting a family. It may even drive some to take their lives. A newly released study linked 60,000 suicides in India over the last 3 decades to climate change.[22] The victims were primarily farmers who lost their land ended up in a vicious circle of loss, debt, and despair.

It has even been said that many climate scientists suffer from *pre*-traumatic stress disorder, because they foresee a future with serious and even catastrophic consequences to our planet and civilization. Many in the field have heard their motives questioned and have been mocked and ridiculed and even threatened for doing their jobs. It is easy to imagine the feeling of powerlessness and frustration over the slow pace at which the world is reacting to the very real threat of climate change and to understand the negative psychological effect those in the know experience.

Health Benefits from the Sustainable Energy Revolution on Health and Climate-Resilient Healthcare

Solving the climate crisis has a number of potential health benefits. First and foremost, the health sector needs to step up to the plate and take the lead on this issue. They must recognize the harmful health effects of climate change and advocate the transition to green energy. The health sector itself is a major part of the economy and is, therefore, responsible for significant emissions. This falls in alignment with the do-no-harm principle; as they are aware of the adverse health effects of burning fossil fuels, they have a moral obligation to become carbon neutral.

The medical profession has addressed many major health issues historically, such as HIV/ AIDs, polio, and the health problems associated with tobacco use. In the case of the latter, they took on major economic interests that parallel those of the fossil fuel industry. *Merchants of Doubt* documents that the fossil fuel industry is, in fact, following the same playbook as big tobacco and even hired the same public relations company to represent them. Their primary objective is to cast doubt on the science behind climate change.

This is done through a web of so-called experts and think tanks that pose as independent analysts or experts on climate change; in reality, these are just talking heads, bought and paid for by the fossil fuel industry to muddy the waters and cast doubt on science. As the *Lancet* commission concludes, the risk from climate change is unacceptable, and the medical profession must counter the false narratives with facts about how climate change affects human health.

Moving toward a fossil fuel-free society means less particle pollution. This is perhaps the biggest health benefit, as this pollution accounts for around three million deaths each year. There are also health benefits for individuals who choose a carbon-neutral lifestyle: If you choose active transportation over driving your vehicle, that exercise will lower your risk for cardiovascular disease, as well as save money and reduce your carbon footprint. The same is true for choosing a climate-friendly diet: Eating less meat lowers the carbon footprint, and it is less expensive to eat healthier. It has been suggested that an individual on vegetarian diet has a 10-12 percent lower risk of cancer, and lower cholesterol and weight loss and control are additional benefits.[23]

---

[22] A. Carleton, Tamma. (2017). Crop-damaging temperatures increase suicide rates in India. Proceedings of the National Academy of Sciences. 114. 201701354. 10.1073/pnas.1701354114.
[23] Lanou, Amy Joy, and Barbara Svenson. "Reduced Cancer Risk in Vegetarians: An Analysis of Recent Reports." Cancer Management and Research 3 (2011): 1–8. PMC. Web. 7 Sept. 2018.

Moving the health sector toward green energy is especially advantageous in parts of the world where there is little or poor electricity. Many clinics in the developing world have no power at all. Installing renewable, off-grid solutions such as solar panels will improve health services immensely. Health systems must be climate resilient, able to withstand the sudden stress of an extreme weather event, so they can function when they are most necessary. Early warning systems should also be put in place, so services can be prepared and operational during heatwaves, floods, wildfires, and other natural disasters. In the U.S., the CDC has formed the "Building Resilience Against Climate Effects" (BRACE) framework, which is designed to address the challenges of climate change to the healthcare system. The purpose is to adapt the healthcare system to meet challenges from climate change. This five-step loop includes the following:

Step 1: Anticipate Climate Impacts and Assessing Vulnerabilities Identify the scope of climate impacts, associated potential health outcomes, and populations and locations vulnerable to these health impacts.

Step 2: Project the Disease Burden Estimate or quantify the additional burden of health outcomes associated with climate change.

Step 3: Assess Public Health Interventions Identify the most suitable health interventions for the identified health impacts of greatest concern.

Step 4: Develop and Implement a Climate and Health Adaptation Plan Develop a written adaptation plan that is regularly updated. Disseminate and oversee implementation of the plan.

Step 5: Evaluate Impact and Improve Quality of Activities Evaluate the process. Determine the value of information attained and activities undertaken.

The healthcare system is a significant part of societal economy, and moving it away from fossil fuels will make a significant, positive impact on emissions. Healthcare professionals have high credibility in society, and this should be used to the fullest to advocate and pioneer the transition toward sustainable energy. The health sector is currently responsible for a substantial amount of pollution, resulting in adverse public health effects. This is hypocritical, unacceptable, and contrary to the medical philosophy of doing no harm, and it must be reversed.

Climate Crisis "Keep/Stop/Start" Actions

Below for your consideration, are several author-suggested thought, behavior, and action items. For a further explanation of the Keep/Stop/Start organizational tool, please see How to Read This Book.

KEEP:

1) *Keep buying food locally.* Farmers markets, community gardens, and locally sourced grocery stores have recently grown in popularity as many people begin to understand the detriments of buying food from far away. In America, on average, food travels roughly 1,750 miles from farm to market and with trucks that average 5.5 miles per gallon, that means that for every 100 pounds of non-locally sourced food that you buy, you are indirectly using about a gallon of fuel.[24] Though this might not sound like much, it certainly adds up. Furthermore, locally-sourced food typically has more nutrients, is less likely to be contaminated,

---

[24] http://www.washingtonpost.com/wp-dyn/content/article/2008/03/07/AR2008030702520.html

and is more flavorful.[25] Overall, eating locally improves your personal health, your community's health, and helps mitigate climate change.

2) *Keep exercising often.* Though your personal time at the gym may not have any direct benefit for the climate, it can help your body become more resilient to the effects of climate change. As described above, there are many tertiary effects of climate change that can negatively impact your health, such as heat stroke, vector-borne diseases, and effects on mental health to name a few. Getting in a regular workout— whether lifting weights, cardio, or yoga—can strengthen your immune system, cardiovascular system, and mental health which will make you all the more resilient as the effects of climate change continue to worsen.

STOP:

Stop believing the first thing you read in regards to health. Other than the debate about what to do about climate change, the debate about what's healthy and what isn't is almost as polarizing. There is so much misinformation about what diets, exercises, and nutritional supplements are best and as such, you need to be vigilant in your research. Furthermore, after researching what affects certain lifestyle changes can have for your personal life, look into how they affect the planet. For example, palm oil has recently stormed the market as an end-all solution to improve your cholesterol and cardiovascular health, however it has simultaneously decimated tropical rainforests and biodiversity around the world. Specifically, from 2001-2010, land-use carbon emissions from palm oil production in Indonesia alone was the equivalent to the emissions of 45-55 million cars.[26]

Another example is that of microbead-based toiletry products. Though these particular products can improve your health through boosting the effectiveness of shampoos, toothpastes, and facial scrubs, the microbeads are made up of fossil fuel-based plastics which not only produce carbon in manufacturing, but pollute the environment through our waterways. I know it can be difficult to determine which things are the healthiest, let alone what is also best for the planet, but a little research can go a long way, and the planet will thank you for it.

START:

1) *Start eating more fruits & veggies* and eating less meat. We all already know how beneficial a balanced diet of fruits and vegetables can be for your personal health, but adding more produce to your diet and reducing your meat consumption can also help mitigate climate change. Aside from the excess methane released from raising animals for food, it is simply inefficient in terms of energy. It takes about five to seven kilograms of grain to produce one kilogram of beef, which requires additional water and energy to produce, manufacture, and transport.[27] According to researchers at UC Santa Barbara, the food system contributes about 30 percent of total U.S. greenhouse gas emissions, with the majority of that coming from animal-based diets.[28] They estimate if Americans transition toward healthier diets, this could contribute up to 23 percent of the U.S. Climate Action Plan goal of reducing greenhouse gas emissions 17 percent below 2005 levels by 2020.[29] So, an apple (and a bunch of kale) not only keeps the doctor away, but keeps the planet cooler as well.

---

[25] http://msue.anr.msu.edu/news/7_benefits_of_eating_local_foods
[26] https://www.ucsusa.org/global-warming/solutions/stop-deforestation/palm-oil-infographic.html#.W5MMd-hKg2w
[27] https://davidsuzuki.org/queen-of-green/food-climate-change/
[28] https://www.sciencedaily.com/releases/2017/03/170308154423.htm
[29] Ibid.

2) *Start paying attention to your mental health.* If there's one thing you've probably learned from this book, it's that all this doom and gloom discussion about climate change can be depressing. Things seem bleak at times, and given the current political climate it looks uncertain that things will improve any time soon. However, it is important to stay positive and hopeful about the future. There are so many immense benefits that can be reaped from acting on the climate crisis, and making ourselves dispirited about the future isn't going to help anything. As such, be cognizant about your mental health. Take time to meditate, do some yoga, or go for a hike—whatever you need to do to stay positive in both your day-to-day life as well as your outlooks about the future of the planet, take care of your mind as much as your body.

3) *Start reading:*

- The previous U.S. Surgeon General Vitek Murty stated, "Climate change will be a serious danger to public health, worse than polio in some respects." This is due to the multifaceted impacts on human health. That administration published a report "The Impacts of Climate Change on Human Health in the United States: A Scientific Assessment," which can be found on the U.S. government website https://health2016.globalchange.gov/, and anyone concerned about human health in regard to climate change is encouraged to read it.

- Even though the current administration has removed some climate-related material from the EPA website, relevant information can still be found on the Center for Disease Control website. The CDC formally established the climate change and public health framework in 2009, leading the effort to identify populations vulnerable to climate change, prevent and adapt to current and anticipated health impacts, and ensure that systems are in place to detect and respond to current and emerging health threats.

- The American Public Health Association declared 2017 the Year of Climate Change and Health, and they offer many resources and events to the public. This area of discussion is receiving more and more attention, as it becomes more and more apparent that we are endangering our own health by allowing climate damage to continue. The WHO has estimated that between 2030 and 2050, climate change is expected to cause an additional 250,000 deaths a year from malaria, malnutrition, heat stress, and diarrhea. The time to act is now, in our daily lives, because those lives may very well depend on it.

~ ~ ~

Kasper Eplov, MD is a graduate of Copenhagen University and currently a General Practitioner trainee and has published in the Danish medical journal on the topic of climate change and health. He has given talks on the subject at several medical conferences. In 2016, he served in an advisory capacity for the Danish government's Advisory Board on Ethics. In June of 2017, he presented the topic "Climate Change and Health" at the Twentieth Nordic Conference on General Medicine. Kasper lives in Haessleholm, Sweden. Read more at: www.climateabandoned. com

Chapter 27

Climate & Health: A Nurse's View

by

Cindy Martinez, R.N.

(To all the nurses who work tirelessly to promote good health and prevent the harmful effects of bad stewardship of our planet.)

"Never doubt that a small group of thoughtful, committed citizens can change the world; indeed, it's the only thing that ever has." – **Margaret Mead**

The silver SUV sped down the highway as the young mother glanced in her rearview mirror, simultaneously keeping her eyes on the road and her 1-year-old in the back seat. Her heart began racing nearly as fast as the car when she noticed a blue hue in her young boy's lips. He audibly wheezed and fought to draw air into his tiny lungs. She wasn't a religious woman, but she prayed to every god she could think of, begging that they would make it to the emergency room in time.

Now, this probably isn't your first thought when you think of the effects of climate change, but more and more adults and children are struggling with climate-related health problems, including severe, life-threatening asthma like this precious baby experienced.

Asthma is caused or triggered by exposure to allergens and/or irritants in the ambient air. Unfortunately, extreme weather events, particularly wildfires and intense flooding, are causing these triggers to become more and more prevalent. How can we human beings slow and even stop the pollution and greenhouse gases that are changing our climate and affecting our health? Even more importantly, how can we prepare for and mitigate the changes we can't prevent? There are some things we can do, and we must make an effort to do them.

The American Public Health Association (APHA) chose climate change and health as the theme of their 2017 annual meeting, with good reason. Executive Director Georges Benjamin believes anyone who doesn't think climate change is a severe problem is only fooling themselves.

Nursing is considered one of the nation's most trusted professions. That puts me and other healthcare professionals in a unique position to inform and improve biological response to the science of climate change. Sure, I may be better versed in anatomy than I am in atmospheric science, but it's still an easy line for me to draw from climate change to human health. Hopefully, in turn, I can help increase public perception and awareness of mitigation and preparation strategies for these changes and reduce the harm to come. At the very least, nurses can bring the voice of health to the conversation.

Climate change affects our health in different ways, and this primarily depends on where you live in the world. Your proximity to oceans or rivers and the weather patterns in your region are big factors. Nevertheless, most Americans seem unaware of the health harms of global warming caused by rising sea levels. According to a recent survey, most have not even taken the time to consider how climate change may affect their health or the health of their loved ones. The Medical Society Consortium on Climate and Health said few Americans, just 32 percent, can name a specific way in which climate change is harming our health. If you're part of that 32 percent, read on; I will enlighten you to quite a few.

The most obvious and direct health effects from changing weather patterns are due to exposure. Extended hot spells, larger and hotter wildfires, floods from changes in precipitation, drought, and more frequent and intense storms take a toll on us. Some less obvious effects include illnesses due to increased air pollution and poor air quality, allergies, heat-related sickness, disease carried by insects, disaster-related injuries, and mental health stressors. Some that may never come to mind are related to lack of water availability, food insecurity, threatened food crops, and nutrient collapse, a hidden issue I'll discuss later.

Like millions of Americans, I watched while Hurricane Irma barreled through Florida. A CNN story about the eighty first responders gathered at the St. Petersburg Fire Rescue Station really resonated with me. For safety reasons, the workers were not allowed to respond to 911 calls, but they were able to see a list of those incoming calls on their computer. Simultaneously, four cardiac arrests and respiratory problems for a 12-year-old occurred. Miami Mayor Tomas Regalado lamented the flooding caused by Irma: "If this isn't climate change, I don't know what is," he said, looking around at his destroyed city. He even referred to the hurricane as the "poster child for what is to come."

Kansas City area resident Craig Wolfe, President of the Heartland Renewable Energy Society, is focused less on health and more on survival, claiming, "You can't be healthy if you don't survive." Energy and the environment are and have always been a big part of his life, and he considers the climate crisis an emergency situation. He says we have a tremendous opportunity to solve the energy crisis, and he is certain things can only improve after that. There wasn't a word for it until Wolfe coined the term *planetary treason,* an act by a person who betrays their duty to protect the survival of all species and civilization by willfully disregarding science for personal or political gain. If that is a crime, I can already think of a few modern-day Benedict Arnolds, and one rhymes with Ronald Grump!

Existing health threats will intensify and new ones will emerge if we don't respect our planet, according to the Climate Reality Project. "That is why climate change is ultimately a human issue, because our health, our livelihoods, and our homes are directly impacted by the changes happening in our environment," they recently stated. In 2014, Pope Francis referred to human beings as "custodians of creation" and insisted that we must safeguard creation and respect the beauty of nature and the grandeur of the cosmos.

In 2017, the World Health Organization (WHO) called climate change a global health emergency, because our physical and mental health are at stake. Furthermore, they noted that climate change affects the social and environmental determinates of health: clean air, safe drinking water, sufficient food, and secure shelter. It already claims tens of thousands of lives a year. In addition, the WHO believes some of the largest disease burdens are climate sensitive, such as under-nutrition, diarrhea, malaria, and extreme weather events that cause injury, trauma, stress, and death.

I have a long-term interest in environmental issues, and it all began on my environmental health class in college. In that class, the professor showed images of large cities smothered with a thick blanket of gray, filthy smog, courtesy of dirty coal ash spewing out of smokestacks. They told us about recycling and energy conservation during the 1970s oil crisis, and frankly, that seemed like the ideal time to cut our reliance on oil and dirty fuels. In my naiveté, I could not imagine why anyone would not want a clean, livable environment. The Clean Air Act has done a lot to clean up our air since it was enacted years ago, but in our current political atmosphere, it's hard to tell what the future of environmental regulations will look like.

Respiratory Disease

Air pollution and air quality-related health effects are the most visual of manmade global warming. As a Kansas City home health and public health nurse for many years, I experienced firsthand the effects of this air pollution on residents in the area. The long, hot summer days there were tough for everyone, but those who suffered the worst were people with chronic respiratory problems. So many times during home visits, I heard wheezing, gasping people utter, "I can't breathe. It's too hot." Their breathing and lung capacity diminished, and they required more oxygen and rescue medications just to get through the hot days. In most cases, if they didn't answer their phone, it meant they were already hospitalized for a respiratory event.

Smog and ground-level ozone, both considered air pollution, intensify as temperatures climb. Both are powerful lung irritants that can trigger asthma attacks and interfere with lung development, among other health problems. Because of their growing minds and bodies, children suffer the most adverse health impacts of ozone and pollution, which contribute to asthma, reduced lung function, and increased sensitivity to irritants and allergens. Furthermore, these cause chest pain, coughing, and nausea, even in healthy children. Climate change increases ozone levels, particularly in urban areas, and respiratory allergens and allergies increase in warmer climates. Studies also show that pollen seasons have become more lengthy and severe. Consequentially, respiratory physicians report seeing greater severity of lung disease and allergies. The increase in allergens also triggers asthma attacks.

Kansas City (KC) is a leader in emergency room visits and hospitalizations due to asthma, with over 11,842 attacks a year. In 2016, the Clean Air Task Force reported that KC is vulnerable to ozone smog in the summer, thanks to oil and gas production. This pollution directly harms our health and speeds climate change. The American Lung Association has called on the U.S. government to phase out the burning of coal, which adds to outdoor air pollution and causes cardiac deaths.

When asked about these issues, Richard Randolph, MD, Chief Medical Officer for Heart to Heart International, refers to a study by the American Lung Association. Based on that study, we should consider the following: The average cost of a simple emergency room visit is $1,000. Every time you fill up a 16-gallon tank of gas and burn through it, you generate $11.85 in healthcare costs. If you happen to get 18 miles per gallon and drive 12,000 miles per year, it would take 2 whole years to generate money enough to send someone to the emergency room.

Heat-Related Illnesses

The most obvious effect of record-breaking warming temperatures is extreme heatwaves, a quiet killer. These happen globally, harming young and old alike. Heat-related illnesses include heat rash, dehydration, heat exhaustion, and deadly heatstroke. In addition, heart and lung problems worsen. The brain is one of the first organs to be affected by the heat, and this can lead to a myriad of problems, including poor judgment. People worldwide swelter and die in heatwaves. Many do not have air conditioning, and those who do are often afraid to use it, as they need to save that money for rent and food. This is yet more evidence that climate change affects our lowest socioeconomic communities first.

The Kansas City Health Department often tweets notifications from the Jackson County medical examiner about possible heat-related death investigations. Many of these fatalities are caused by the urban heat island effect. This comes as no surprise, since several recent studies have pegged Kansas City as a textbook example of this problem. Why? Because paved streets, concrete, tar roofs, brick-and-mortar buildings, and asphalt all soak up sunlight during the day. The heat from that sunlight slowly radiates back into the air, which can push temperatures up by twenty degrees in the urban core. As people struggle to cool their aging homes, energy costs rise. When air-conditioning units are cranked up, we also see increases in ozone levels, bringing even more risk to those with respiratory and other health problems. Not only that, but increased heat produces more smog and pollen. In fact, more people die in Kansas City from heat-related conditions than from all other extreme weather events. The body's ability to cool itself through perspiration is compromised by the heat, especially in those who are already ill, injured, or on certain medications.

One elderly man suffering from chronic obstructive pulmonary disease (COPD) sticks out in my memory. On hot days, I found him sitting on his couch, with his oxygen tube in place. He didn't want to move about, as he knew that would require the use of his inhaler, and he feared running out of medication. He wasn't alone, as many of the elderly respiratory patients I treated struggled with repeated hospitalizations and emergency room visits due to COPD or asthma caused by poor air quality and hot conditions. Their other existing health issues, like cardiovascular disease or chronic bronchitis, also worsened during heatwaves.

Climate change literally adds fuel to the fire, creating hotter and dryer conditions. This, in turn, results in more droughts and wildfires, like those that threaten the western United States. The National Interagency Fire Center reports 2017 has one of the worst fire seasons they've ever seen, and over two million acres of land have succumbed to burning. The California wildfires are raging again at the time of this writing, filling skies with billowing smoke and ash and leaving the landscape looking like some horrible scene out of the Book of Revelations. Whether you live in California or not, this is likely to affect you in some way. Do you enjoy a crisp Napa Valley pinot grigio or, perhaps, a smooth cabernet sauvignon? Truth be told, wine country is going up in smoke. Governor Jerry Brown called this year's fires one of the greatest tragedies California has ever faced, leaving many dead and entire subdivisions incinerated. Any wildfire smoke triggers coughing, asthma flare-ups, and heart attacks, and these affects can be felt for miles downwind of the wildfires.

At a training conference in Denver, Colorado in 2013, we were warned not to walk outside too much due to the nearby wildfires, especially those prone to respiratory problems. Even downtown in the mile-high city, the air was filled with fine soot and ash, and the sky held an ominous brown hue.

Weather-Related Disasters

Extreme weather events cause disaster-related injuries, trauma, stress, and death, and climate change only increases these. Heatwaves, rain bombs, storms, droughts, wildfires, and torrential downpours, as well as stronger and more frequent hurricanes, tornadoes, and floods, are catastrophic to all live. Extreme precipitation and storm surge cause extreme flooding. Unfortunately, the risk of injury, death, and drowning increases due to population growth in coastal and urban areas. On the other hand, the lack of rain causes severe droughts, which lead to dehydration and death.

The devastation caused by Hurricanes Harvey and Irma are now well recorded. Irma is the strongest hurricane on record in the Atlantic basin, outside the Caribbean and the Gulf of Mexico, according to the Climate Reality Project. There was much damage done to property, of course, but perhaps even more was done to human health.

Extreme floods also cause drowning and major medical concerns, especially when patients with chronic conditions are unable to reach hospitals and may be left without electricity, food, or water for days. Flooding from heavy downpours eventually leads to the growth of mold, which increases allergy problems and worsens asthma. Wading in floodwaters leave people vulnerable to stepping on something or picking up viruses or bacteria, since that water eventually becomes a toxic mix of organisms, chemicals, and waste.

Dr. Randolph was recently stationed in the Houston, Texas area in the aftermath of Hurricane Harvey. He and many others treated people on dialysis with chronic health problems like COPD, heart failure, asthma, and respiratory illness. They also had to treat people with everyday emergencies and other health concerns resulting from water and food contamination, infectious disease, wounds, mold, mosquitoes, and mental distress.

Dr. Randolph said the power of the storms is magnified by warmer surface temperatures. In fact, the surface waters in the Gulf of Mexico were two degrees warmer than average, and the surface air temperature was five degrees cooler. This dangerous combination pushed moisture into the air, thus supercharging Hurricane Harvey and generating hard rain. The same thing happened in the Philippines in 2013, with Typhoon Haiyan. Dr. Randolph was stationed with Heart to Heart International there during that disaster, and he reported that similarly to the case with Harvey, the surface water was 2 degrees warmer than the 20-year average, and the storm sustained winds of 200 miles per hour and wind gusts of 220.

There is even an increase in disease and health threats when floodwaters recede. Doctors must watch out for flesh-eating bacteria, cancer-causing chemicals, and mold. The Texas Gulf Coast is home to many oil refineries and chemical plants. Harvey flooded or damaged more than fifty of these, dumping pollutants into the air and water and further complicating the health situation as a result.

Insect-Borne Illnesses

Besides flooding caused by powerful storms like Harvey and Irma, which are indirectly worsened by climate change, there is also a marked increase in vector-borne diseases in warming climates. Vectors include mosquitoes, ticks, fleas, and kissing bugs (also known as assassin bugs or vampire bugs). A warming world alters and increases the range and breeding season of many of these disease-carrying insects. The Climate and Health Assessment says fourteen vector-borne diseases in the United States are public health concerns. West Nile virus is the number-one disease spread by insects in Kansas and the U.S. Yes, that deadly disease from Africa has migrated all the way here to maintain a comfortable living, all because of a warming climate.

In June of 2017, the CDC reported two types of mosquitoes that carry Zika and other viruses, right into the heartland of the United States. Mosquito researchers believe the diseases they transmit are soon to come, but that is not the worst of what bugs can deliver.

The U.S. Center for Disease Control estimates that 300,000 people living in the U.S. have Chagas disease, mostly in California, but in Kansas? Yes! According to public health officials Dr. Randolph has already talked to, the disease has been acquired by people in Kansas. In Central and South America, this condition is the second-leading cause of heart failure, as it results in life-threatening heart damage if it's not treated early. Chagas is a tropical parasitic disease spread by blood-sucking triatomine, kissing bugs. In spite of what may appear to be a cute nickname, there is no love in these creatures, which bite exposed skin around the lips and mouth. However, the bite is not what causes the infection; rather, the parasite (*Trypanosoma cruzi*) in the feces can enter the person's body through the eyes, nose, mouth, or the wound from the bite itself. The lips or mucous membranes swell as a first symptom, and chronic infection causes enlargement of the heart ventricles of the heart, leading to cardiac failure. As of 2017, no vaccine had yet been developed for Chagas. Treatment requires a special type of anti-parasitic drug, since it's not a bacteria. The parasites are small multi-cellular animals, similar to those found in malaria.

Dr. Randolph added that experienced doctors fail to even look for Chagas in Kansas, since it's so new to the area. This is merely one example of the numerous vector-borne diseases spreading into North America due to the warming climate. Zika, Dengue fever, West Nile virus, Lyme, and malaria are also part of this growing list.

Scientists predict that vector-borne illnesses will increase as pathogens move into previously uninflected areas. In their new book, Enviromedics: *The impact of Climate Change on Human Health*, Physicians Jay Lemery and Paul Auerbach explain that global warming is not causing new disorders but is spreading old disorders and making them worse.

Mental Health

Climate change is not only linked to heat-related illness and death, extreme weather events, and food insecurity because of contaminated water and food. It can also be linked to plenty of mental health issues, including trauma, anxiety, and depression. Exposure to natural disasters naturally creates trauma. The loss of property, possessions, jobs, and loved ones causes extreme stress and depression. The psychological impact of climate change include *pre*-traumatic stress disorder, a term coined by Lise Van Susteren, a Climate Reality leader and psychiatrist in Washington, DC. Those who witness the mounting toll of climate change, either in person or in the news, are affected by this disorder. Obstetricians also talk about the frightening risk of the Zika virus for pregnant women in the mainland U.S., as it causes severe birth defects.

Among the lost loved ones are our furry friends, and this can have a deep psychological impact on pet owners. The loss of support from pets is associated with depression, acute stress, anxiety, and post-traumatic stress. Studies show an increase in these impacts after Hurricane Katrina. Pets provide nonjudgmental support, a buffer against physical and mental health problems, and even those who lost a goldfish or a hamster can likely relate to the pain of such a tragedy

The Medical Society Consortium on Climate and Health says extreme weather events cause many people to experience stress and serious mental health problems including PTSD, depression, and suicidal tendencies. Katrina flooding victims had a 47 percent higher mortality rate than usual, and 25 percent of them required mental health counseling in the aftermath. These disasters are also associated with increases in alcohol or drug abuse as survivors struggle to cope with all they have lost.

I attended "Community Discussion on the Health Impacts of Climate Change," sponsored by the Climate and Energy Project and the League of Conservative Voters in Kansas City in August of 2017. The Climate and Energy Project is a Kansas-based, nonpartisan organization working to find practical solutions for a clean energy future. One of their noteworthy goals is to address the impact of climate change on public health in the heartland. They hope to increase community resilience and encourage wind and solar energy. CEP Assistant Director Rachel Myslivy said 70 percent of the U.S. supports setting strict CO_2 emission limits on existing coal-fired plants to reduce global warming and improve public health.

In one tragic tale at that discussion, a mother had to file bankruptcy because her child required continuous emergency room trips for asthma treatment. Another shared her struggles with chronic lung disease contracted during a recent business trip; she didn't want it to spread to anyone else because she was so debilitated by it, and that was why she became involved with the Climate and Energy Project in the first place.

The Climate and Energy Project encourages people to join the conversation and tell their stories about how climate change has affected their health or the health of their loved ones. In this way, they gain support for solutions. The majority of people in attendance were alarmed or concerned about global warming and wanted to know what they could do. Many Americans unfortunately still see climate change as a distant concern, and they don't feel it can happen to them, so awareness is key. It does not seem like a real, immediate threat like someone pointing a gun at you until it actually makes you sick. There is an obvious disconnect between the effects of climate change, and many people think the effects are decades away, but this couldn't be further from the truth.

At the discussion, we learned that few people are aware of the groups of Americans most likely to be harmed by climate change. Low-income communities, the elderly and young, pregnant women, people with chronic illnesses and allergies, student athletes, and communities of color are the most at risk. In fact, African-American children are three times as likely to be admitted to the hospital and twice as likely to die from an asthma attack than Caucasian American children. It seems there's little public discussion about these harms, and doctors are in a great position to share the health consequences.

As briefly mentioned already, disasters and rising temperatures also cause food insecurity and impact global food supply. Consider the orange crops damaged by Hurricane Irma in Florida. Many say the destruction is the worst they have ever seen, with up to 50 percent of the citrus crop damaged, as well as sugar cane and vegetable crops. This is not good, since the Sunshine State is a key source of fresh fruits and vegetables for our country in the winter. Droughts also cause crop loss, leading to more malnutrition, and of course children suffer disproportionately.

As if this is not precarious enough, a recent found that the atmosphere is literally changing the food we eat, for the worse. In Helena Bottemiller Evich's Politico piece, "The Great Nutrient Collapse," she discusses her discovery that when plants get too much CO_2 from the atmosphere, it affects the quality of those edible plants, rendering them less useful to us from a nutritional standpoint. The junk food effect appears to be happening in fields and forests around the world. Imagine making a healthy choice, an apple instead of a bag of chips, only to find that they have the same nutritional content. An apple a day may not keep the doctor away for long, because that's the world we're heading for.

Irakli Loladze, the focus of that interview for that written piece, stated that many of our most important foods are becoming less nutritious, with increased carbohydrate and sugar content and reduced mineral, vitamin, and protein. The increasing CO_2 causes more plant growth, but size doesn't matter! In other words, the quality of the plants has dropped, and no one knows what will happen to plant life in the long term. We do know that plants are a crucial source of protein in the developing world. In the U.S., some researchers say the increased carbohydrates and sugars in plants will contribute to obesity, diabetes, and cardiovascular disease, which are already nearing epidemic levels today.

In his *Drawdown*, author Paul Hawkins says the most comprehensive plans ever proposed to reverse global warming are technology-based solutions. After Hurricanes Harvey and Irma, Neil deGrasse Tyson, a scientist and astrophysicist, asked what would it take for people to recognize that we can all benefit from learning objective truths about the natural world. Fifty inches of rain in Houston? A hurricane the width of Florida? You'd think somebody would pay attention to that!

Our health is at risk because our climate is changing, but there are some things we can do to slow or halt these issues. You, as an individual, can make a difference in solving our climate crisis. According to ecoAffect, almost everything we do, from what we drive to what we eat to how we talk about climate change, affects the people around us.

The positive choices we make can help protect our personal and public health. Anyone can promote knowledge about the links between climate change and health, but this is especially true for students, nurses, doctors, and other healthcare professionals. We must speak on the issues that affect health, the causes of climate change, such as fossil-fuel emissions, and disaster preparedness. Nurses can and should play a key role in educating the public about climate change.

Demand that lawmakers use their leadership to convey the urgency of this global disaster. Dr. Randolph agrees with me that Hurricanes Harvey and Irma should be turning points that encourage us to talk about these issues, that they should at least create opportunity for conversation because many more people experienced and saw what happened in that recent onslaught, and several are still suffering from them.

We cannot turn our backs on the people who lost their lives, their homes, and their families. We cannot pretend it didn't happen or that it won't happen again. Let's get together and move toward a solution for ourselves and our families. We can all be part of the solution, but we must act now. Below, you will find some suggestions of things you and those in your community can do to reduce climate change and lessen the adverse effects on human health.

Climate Crisis "Keep/Stop/Start" Actions

Below for your consideration, are several author-suggested action items. For a further explanation of the Keep/Stop/Start organizational tool, please see How to Read This Book.

KEEP:

1) *Keep the budget of the Environmental Protection Agency intact.* Contrary to popular belief, the primary mission of the EPA is to protect human health and the environment, not to promote fossil fuel development. According to the League of Conservative Voters, any cuts to the EPA will devastate efforts to address climate change and protect our clean air and water, so these budgetary cuts can be seen as a direct threat to endangered species and public health.

2) *Keep the Clean Air Act intact,* protected, and enforced. The American Lung Association says the U.S. has seen air quality improvements every year since this act began. Abandoning clean air laws now will only exacerbate the health problems for the chronically ill. These illnesses force otherwise productive members of society to miss school and work and spend more days in the hospital. The quality of life decreases, while stress increases. Don't become complacent about the air because it is less polluted due to the Clean Air Act.

3) *Keep accelerating our transition* to clean, renewable energy like solar and wind. This will increase clean energy jobs. Support renewables and tell your power company and governor that you demand more renewable energy sources like wind and sunshine in your electricity mix.

STOP:

1) *Stop Congress and the current administration from dismantling* and cutting the budget of the EPA. According to the League of Conservation Voters, any cuts to the EPA would devastate efforts to address climate change and protect our clean air and water and are a threat to endangered species and public health.

2) *Stop denying climate change.* Denying climate change puts people in harm's way. Rising sea levels and stronger storms are real and unless we take steps to keep fossil fuels in the ground and plan for a changing world, more people will lose their homes or their lives, Jamie Henn of 350.org said. What can you do about climate change? Go to 350.org for more solutions to cut your own global warming pollution.

3) *Ask Congress to work to find real solutions* to stop climate change and the further destruction of our planet. Sign petitions or call your members of Congress to make your voice heard. You can make a real difference in solving our climate crisis. According to ecoAffect, almost everything we do, from what we drive to what we eat to how we talk about climate change, affects the people around us. The positive choices we make can help protect our personal and public health. Anyone can promote knowledge about the links between climate change and health but, especially students, nurses, doctors and other professionals. Health professionals must speak to the issues that affect health, the causes of climate change, such as fossil-fuel emissions, and disaster preparedness. Nurses can play a key role in educating the public about climate change.

4) *Stop or limit your own contribution to carbon pollution* by eating less red meat, insulating your home, and driving less. MOMs clean Air Force says moms and dads can be engaged citizens. Tell the President, Congress, your mayor—- and anyone else who will listen that you are deeply concerned about climate change, and they should be too.? Demand that lawmakers use their leadership to convey the urgency of this global disaster. I asked Dr. Randolph if Hurricanes Harvey and Irma will be turning points when we talk to people about climate and health? He said they should at least create an opportunity for conversation because many more people experienced and saw what happened with the recent trio of hurricanes. Let's not turn our backs on the people who lost their lives, their homes, their families and pretend it didn't happen or that it won't happen again. Let's get together and move toward a solution for ourselves and our families. We can all be a part of the solution. Spread the word.

START:

1) *Start supporting community solar farms*, if you aren't supporting them already. Last Saturday, I toured one of the first ones that sprang up in Kansas City, near Independence, Missouri, operated by Independence Power and Light and owned by MC Power. The tour of the amazing 3-acre, south-facing farm was sponsored by Heartland Renewable Energy Society. Homeowners and renters share in the benefit of solar energy, even if the panels are not on private roofs. Presenters noted that Missouri's frequent rains are perfect to clean the more than 11,000 solar panels, and there are plans in place to double the size of the farm in the near future.

2) *Start electing politicians who care about solutions* to our climate crisis and are working toward them. Remove climate deniers from our government. This may seem easier said than done, but four years won't last forever.

3) *Start using EPA air quality flags in schools*, from the AirNow Program.

4) *Start green teams* in healthcare and business settings.

5) *Start joining campaigns* to strengthen EPA efforts to reduce carbon, soot, ozone, and methane pollution.

6) *Start using your unique voice* as a nurse or other healthcare professional to motivate and inform people about climate change and its effects on health.

7) *Start doing what you can personally* to change to renewable energy. Make your home more energy efficient; for advice, visit energy.gov/public-services/homes. Make lifestyle changes that will allow you and Planet Earth to heal.

8) *Start saving money by saving energy*. Enjoy better health and a healthier bank account by eating locally grown food and biking or walking instead of driving, which reduces carbon pollution. Encourage healthy food initiatives.

9) *Start or continue to support* the U.S. Climate Alliance and encourage state and local governments to do the same. As of this writing, nine states—California, Connecticut, Hawaii, Massachusetts, New York, Oregon, Rhode Island, Vermont, and Washington—have joined the alliance, vowing that they will abide by the Paris Climate Agreement and urging other states to follow suit.

10) *Start demanding* efficient buildings, which reduce energy waste.

11) *Start buying products made from recycled* materials and recyclable products and take advantage of recycling programs in your area. The more you recycle and compost your garbage, the less smoke goes into the atmosphere. Less smoke equals less asthma and lung cancer.

12) *Start planting trees.* A Rotary Club president in Kansas started a program that requires every member to plant one tree by Earth Day, so you can also set this goal for yourself and your family.

13) *Start driving less*, relying on less heating and cooling in your home, and eating less red meat. Meat production is water, oil, and greenhouse-gas intensive. Moving toward a more plant-based diet reduces these gases, as does buying electric vehicles and choosing local and organic produce.

14) *Start contacting decision makers* about important issues like air pollution, water contamination, and hazardous chemicals that threaten your family's health and safety. Ask congress to find real solutions, to stop climate change and any further destruction of our planet. Sign petitions or call your elected officials to make your voice heard. Tell them they must condemn President Trump's climate denial and draw the links between climate change. If they need evidence, talk to them about three monster hurricanes in two weeks, the wrath of Harvey, Irma, and Maria!

~ ~ ~

Cindy Martinez, **R.N.** is a registered nurse with a background in public and home health. She became an environmental activist and testified at hearings on clean air, water, and in support of wind energy. Cindy is the author of two mystery/suspense novels, *River Stalk*, a Library Journal SELF-e Select choice in 2016; and *Mommy's Missing*, a Global eBook honorable mention winner for suspense fiction in 2014. In August of 2016, she completed Climate Reality Project leadership training. Cindy lives in Kansas City, Missouri. Read more at: www.climateabandoned.com.

Chapter 28

Climate Crisis & Religion

by

Peter M. J. Hess, Ph.D.

"It is right to save humanity. It is wrong to pollute this earth. It is right to give hope to the future generation."
– Vice President Al Gore

Long before I glimpsed flames through the surging smoke, I heard the fire roaring down Boggs Mountain. The sound was unlike anything I'd ever heard in my life, or hope ever to hear again: it was like the whine of a jet engine at full throttle superimposed on the thunder of Niagara Falls.

Imagine an unearthly roar with its own low musical tone, punctuated at intervals by the loud booms of exploding propane and automobile fuel tanks. Conjure up the sight and smell of billowing clouds of smoke ten thousand feet high, so thick that they blot out the sun, turning day into night.

Picture the panic and chaos of thousands of cars and trucks and horse trailers fleeing down curving mountain roads while helicopters and air tankers dump paltry loads of pink retardant in a futile attempt to slow the flames. It called to mind an apocalyptic scene in the Book of Revelation: "The first angel sounded his trumpet, and there came hail and fire mixed with blood, and it was hurled down on the earth. A third of the earth was burned up, a third of the trees were burned up, and all the green grass was burned up." (*Revelation* 8:7)

Only five days earlier in 2015, my family and I had enjoyed a peaceful Labor Day holiday at our home in Cobb Valley, California, observing how a multi-year drought had contributed to the death-by-bark beetle of about a third of the Ponderosa pines in the region. We knew the chaparral was very dry that year, with a moisture level at only 45% of normal (it is considered critical when brush moisture drops below 65%). We were discussing the effects of climate change along the Pacific coast, particularly the desiccation of the American Southwest. Our conversation was interrupted by a flurry of excitement Monday afternoon: a small fire erupted north of us and was quickly and efficiently contained at less than an acre.

A week later the fire was quite different. September 12th dawned nervously to a fitful, erratic wind, and soon developed into a living hell in the course of one hot, gusty, roaring, frantic, terrifying afternoon. The California Department of Forestry and Fire Protection tweeted at 1:15 p.m. that "a small brush fire two acres in extent with a moderate rate of spread" had ignited on High Valley Road just north of my family's property. I was away and attempted to rush to Cobb to activate our wildfire protection system, consisting of a high-pressure pump, five hundred feet of fire hose, and 15,000 gallons of water. However, I never made it to Cobb. In thirty minutes, freak meteorological conditions propelled that small blaze into a raging inferno with winds gusting up to sixty miles per hour.[1]

Stopped at a road block, I helped water down a family's roof while they hastily loaded two vehicles with a few prized possessions. When I felt radiant heat on the back of my neck I turned around to see through the trees a gigantic wall of flame a mile and a half away. Relentlessly it bore down Boggs Mountain toward us, flames flaring up massive pine and fir trees as if they were twigs in a bonfire. The deputy sheriff announced through his loudspeaker that we had to evacuate the community or risk being trapped and killed. Over the course of nearly a week the 2015 Valley Fire burned fourteen hundred structures in four communities, most of them family homes

---

[1] Kevin Schultz, "Hurricane Linda may have hastened Valley Fire," San Francisco Chronicle, September 27, 2015. https://www.sfgate.com/bayarea/article/Hurricane-may-have-hastened-Valley-Fire-6532084.php

— including ours. A handful of people and hundreds of pets died in the conflagration, and an estimated seven million trees were killed along with countless wild animals.[2]

This excruciating experience is matched every year by increasingly catastrophic wildfires in other states and in countries around the globe: Australia and Canada, Chile and France, Portugal and South Africa. Losing people and homes and entire communities to a disaster - whether to a wildfire or a tornado, to a major flood or a devastating hurricane - gives human faces to the varying effects of the climate crisis. To be sure, we cannot prove conclusively that the Valley Fire or any other particular conflagration is the direct result of climate change, as there are too many variables involved for us to be certain. The same caution must be exercised with respect to the frequency or severity of hurricanes, typhoons, tornadoes, and other storm-related weather events. Although no particular hurricane can be ascribed to climate change, greater atmospheric energy can on average be expected to increase typhoon frequency and intensity.[3]

Climate scientists look at long-term patterns such as the six-year drought in the Southwest, and conclude that the wildfires we are now experiencing reflect the predictions of the most cogent models of what we might expect on a warming planet.[4] The consensus of climate experts - well established now for a decade - is that anthropogenic (human-caused) global warming and consequent climate change are now part of human reality. Climate change denialists abound, of course, but fortunately, in response to them, the website "Skeptical Science" analyses and disposes of a comprehensive list of 197 false arguments.[5]

The Relevance of Climate to Religion

What is the relevance of climate to religion? As a Roman Catholic scholar of the relationship between theology and science, I would say "all the relevance in the world!" Every human religious tradition emerged during a relatively benign period of climatic stability following the end of the last ice age, about 11,700 years ago. Long-standing confidence in a benevolent climate was expressed by German pietist poet Matthias Claudius (1740-1815) in lines that became England's most popular harvest festival hymn:

> *We plough the fields and scatter the good seed on the land,but it is fed and watered by God's almighty hand;he sends the snow in winter, the warmth to swell the grain,the breezes and the sunshine and soft refreshing rain.[6]*

Nature was idealized as a stable and dependable order conducive of thankful worship. The Book of *Genesis* set Adam and Eve in a garden of perfection, with an ideal climate and plants laden with fruit ripe for the eating.

Of course, this ambient gentleness was only imagined — in reality, *Homo sapiens* never did inhabit such an idyllic Eden. Our dynamic environment is replete with floods, avalanches, and landslides during the rainy season and with heat waves, droughts, and wildfires in the dry time of year. But now we are entering a period of climatic instability that is unprecedented in human experience — a key point that many people miss. Of course, the terrestrial climate has changed over millions of years, but never has it happened so drastically or during a time when humans have occupied nearly every biological niche of the globe, where we experience intimately its manifold effects. Climate change of this rapidity and magnitude is driven not so much by natural cycles as by the high-energy-consumptive lifestyle of a human population ballooning beyond sustainability.[7]

---

[2] For a first-hand account see http://www.huffingtonpost.com/peter-m-j-hess-phd/wildfires-angels-and-the-hand-of-god_b_8150728.html
[3] http://www.climatecentral.org/news/warming-increases-typhoon-intensity-19049. Some extreme weather and climate events have increased in recent decades, and new and stronger evidence confirms that some of these increases are related to human activities.
[4] https://www.climate.gov/news-features/featured-images/risk-very-large-fires-could-increase-sixfold-mid-century-us
[5] See http://www.skepticalscience.com/argument.php
[6] http://www.stmartinsmethodist.org.uk/revs-blog/the-story-behind-the-hymn-we-plough-the-fields-and-scatter/
[7] Antony Barnosky and Elizabeth Hadly offer an engaging if sobering discussion in *Tipping Point for Planet Earth: How Close are we to the Edge?* (New York: St. Martin's Press, 2015).

Religion and theology should play a prominent role in our accelerating conversation about climate, because billions of religious believers (along with everyone else) are at risk of devastation wrought by rapid global change. Governmental or corporate responses alone are not sufficient to address this complex problem, and it is encouraging to see that many religious leaders now understand the irreparable harm anthropogenic climate change is causing, and are rethinking their theologies, ethics, and programs of action accordingly.[8]

Religious Voices: A Representative Sampling

Virtually all major religious traditions are aware of the seriousness of climate change and most have issued position statements about it. Readers interested in a more comprehensive collection of such statements are directed to two useful on-line repositories: "Interfaith Power and Light: A Religious Response to Global Warming,"[9] and "The Forum on Religion and Ecology at Yale."[10] Here is a spare but representative sampling of religious views on the subject.

Indigenous Religious Voices

Due to their dependence upon and close relationship with the environment and its resources, Indigenous communities are among the first to face the direct adverse consequences of climate change. In 2009 the United Nations Permanent Forum on Indigenous Issues issued a paper on "Indigenous Peoples and Climate Change":

> *While climate change may still be a distant threat for some people it is already a grim reality for many indigenous and local communities, especially those on the three regions mentioned....Climate change has a harmful effect on biological diversity and the related knowledge, Innovations and practices of indigenous and local communities. Traditional knowledge is an inseparable part of indigenous and local communities' culture, social structures, economy, livelihoods, beliefs, traditions, customs, customary law, health and their relationship to the local environment. It is the totality of all such elements that makes their knowledge, innovations and practices vital in relation to biological diversity and sustainable development.[11]*

This descriptive summary of the impact of climate change was expressed in spiritual language in 2015 at the United Nations Convention on Climate Change, at which "first nations" representatives published the *Indigenous Elders and Medicine Peoples Council Statement*, concerned that "the sacred has been excluded from all our discussion and decisions":

> *We, the original care takers of Mother Earth, have no choice but to follow and uphold the natural law, which sustains the continuity of life. We recognize our umbilical connection to Mother Earth and understand that she is the source of life, not a resource to be exploited. We speak on behalf of all creation today, to communicate an urgent message that man has gone too far, placing us in the state of survival.[12]*

This theme was endorsed in the concluding message of the World Council of Churches from the "Conference on Just Peace with Earth" (Iceland, October 2017), which encouraged the recognition of the wisdom of Indigenous Peoples "who have deep and longstanding traditional knowledge of the environments that are their ancestral

---

[8] See the essays in Richard W. Miller, ed., *God, Creation, and Climate Change: a Catholic response to the Environmental Crisis* (Maryknoll, New York: Orbis Books, 2010).

[9] http://www.interfaithpowerandlight.org/

[10] http://fore.yale.edu/climate-change/articles-on-religion-and-climate-change/

[11] Interagency Support Group on Indigenous Peoples' Issues: Collated paper on Indigenous Peoples and Climate Change (2008), https://www.un.org/esa/socdev/unpfii/documents/2016/egm/IASG-Collated-Paper-on-Indigenous-Peoples-and-Climate-Change.pdf

[12] http://spiret.org/wp-content/uploads/2015/11/Formal-Statement_IndigenousEldersandMedicinePeoples_COP21.pdf

homelands." The global religious community should safeguard such spirituality and wisdom "for the well-being of all created life and the earth" for the generations to come.[13]

Jewish Voices

Judaism has long had what might be called an "ecological conscience." In close contact with flocks, wild animals, and the desert wilderness surrounding small cities, pre-exilic Hebrew culture was morally attuned to the ecosystem in which it was embedded, often articulated in the Psalms:

Praise the LORD FROM THE EARTH, YOU GREAT SEA CREATURES AND OCEAN DEPTHS, fire and hail, snow and frost, stormy wind fulfilling his command!

Mountains and all hills, fruit trees and all cedars! Wild animals and all cattle, creeping things and flying birds! (Psalm 148:7-10)

Israeli scholar Noah Efron notes that although there is relatively little in the Jewish tradition that finds intrinsic religious value in nature itself, "those who feel that their Judaism requires of them concern for the environment see this responsibility as part of their covenantal relationship with God."[14] In Israel's prophetic period (ninth to sixth centuries BCE) the Jewish prophets brought to bear a strong ethical consciousness critical of social and political injustice. Judaism is thus primed to take a leadership role in addressing the forces and structures underlying anthropogenic climate change.[15]

The Coalition on the Environment and Jewish Life (COEJL) is a network of organizations whose members include leaders from across the political and religious spectrum. COEJL has established a goal of reducing by 2050 the Jewish community's contribution of greenhouse gases by 83% of the 2005 level, encouraging the greening of synagogues, homes, and buildings: "We are God's caretakers for the earth and...destroying the conditions for much life on earth violates this duty of stewardship.[16]

The Reform wing of Judaism is represented by the Religious Action Center, which in 2009 passed a resolution on climate change and energy, supporting among other things tougher fuel economy standards and "cap-and-trade" policies to reduce greenhouse gases. COEJL recognizes the weight of our responsibility for addressing climate change:

We weep at the heavy burden that climate change imposes on the world's poor, we mourn it's impact on the diversity of God's creations, we tremble at the harm we impose upon our own descendants – and we are alarmed by our own vulnerability, here and now.[17]

Patriarch Noah and his life-saving ark from ancient Jewish myth will not be around to rescue us from the coming very real inundations brought about by rising sea levels. Tackling this awesome task will be up to present and future human beings.

---

[13] World Council of Churches, "Conference on Just Peace with Earth" (2017) https://www.oikoumene.org/en/resources/documents/wcc-programmes/final-message-of-the-wcc-conference-on-just-peace-with-earth-iceland-october-2017.

[14] Private Correspondence, Nov. 2, 2017.

[15] Stephen A. Jurovics, http://jewcology.org/2017/02/seasons-in-the-era-of-climate-change/. Jewcology.com is an excellent resource for a wide range or of writings on Judaism and climate change.

[16] The Jewish Climate Initiative, http://www.jewishecoseminars.com/ethics-basics/

[17] http://multi.jewishpublicaffairs.org/coejl/resources/jewish-energy-covenant-campaign-declaration/

Christian Voices

Roman Catholicism has a long-standing involvement with environmental matters, recognizing as it does that the ecological health of the planet is essential to the well-being of society.[18] The United States Conference of Catholic Bishops has taken a leadership role, launching in 2006 the Catholic Climate Covenant, part of the USCCB's Environmental Justice Program.[19] In 2009 the coalition developed the St. Francis pledge (named for the medieval saint and patron of animals). Those taking the pledge undertake to:

- Pray and reflect on the duty to care for God's creation.

- Learn about and teach the causes and moral dimensions of climate change.

- Assess how we contribute to climate change by our own energy use, consumption, waste, etc.

- Act to change our behaviors to reduce our contribution to climate change.

- Advocate for Catholic principles in climate change policies, especially as they impact the poor and vulnerable.

Thousands of individuals and organizations have taken the pledge, and the Franciscan Action Network translates this into practice by offering resources for incorporating climate change awareness into church worship by way of prayers, homiletical themes and liturgical music.[20] Pope Francis is both leading the Catholic Church on the issue of climate change and reaching out to other faith traditions. This is likely a contributing factor to why approximately 7 in 10 Americans view Pope Francis favorably, regardless of religious ideology.[21] In 2015 he issued his influential encyclical letter *Laudato Si': On Care for Our Common Home*, the first ever encyclical about climate change.[22]

The Orthodox communion likewise is developing strong initiatives on climate change, led by Ecumenical Patriarch Bartholomew of Constantinople. Prior to the international meeting resulting in the Kyoto Protocol (1997), Bartholomew issued this statement:

> To commit a crime against the natural world is a sin. For humans to cause species to become extinct and to destroy the biological diversity of God's creation… for humans to degrade the integrity of Earth by causing changes in its climate, by stripping the Earth of its natural forests, or destroying its wetlands… for humans to contaminate the Earth's waters, its land, its air, and its life, with poisonous substances… these are sins… It is certainly God's forgiveness for which we must ask, for causing harm to His Own Creation.[23]

Pope Francis and Patriarch Bartholomew jointly issued a statement in 2017 on the World Day of Prayer for Creation:

> Our propensity to interrupt the world's delicate and balanced ecosystems, our insatiable desire to manipulate and control the planet's limited resources, and our greed for limitless profit in markets

---

[18] Drew Christiansen and Walter Grazer, eds., *And God Saw that it was Good: Catholic Theology and the Environment* (United States Catholic Conference, 1996).
[19] http://www.catholicclimatecovenant.org/
[20] https://franciscanaction.org/
[21] http://www.pewresearch.org/fact-tank/2017/01/18/favorable-u-s-views-pope-francis/
[22] http://w2.vatican.va/content/francesco/en/encyclicals/documents/papa-francesco_20150524_enciclica-laudato-si.html
[23] https://www.patriarchate.org/patriarchal-documents/-/asset_publisher/2IzbCNORLysD/content/address-of-ecumenical-patriarch-bartholomew-at-the-environ-mental-symposium-saint-barbara-greek-orthodox-church-santa-barbara-california?inheritRedirect=false

— all these have alienated us from the original purpose of creation. We no longer respect nature as a shared gift; instead, we regard it as a private possession.[24]

Other Christian denominations have been equally forthright. The worldwide Anglican Communion issued a powerful statement on climate change at a meeting of bishops[25] held in 2015 in Cape Town, South Africa:

We believe that the problem is spiritual as well as economic, scientific and political, because the roadblock to effective action relates to basic existential issues of how human life is framed and valued, including the competing moral claims of present and future generations, human versus non-human interests, and how the lifestyle of wealthy countries is to be balanced against the basic needs of the developing world. For this reason the Church must urgently find its collective moral voice.[26]

The bishops note that it is crucial to invite into the conversation the voices of indigenous people who are most at risk, women who make up the majority of the world's poor people, and younger generations who will inherit a climate in shambles. The Evangelical Lutheran Church in America (ELCA) likewise notes that in our role as Christian believers we have an obligation to the whole world:

With the reach of our contemporary human knowledge and the power we employ in new technologies, this responsibility in terms of caring for creation now includes the global future itself.[27]

As of 2016 twenty of the ELCAs synods had passed climate change resolutions. Lutheran theologian James Martin-Schramm claims that the "climate question" is the most important issue facing the church today.[28]

The Church of the Brethren situates climate change in the context of justice, noting structural imbalances:

The industrialized nations, representing less than 20 percent of the world's population, are responsible for 75 to 80 percent of the annual greenhouse gas emissions. Yet those who live in poor and developing countries will be most seriously affected by global warming."[29]

The United Methodist Church's *Book of Discipline* (2016) places a priority on changing economic, political, social, and technological lifestyles "to support a more ecologically equitable and sustainable world leading to a higher quality of life for all of God's creation."[30] The Presbyterian Church (PCUSA) recognizes a diversity of approaches ranging from bringing local groups to awareness of their congregational impact on climate, to partnering with environmental organizations working to reduce the carbon footprint of society.[31]

An energetic movement driven by religious conviction is Young Evangelicals for Climate Action (YECA), whose members are spiritual, scientifically aware, and politically savvy. YECA are convinced that the climate crisis is both a pressing challenge to justice and freedom, and a profound threat to "the least of these" with whom

---

[24] 1 September 2017. https://w2.vatican.va/content/francesco/en/messages/pont-messages/2017/documents/papa-francesco_20170901_messaggio-giornata-cura-creato.html

[25] Anglican Bishops for Climate Justice; the Cape Town meeting was chaired by Dr. Thabo Makgoba, Archbishop of Cape Town and Chair of the Anglican Communion Environmental Network.

[26] http://www.interfaithpowerandlight.org/2015/03/17-anglican-bishops-from-all-six-continents-have-called-for-urgent-prayer-and-action-on-the-unprecedented-climate-crisis/

[27] http://download.elca.org/ELCA%20Resource%20Repository/Climate_Change_Issue_Paper.pdf

[28] James B. Martin-Schramm, "Bonhoeffer, the Church, and the climate question," in *Eco-Reformation: Grace and Hope for a Planet in Peril*, ed. Lisa Dahill and J.

[29] http://www.webofcreation.org/ncc/statements/cob.html

[30] United Methodist Church, Book of Discipline (2016), 161D

[31] https://www.yaleclimateconnections.org/2012/04/presbyterians-and-climate-change/

Jesus identifies in Matthew 25:40. Partly in reaction to obstinate deniers of climate change,[32] YECA strongly believe that God is calling people of the millennial generation to take action toward overcoming the climate crisis:

For us, this means living as good stewards of God's creation, advocating on behalf of the poor and marginalized, supporting our faith leaders when they stand up for climate action, holding our political leaders accountable for responsible climate policies, and mobilizing our generation and the larger church community to join in."[33]

Islamic Voices

Muslims comprise about 20% of the earth's population, and Muslim-majority countries are among the nations at greatest risk either from flooding or from drought and famine. The International Islamic Climate Change Symposium in 2015 issued the Islamic Declaration on Global Climate Change. Based on the essence of Qur'anic ethics ("knowledge of creation"), the declaration calls on Muslims

to tackle habits, mindsets, and the root causes of climate change, environmental degradation, and the loss of biodiversity in their particular sphere of influence, following the example of the Prophet Muhammad (peace and blessings be upon him), and bring about a resolution to the challenges that now face us."[34]

Allah admonishes believers in the Qur'an, "Do not strut arrogantly on the earth. You will never split the earth apart, nor will you ever rival the mountains' stature." (Qur'an 17:37). As noted by Professor Seyyed Hossein Nasr,

In the Islamic point of view, nature is alive. It's conscious. It follows God's laws. And what we're doing is breaking those laws in the name of our own earthly welfare, and now we're destroying the very habitat that God created for us.[35]

Religious Voices in Other Traditions

The Hindu Declaration (issued in 2015) recognizes climate change as being among the things that create pain, suffering, and violence:

We must base our response to climate change on a number of central principles, expanding on the truism that the Divine is all and all life is to be treated with reverence and respect: Internalizing vasudhaiva kutumbakam (the family of Mother Earth), promoting sarva bhuta hita (the welfare of all beings), and acting with an understanding of karma and the cycle of birth, death, and rebirth.[36]

Hindus strive for ahimsā, to minimise the harm we cause through our actions in our day-to-day lives: "When we embody this...we become servants of the Divine, with all our actions, including those in protection of the world around us and all the beings therein, becoming acts of worship."

The Association of Soto Zen Buddhists addressed climate change in 2016, placing it in the context of the Buddhist tradition of living close to nature, of stepping lightly and mindfully on the earth. Buddhism's Four Noble Truths provide the framework for diagnosing our situation and formulating appropriate guidelines: "Our

---

[32] E.g., The Cornwall Alliance for the Stewardship of Creation http://cornwallalliance.org/, publishers of the book and video series, *Resisting the Green Dragon* (2011). Cornwall reject the scientific consensus that human activities are largely responsible for the climate change we observe.

[33] https://www.facebook.com/newsyvideos/videos/10155205389703775/?hc_ref=ARTMWMVEg1wo_zQcvQWac1ChQgMyuJzC_JyyxZJ4a3Q-27rqn8_SzoM3bxGa-T55oPJk

[34] http://www.ifees.org.uk/wp-content/uploads/2016/10/climate_declarationmMWB.pdf

[35] https://www.yaleclimateconnections.org/2016/05/islamic-declaration-on-climate-change/

[36] *Hindu Declaration on Climate Change* (2015), http://www.hinduclimatedeclaration2015.org/english.

ecological emergency is a larger version of the perennial human predicament. Both as individuals and as a species, we suffer from a sense of self that feels disconnected not only from other people but from the Earth itself.[37]

The Ba'hai International Association perceives humanity as a unified whole, "infinitely differentiated in form and function yet united in a common purpose which exceeds that of its component parts." Climate change affects us as a disease affects a body.[38]

Religious Responses to Climate Change

Building upon this sampling of formal statements by world religious traditions, the final section of this chapter sketches four concrete responses that religious believers can make to the unfolding climate crisis: addressing the two main drivers of climate change, responding to crises in ecological ethics, and offering a witness of hope.

Addressing Overconsumption

Anthropogenic climate change has two primary drivers: overconsumption of resources and overpopulation of Earth by one species: *Homo sapiens*. Together these factors account for the exhaustion in a few centuries of fossil fuels laid down over hundreds of millions of years. Together they are responsible for deforestation, habitat destruction, extinctions of species, skyrocketing levels of CO_2 and methane, coral reef bleaching, melting glaciers, rising sea levels, inundation of coastal cities and island nations, accelerating ecological refugeeism around the globe, and countless other side-effects listed in this book that I may have missed. There are two aspects of overconsumption that are of particular relevance to religion.

First, if we do not reduce our consumption to a level sustainable by the annual energy input from the sun, every human aspiration, every value we hold dear, every religion we cherish, will vanish along with the human species. Can we change our consumptive habits? Replacing fossil fuels might seem daunting, but consider the context: the annual worldwide average energy consumption is roughly eighteen terawatts, the equivalent of about five billion barrels of oil per year.[39] Although this may seem like a lot, the Earth receives an average of about 84 terawatts of solar energy *per day*.[40] Granted, the majority of this solar radiation falls upon oceans and uninhabitable regions of the planet, but the key point is that it is indeed possible to think about and plan for a post-fossil fuel civilization.

David Sears notes that all religious traditions include moral injunctions against wasteful overconsumption:

> A spiritually attuned person will recognize that every creature is essentially bound up with every other creature, and that we share a collective destiny. Thus, our most fundamental attitude should be one of compassion, not acquisitiveness or aggression. This ethic applies toward all levels of creation.[41]

Religious traditions counsel the spiritual values of simplicity, caring for the less fortunate, and working for social justice. For religious leaders the challenge is to move those who are accustomed to an energy-intensive lifestyle toward a simpler way of life that can help us mitigate the dumping of greenhouse gases into the atmosphere.

Second, consumption as measured by carbon footprint is an issue of distributive justice, with consumption varying widely across the human species. Some groups exert a much greater ecological impact than do others,

---

[37] http://szba.org/a-western-soto-zen-buddhist-statement-on-the-climate-crisis/

[38] https://www.bic.org/statements/seizing-opportunity-redefining-challenge-climate-change.

[39] https://www.zmescience.com/ecology/climate/how-much-renewable-energy/

[40] http://sunposolar.com/solar-basic/basics-of-solar-energy/

[41] http://www.chabad.org/library/article_cdo/aid/255521/jewish/Ecology-and-Spirituality-in-Jewish-Tradition.htm

and often the latter communities suffer the climate effects more seriously.[42] People living at a subsistence level exercise a minimal impact upon the natural environment; at the opposite end of the spectrum are people in developed countries who consume resources at a much greater and unsustainable rate.[43]

Consider that if everyone on Earth consumed as much as the average American, we would need approximately 4.1 Earths to support our global population. Contrarily, if everyone on Earth consumed as little as the average Ugandan, the world could sustain life on an area less than two-thirds the size of Earth.[44] Ethically speaking, how much in the way of luxury or conveniences are people in developed nations obligated to forego? How much are people in developing nations obligated to give up before enjoying the fruits of advanced civilization? Who determines what level of consumption is fair and just in different regions of the world?

Let's take one concrete example of how religious communities can think about and combat a crucial aspect of overconsumption: our waste of food resources. In Eastern religions and philosophies such as Hinduism, Jainism, and Buddhism, modest consumption of food — and in some cases abstention from eating meat at all — is a time-honored element of the tradition. Fasting is equally important in the Abrahamic religions: in Judaism self-denial of water and food serves as a ritual expression of remorse, submission, and supplication; it is penance for both the individual and the community.[45] Fasting plays a similar penitential role for Muslims during the holy month of Ramadan, and for Christians during the forty days of Lent leading up to Easter. Ecological theologian Norman Wirzba urges,

"People should feast so they do not forget the grace and blessing of the world. They should fast so they do not degrade or hoard the good gifts of God. In short, we feast to glorify God, and we fast so we do not glorify ourselves."[46]

Acutely aware of the need for a drastic reduction of our human footprint on earth, religious leaders and congregations are turning to scriptural, ethical, and ritual resources within their traditions to preach and live this new consciousness about food and fasting.[47]

Addressing Overpopulation

The second primary driver of climate change is overpopulation. In the last 150 years the human population has ballooned from one billion to 7.6 billion[48] due to improved agricultural output, decreased infant mortality, and increased human lifespan. To be sure, a decrease in starvation and other forms of premature death is in itself good and worthy of celebration. But sooner or later a constantly growing human population will compromise Earth's delicate ecological balance, ensuring that humans out-compete most other species for water, food, and habitat.[49] Accelerating population growth has created a negative feedback loop: more humans demand more energy and therefore push the exploration for and extraction of more oil, coal, and natural gas. This increased availability of energy in turn supports more population growth.

But addressing the issue of overpopulation is fraught with theological peril, as it touches upon an issue regarded in many traditions as sacred: the "miracle" of human reproduction. In the Hebrew creation story God blessed Adam and Eve and said to them, "Be fruitful and increase in number; fill the earth and subdue

---

[42] Katharine Hayhoe discusses this problem, citing the Global Humanitarian Forum who report that climate change claims 300,000 lives annually in developing countries. See "At the intersection of belief and knowledge: climate science and our Christian faith," in Mallory McDuff, ed., *Sacred Acts: How Churches are Working to Protect Earth's Climate* (New Society Publishers, 2012), 84.

[43] For an extended discussion of how unsustainable high consumption is, see the excellent Worldwatch Institute book *Is Sustainability Still Possible?* ed. Eric Assadourian and Tom Prugh (Washington: Island Press, 2013). It is available as a PDF here: https://islandpress.org/book/state-of-the-world-2013.

[44] http://www.bbc.com/news/magazine-33133712

[45] See the discussion of feast and fast days in the "Jewish Virtual Library" https://www.jewishvirtuallibrary.org/fasting-and-fast-days

[46] Norman Wirzba, Food and Faith: a theology of Eating (Cambridge University Press, 2011), 137.

[47] "Fast for the Climate," Climate Action Network International, http://www.climatenetwork.org/fastfortheclimate.

[48] As of June, 2018. Source: https://www.census.gov/popclock/

[49] Peter M. J. Hess and Richard J. McDonald, "Negotiating a human future: evolution, population and ethics at the end of affordable oil," in *Science, Wisdom and the Future: Humanity's Quest for a Flourishing Earth* (Santa Clarita, CA: Collins Foundation Press, 2012), 303-304.

it." (*Genesis* 1:28). Jews and Christians have long taken this text as a mandate to increase their populations; similar support for more-than-replacement reproduction exists in other religions. The mandate is historically understandable: The Bible was written at a time when humans had virtually no control over the vectors of drought, famine, epidemic disease, and other forces that threaten communities, and people experienced an urgent need to grow their populations. However, we have succeeded many times over by now in "filling the earth and subduing it," and our proliferation as a species is coming back to haunt us and our planet mates. With a human population fourteen times greater than the roughly 500 million people living on Earth at the time *Genesis* was written, I submit that the mandate to "be fruitful and multiply" has been fulfilled many times over, and should now be shelved indefinitely.

In his otherwise excellent encyclical letter *Laudato Si' Señor: On Care for Our Common Home* (2015), Pope Francis does not address human overpopulation as a critical driver of accelerating climate change.[50] Of the sixteen occurrences of the word "population" in *Laudato Si'*, only three are relevant, and all are found in paragraph 50.[51] The first instance is a flat-out denial of the problem: "demographic growth is fully compatible with an integral and shared development." *Laudato Si'* ignores the fact that there is a biological limit to Earth's carrying capacity for humans, just as there is for every other species. The second use of "overpopulation" correctly points to the arrogance and danger of ignoring excessive consumption: to blame climate change "on population growth instead of extreme and selective consumerism on the part of some, is one way of refusing to face the issues."

However, the third reference completely fails to reflect the basic fact of biological equilibrium, and reduces the matter to "imbalances in population density" that lead to "complex regional situations." What *Laudato Si'* neglects to acknowledge is that while we have temporarily removed some of the natural constraints on our food supply, eventually our excess population beyond carrying capacity will be pitilessly trimmed by the factors of famine, disease, ecological refugeeism, and brutal wars over water, energy, land, and resources.

For a pro-life position this is a serious moral problem. Catholic missionary Sean McDonagh has asked, "Is it really pro-life to ignore the warnings of demographers and ecologists who predict that unbridled population growth will lead to severe hardship and an increase in the infant mortality rate for succeeding generations? Is it pro-life to allow the extinction of hundreds of thousands of living species which will ultimately affect the well-being of all future generations on the planet?"[52]

Responding to Ethical Crises: The Case of Ecological Refugeeism

Climate change brings to the fore numerous issues for religious and ethical reflection.[53] Here we have time for only one example: ecological refugeeism. As oceans rise, low-lying infrastructures will begin to flood, ranging from villages to airports, highways, boats docks, cargo loading facilities, passenger terminals, and entire cities. From Fiji to Cape Town to Miami, all coastal areas are under threat from rising waters, but the "canaries in the coal mine" are communities just a meter or two above sea level. Examples include Tuvalu in Polynesia and the Iñupiak community of Kivalina in the Northwest Arctic Borough of Alaska.[54] The poster child of climate-related sea level rise is the Republic of Kiribati, an island nation in the south-central Pacific Ocean. The human population of about 120,000 is dispersed over thirty-three coral atolls spread over 1,351,000 square miles of

---

[50] Peter M. J. Hess, "Is Pope Francis Pro-life? The Perplexing Silence of Laudato Si' on Human Overpopulation," at http://www.huffingtonpost.com/peter-m-j-hess-phd/is-pope-francis-prolife-t_b_10709504.html

[51] Pope Francis, *Laudato Si Senor*, Vatican website: http://w2.vatican.va/content/francesco/en/encyclicals/documents/papa-francesco_20150524_enciclica-laudato-si.html

[52] Sean McDonagh, *The Greening of the Church* (New York: Maryknoll, 1990), 65.

[53] James Garvey explores an ethical framework for discussing such issues in *The Ethics of Climate Change: Right and Wrong in a Warming World* (New York: Continuum Publishing, 2008)

[54] http://www.dw.com/en/climate-change-a-village-falls-into-the-sea/a-18717942

ocean. The average elevation of Kiribati is two meters, equal to the two-meter rise conservatively predicted by some for the end of this century.[55]

When considering the ethical questions raised by threats to these populations, it is helpful to examine a popular thought experiment by Peter Singer, "The Drowning child and the expanding circle."[56] Imagine you are walking past a shallow pond and see a child drowning. Most people would agree that you would feel obligated to jump in and try to save the child. But what if you had the same knowledge and concern for the drowning child but you were ten miles away — would you have the same obligation to save her? What if the child were 100 miles away? Most people would still answer "yes." According to Singer, very few people "challenge the underlying ethics of the idea that we ought to save the lives of strangers when we can do so at relatively little cost to ourselves."[57] In the case of climate change, we have the knowledge that our actions on one side of the globe can affect our fellow humans on the other side, but how many of us are actually willing to change our beliefs or lifestyles?

To bring this back to the threat to Kiribati: Ethically speaking can people in developed nations idly sit by while Kiribati and hundreds of island cultures are erased from the Earth? In order to preserve a culture, a language, and a maritime way of life with all its ancient customs and ceremonies and spiritual beliefs, Kiribatians need a place to live where they will not be broken up and dispersed. Settling some Kiribatians in New Zealand and some in Australia and some in the United States might save them from drowning, but it would be the end of them as a socially cohesive culture with an ancient and treasured way of life. Does the human community have an obligation to preserve Kiribatian culture as a coherent whole?

It is not only island dwellers who will be affected by climate refugeeism. People seeking to move from Africa to Europe, from Bangladesh to India, or from Indonesia to Australia will be one of the central moral issues of the twenty-first century. In 2013 an overcrowded fishing vessel sank in the Mediterranean near the small Italian island of Lampedusa, killing more than 360 people.[58] India has built a 2,100-mile-long fence along its border to try to keep Bangladeshis from entering, and desperation may drive people to violence.[59]

Do people who have profited most from the industrial development that created global warming have an obligation to step up in support of peoples in need? Are countries morally obligated to welcome refugees from nations that border them, but not obligated to take in refugees from more distant nations? Upon which nation is it incumbent to offer safe haven for the population of a gradually submerging island nation in the Pacific? Religious communities have under our cultural belts centuries of ethical reflection about human rights and responsibilities, but we will need to employ every tool possible in conducting honest and deeply reflective conversations about the issue of huge numbers of people desperately seeking to escape catastrophe and settle in a new homeland.

Offering a Prophetic Witness of Hope

A fourth area in which religious leaders can play a crucial role is offering a witness of hope.[60] What do you say to your children when you realize that their experience of the world will be quite different from yours? How do you negotiate a delicate conversation with your teens about whether or not they should have children of their own?[61]

I have grounded my two sons in the knowledge that the universe is ancient, dynamic, ever-evolving, and incomprehensibly vast. Change is an essential part of our world, without which neither humans, nor plants and

---

[55] http://www.climatechangenews.com/category/south-america/

[56] Peter Singer, "The Drowning Child and the Expanding Circle" In *New Internationalist*, 1997. https://www.utilitarian.net/singer/by/199704—.html

[57] https://www.utilitarian.net/singer/by/199704—.html

[58] Peter M. J. Hess, "Climate change: upping the ante for environmental refugees," https://ncse.com/blog/2013/12/climate-change-upping-ante-environmental-refugees-0015256

[59] http://www.nytimes.com/cwire/2009/03/23/23climatewire-a-global-national-security-issue-lurks-at-ba-10247.html?pagewanted=all

[60] Edward P. Echlin develops this theme in *Climate and Christ: a Prophetic Alternative* (Dublin: Columba Press, 2010, 84-115).

[61] See Peter M. J. Hess, "Children, Climate Change, and Hope in Uncertainty," http://godandnature.asa3.org/column-clearing-the-middle-path1.html

other animals, nor the geographical features that make our world exciting and beautiful, would exist at all. The issue of abrupt climate change carries a great deal of uncertainty about its extent and significance. How will we estimate the likely magnitude and effects of climate change? How will global warming affect people in different parts of the world as sea levels rise? What **can** we do about it? What **should** we do about it? Where do our moral obligations lie? How do we put into practice Jesus' admonition to "love the Lord your God with all your heart and with all your soul and with all your mind, and love your neighbor as yourself"? (Matthew 22:37-39).

There is reason to hope. We share with our non-religious sisters and brothers the conviction that biodiversity and ecosystemic integrity are profoundly important. In the face of wide-ranging and potentially catastrophic threats, we must remain confident that life on Earth will find a way forward as it has for 3.5 billion years. As religious believers, we ground our hope in the conviction that life in the universe has meaning and purpose. A decade ago, the United States Conference of Catholic Bishops issued a prescient statement about climate change:

> As people of faith, we are convinced that "the earth is the Lord's and all it holds" (Psalm 24:1). Our Creator has given us the gift of creation: the air we breathe, the water that sustains life, the fruits of the land that nourish us, and the entire web of life without which human life cannot flourish. All of this God created and found "very good." We believe our response to global climate change should be a sign of our respect for God's creation.[62]

The silver lining of the cloud might include the fact that exciting careers are emerging in alternative technologies, sustainable industries, and environmental engineering. People who have hope can think seriously about realistic and practical solutions, and can re-envision how life might be even better without our current profligate use of resources.[63]

Conclusions

All who accept the reality of global climate change have some personal experience that brings the reality home to them: powerful hurricanes in the Caribbean, melting glaciers in Greenland, droughts in the Middle East, permafrost melt in Siberia. My context of understanding is the increasingly severe wildfires in the American West.[64] If our civilization and our fellow earthlings are to survive far into the future, the unreflective dogma that consumption, industrial expansion, and human population growth can go on forever on a finite planet must be changed. David Brower used to refer to the default assumption of infinite growth on a finite planet as "stark raving madness."[65]

The world's religious communities have a vital stake in this endeavor, for climate change carries major challenges for every aspect of human existence. Issues of international justice confront us in the differential ability of rich and poor nations to deal with inevitable sea level rise, as for example, in the comparison between Florida and Indonesia. Intergenerational justice requires that we not pillage the planet relentlessly, but that we leave it flourishing for thousands of generations to come. And ecological justice requires that we consider not merely the needs of our species, but the integrity of creation, the global ecosystem as a whole.[66]

A million years from now Earth will still orbit the sun. The future diversity of life on our planet is in the hands of two generations: the one now in control of society and the generation of our children. Our actions during the next thirty years will determine the kind of world we leave to future humans and to the plants and

---

[62] http://www.usccb.org/issues-and-action/human-life-and-dignity/environment/global-climate-change-a-plea-for-dialogue-prudence-and-the-common-good.cfm
[63] Hayhoe, "At the intersection of faith and knowledge", 84-86.
[64] https://insideclimatenews.org/news/18102017/california-wildfires-global-warming-drought-wind-climate-change-fire.
[65] John de Graaf, "Memories of a conservation giant," *Sierra Magazine* (2012). See also Terra S. Rowe, "Grace and climate change," in Dahill and Schramm, eds., *Eco-Reformation*, 255.
[66] See Jame Schaefer, "Environmental degradation, social sin, and the common good," 69-87 in Richard W. Miller, ed., *God, Creation, and Climate Change: a Catholic response to the Environmental Crisis* (Maryknoll, New York: Orbis Books, 2010).

animals with whom we share our evolutionary history. Let us make our decisions wisely, using every tool at our disposal — scientific, historical, economic, political, ethical, and religious. With nature threatened on every side by the rapacious demands of an ever-expanding human population, it is up to our species alone to safeguard the vision articulated in Psalm 148, that all of nature should praise God, from the highest heavens to the depths of the ocean:

Let them praise the name of the LORD, for at his command they were created, and he established them for ever and ever—he issued a decree that will never pass away.

Climate Crisis "Keep/Stop/Start" Actions

Below for your consideration, are several author-suggested thought, behavior, and action items. Action nourish inspiration! For a further explanation of the Keep/Stop/Start organizational tool, please see How to Read This Book.

KEEP:

Keep informing yourself about the relationship between religion and climate change. Worldwide, more than eight in ten people identify with some sort of religious ideology. By emphasizing the importance of environmental stewardship of the natural world, the faith community can be mobilized to join the resistance to the climate crisis.

As described above, each religion has a nuanced view of climate change. But I hope it is now clear that nearly everyone agrees that there is a moral obligation on the part of humans to take care of the natural environment, as this has cascading effects on social and economic equity as well. The faith community is one of the most resilient, steadfast, and motivated of human communities, and cannot be ignored in our efforts to solve the climate crisis.

STOP:

Stop allowing extremists to distort your view of religion. Regardless of which religion you examine, there will likely always be an extremist on one end of the spectrum who tries to steal the attention away from the responsible community. The news media know that people are fascinated by outlandish stories and they cover such extremes to excess. This has become evident in recent years as Western media focus on terrorism perpetuated by religious zealots.

As a result, the public has formulated a distorted view that promotes fear, anger, and suspicion of unfamiliar religions. The overwhelming majority of religious believers do not engage in or condone behaviors exemplified by extremists. We should resist glamorizing those aspects of religion that are divisive, and focus rather on addressing in a united chorus the global concerns we all share.

START:

1) *Start stepping out of your comfort zone and engage with other communities.* Differences in religious ideologies have sparked conflict for centuries, but it is now crucial to put these differences aside and work toward a common goal. The effects of climate change will not discriminate between Christians, Muslims, Jews, Buddhists, Hindus, and atheists. Empathy and understanding are essential instruments if the global community is ever to come together to combat climate change effectively and ensure environmental and social justice for all.

2) *Start engaging with faith communities to learn about their unique perspectives, teachings, and actions with respect to climate change.* To understand what solutions will be the most effective, we must examine what has and has not worked in different communities around the globe. Since there is no single correct strategy for combatting climate change, plural approaches can and should be evaluated and adopted where appropriate. The contrasting worldviews, cultural traditions, and geographical contexts of different religious communities offer us a rich panoply of resources on which to draw as we craft our collective human response to the climate crisis.

~ ~ ~

Peter M.J. Hess, Ph.D. is a Roman Catholic theologian who specializes in issues at the interface of science and religion. He has served as international director for the Center for Theology and the Natural Sciences (CTNS) in Berkeley, California and director of Outreach to Religious Communities with the National Center for Science Education (NCSE) in Oakland, promoting dialogue in the areas particularly of evolutionary biology and climate change. Peter Hess earned an M.A. in philosophy and theology from Oxford University and a Ph.D. in historical theology from the Graduate Theological Union in Berkeley. He is co-author of *Catholicism and Science* (Greenwood Press, 2008) and writes on the religious and ethical implications of a growing human population in light of climate change. Peter lives in Berkeley, California. Read more at: www.climateabandoned.com.

Chapter 29
Climate Crisis & Sports

by

Brij M. Singh

"Hockey captures the essence of Canadian experience in the New World. In a land so inescapably and inhospitably cold, hockey is the chance of life, and an affirmation that despite the deathly chill of winter we are alive." **– Stephen Leacock**

Across the harsh Canadian winters, with temperatures consistently below freezing and snow adrift for months at a time, thousands rally around a sport as inhospitable as the barren winter: ice hockey. Whether in large metropolises, such as Toronto or Vancouver, or small remote villages from Dawson's Creek, British Columbia to Moose Factory, Ontario, every weekend in the winter months is full of the same: families, communities, and ultimately, memories being made in one of Earth's most inhospitable conditions. Some of Canada's most famous emerged from these outdoor paradises, such as Wayne Gretzky[1] or Jordan Eberle[2], who often speak of their fondest memories on outdoor rinks, learning and developing the skills that lead to professional careers.

Stories of sport uniting communities is not exclusive to Canada. Football, or soccer in North America, is similar in its ability to connect around the globe. The roots of these lifelong passions come from humble beginnings, whether on the streets of Nairobi or Naples. Moreover, the impact sport can have on lives is second to none. The modern-day churches of the twentieth century have become stadia and arenas, with gospels singing praises of players and coaches rather than holy figures. These arenas, just like the churches they have replaced, drive a hard bargain for membership, with high prices for entry, food, drink, and merchandise.

With a deep-seated passion at a personal level, a worldwide experience that can be considered holy, and the ability to capture billions of dollars in revenue, it is truly a wonder that sport has not prepared for the immediate climate crisis. Like many other industries mentioned, sport has taken baby steps to prepare for an uncertain future, implementing programs to try to mitigate their ecological foot. However, without major fundamental changes for sport as an industry, sport could lose an advantage to positively alter the global landscape, literally. In the next few pages, I will outline some of the sports industry most egregious offenses towards abandoning the climate, while also pointing out a few of the successes, failures, and possible areas to begin the healing process. The sport industry holds a unique place of change, and hopefully, with the right guidance nurturing, it can play a pivotal role in righting the climate crisis.

(Not Enough) Winter Sports in Crisis

At this current time, the run-up to the 2018 Winter Games in PyeongChang are well underway. The world's most elite athletes in a wide variety of sports are en route, along with their coaches, trainers, and sponsors. With billions invested into the athletes, fans spending thousands on travel to South Korea, and television companies bickering for the rights to air the games in their respective countries, little could be left to chance. However, the largest question mark is not on the athletes' performance or which country departs the Korean Peninsula with the most silverware, but rather whether the Games go on as planned.

---

[1] Wayne Gretzky is one of the most accomplished ice hockey of history and is often referred to as the greatest of all time. His records for goals, assists, and awards are routinely characterized as "unbreakable"

[2] Eberle is a household name for many young Canadians; his goal for Team Canada with six seconds remaining against Russia in 2009 is often referred to as "The Goal"

On all of these sports, whether it be skiing, ice hockey, figure skating, or even curling, cold weather and possibly snow is necessary for the athletes to compete. With the current rises in global temperatures, this seeming non-issue becomes a major point of concern in the lead-into the games. Eight years ago, in Vancouver, airlifted blocks of snow and straw were brought in an attempt to mask the unseasonably warm conditions on the slopes, with the Vancouver Games Organizing Committee even employing the use of the Fire Brigade and Army helicopters.[3] According to a Guardian report, temperatures were 38°F higher than typical for the British Columbian February, throwing a curveball that no amount of TV money could solve preemptively. Not only is this an issue to deal with for organizers, where billions of dollars have been invested into the games running on a tight schedule, it poses a safety risk for athletes who train and compete on ideal conditions year-round. Athletes demand pristine conditions, especially when their entire life's work is judged in a matter of minutes, if not seconds, in pursuit of an Olympic Gold Medal[4]. With athletes unhappy, combined with pressure from television companies to keep schedules, along with sponsorship obligations to not cancel or delay events, no wonder the organizing committees utilize all their available resources to fight the changing climate.

Alaska, arguably the state most renowned for their cold-weather hostility, also endured drastic climate changes which in turn affected their most global sport, the Iditarod Sled Dog race. This epic race pits mushers alongside sixteen sled dogs through a journey from Anchorage to Nome, a distance of 352 miles, covered entirely via sled dogs. This event is often hailed as the last "great race" on earth, due to its hostility and duration, alongside the showcasing of snow-loving Huskies in their finest form. In 2016, Alaska held its first snow-less February day in its largest city, Anchorage.[5] As such, the start of the Iditarod was marred with cars packed with snow headed for the trails in an effort to preserve the race schedule. Anchorage received a pithy 7.9 inches of snow since winter began in December; yearly averages predict over 60 inches in the same timeframe.

The ability to host the Winter iteration of the Olympics has become a very small circle of candidate cities and countries. Past Olympic sites, such as Sarajevo, Bosnia and Herzegovina and Sapporo, Japan, do not have the climate to re-host the games in the future. In fact, according to a report from the University of Ontario listed 9 of the previous 21 Winter Olympic sites might "never be cold enough to host the Games ever again"[6] with climate being the largest affecter. With the 2022 Winter Games set to be hosted at the 2008 Summer Games site of Beijing, China, questions of winter fitness are looming larger than ever on whether the Winter Olympics will return to the moderate climates of the early twentieth century.

Canada's ice hockey past is also at risk due to the rise in global temperatures. Between 1951 and 2005, Canada's average temperature rose 40°F, a change so drastic that outdoor skating could be history. Ottawa's Rideau Canal, a stretch of canalway that freezes over for the winter, is under threat as well: "With current emission trends, the canal's skating season could shrink from the previous average of nine weeks to 6.5 weeks by 2020, less than six weeks by 2050 and just one week by the end of the century."[7] Current greenhouse gas trends could seriously endanger one of Canada's most treasured activities, and steal moments of joy, happiness, and culture from future generations of Canadians.

Too Hot to Handle

Winter is not the only season affected by the climate crisis. The summer sports have also required serious adjustment to ensure the safety of the players and fans alike. No example of these adjustments can be seen better than in the first major sporting event each calendar year, the Australian Open Tennis Tournament. After

---

[3] https://www.theguardian.com/sport/2010/feb/10/vancouver-lacks-snow

[4] Olympic Gold is often the standard for elite athletes; many consider it the peak of athletic achievement regardless of era or sport.

[5] https://climatenexus.org/climate-issues/climate-change-and-sports/

[6] Source: http://www.independent.co.uk/sport/football/news-and-comment/sport-climate-change-football-golf-cricket-become-part-of-the-solution-problem-leeds-university-a8201986.html

[7] https://www.straight.com/news/344961/david-suzuki-future-outdoor-ice-rinks-risk

consulting with players, the 1998 iteration was the first to institute an extreme heat policy, where no match can take place when the temperature is above 35°C. This was designed with players' and fans' heat in mind, as the games during midday would regularly result in heat strokes and cramps for players and fans alike. In 2014, the Aussie Open organizers faced significant criticism when upgrading the heat policy to 104°F, after determining that games were regularly played above 95°F, and caving to external pressure on limiting stoppages for more fluid play. Temperatures that year hit over 107.6°F, with 970 fans being treated for heat exhaustion on one day alone, setting a new tournament record.[8] While normal for the Australian Open to be a hot event, an extreme heating policy is only one example of the effect global warming has had on the sporting world.

These conditions were not only matched in Australia. The 2015 US Open Tennis Championships also faced comparable high levels of high heat and humidity. Shockingly, there is an extreme heat policy for women players, but not men.[9] Of the sixteen retired players at the 2015 US Open, fourteen were male, with many players citing the high dew point and consistently high temperatures as the reason for retirement.[10]

While tennis has fared poorly over the rise in global temperatures, American football has also seen its share of rule changes and record statistics regarding heat, especially at lower levels. High school football, one of the largest activities for men during their teenage years, has experienced large changes to the rules regarding their typical offseason, summer-months conditioning. In Georgia, where the temperatures often reach above 86°F for most practices, extreme heat policies are the latest to limit practices in both time and padding, with high humidity often yielding shorter practices and fewer pads. A study[11] by Andrew Grundstein of the University of Georgia that "deaths of high school football players due to heat nearly tripled from 1994 to 2009 compared to the previous 15 years"[12] Maintaining grass fields for football also has become more expensive due to the increasing temperatures, with many short-funded schools often turning to rubber-filled artificial turf to decrease maintenance costs. However, this only exacerbates the problem of heat-related illness, as black rubber filler keeps field temperatures higher for longer, increasing the likelihood of a heat-related injury. This positive feedback loop only increases the damage the rising temperatures has on young athletes, in turn fueling the effects the climate crisis has on sport, putting athletes at risk due to no fault of their own.

Marathons, an activity known for its endurance, could see the finish line prematurely as rising temperatures threaten marathoners worldwide. Faster times are often run when the temperatures are just above freezing at 40°F. An increase to a typical spring day in the mid 60's °F could lead to an increase of about 12 minutes[13] for the average finisher. Elite finishers could add up to four minutes if temperatures were in the 70s, which in the sport of competitive marathoning, is an eternity. The New York City Marathon, one of the world's most prestigious, is experiencing the heat as well. According to the New York Times, "Over the last 50 years, the temperature has exceeded 60 degrees on just 5 percent of the days during the first week of November, when the New York race is typically held. By 2050, this is projected to rise to 18 percent of days, and by 2090 it is expected to be 38 percent." These changes could mean the world's fastest runners chasing elite times could find themselves in cooler climates, such as Scandinavia or Canada, rather than the current prestige locations of New York and London. The Berlin Marathon is often seen as one to attempt a world record pace, with seven of the ten fastest marathons ever recorded occurring in the German capital. However, when Dennis Kimetto of Kenya came to Berlin in 2017, his attempt fell short five kilometers in, as the rain and humidity set in, eventually discouraging

[8] http://www.skynews.com.au/topstories/article.aspx?id=942452
[9] The Women's Tennis Association, or WTA, has a rule for extreme heat, whereas the Association of Tennis Professionals, or ATP, for men has no such policy. All major tennis tournaments are under the rulesets of the WTA or ATP, rather than individual tournaments.
[10] http://www.newsweek.com/heat-humidity-knock-players-us-open-federer-murray-369194
[11] https://www.ncbi.nlm.nih.gov/pubmed/21161288
[12] https://www.scientificamerican.com/article/is-climate-change-making-termeratures-too-hot-for-high-school-football/
[13] https://www.nytimes.com/2017/11/03/upshot/your-race-against-time-how-climate-affects-the-marathon.html

Kimetto of the world record pace. With marathons, timing is everything, and seeing the world's best chasing times in November, December, or January might become commonplace if the climate crisis cannot be reined in.

In a rather bizarre case of fighting the heat, the 2022 FIFA World Cup has looked towards artificial clouds as a method to keep outdoor games cool. Awarded the largest football (soccer) tournament in the world, Qatar, a small coastal nation in the Arabian Peninsula, has looked to negate the 104°F summer average typical for the Middle East by investing in artificial clouds[14]. In 2015, the British firm Arup presented a prototype for a proprietary, suspended cooling "cloud" which could decrease temperatures to nearly 73.4°F, allowing the World Cup to proceed on its typical summer schedule.[15] This prototype was successful in decreasing temperatures in a 500-capacity mock-stadium, but skeptics are still prevalent in its effectiveness in decreasing temperatures for a 50,000-seat ground typical for the World Cup. While the effect of the Qataris is noble in an entrepreneurial setting, the investment into the technology must seem misguided. With such a disruption to the footballing world already[16], it's no wonder many fans around the world are calling for a more reasonable approach.

America's pastime of baseball has also fallen to the might of rising global temperatures. However, unlike other sports who have failed to prepare, baseball has taken steps to mitigate the damage done to their prized athletes. In 2010, the Minnesota Twins moved across town to their new ballpark, the open-aired Target Field susceptible to the local climate, leaving the Hubert H. Humphrey Metrodome, a domed stadium where temperatures were always between 68-72°F, the athletic trainers prepared players by increasing hydration, monitoring players via body weight measurements, and providing IV saline treatments in the case of dehydration.[17] This preparation, along with water-misting fans and air-conditioned clubhouses provide some heat relief. Organizations have also begun to provide services for fans, such as free water, hydration stations, and increased medical staff on hot days. These changes have patched baseball for a while, but while temperatures continue to rise, the threat of a midsummer break to avoid the heat might seriously endanger the national pastime.

Baseball also faces threats from an invasive species which troubles their famous bats. White ash wood has been used around professional baseball for over a century, as its hardiness and stability prevent the bats from shattering on contact with the 90 mile-per-hour fastballs pitchers throw. However, an invasive species of beetle, the Emerald Ash Borer, is threatening major bat producers, such as Rawlings or Louisville Slugger. The Borer, which is an invasive species native between the Midwest and the Northeast, "chews layers of healthy bark into pulp"[18], destroying the wood that would become a healthy baseball bat. According to Rawlings, 70% of their 300,000 bats are made of white ash, the preferred snack of the Emerald Ash Borer. Luckily, this won't kill baseball as we know it, it will rather force other, less ideal wood types, such as maple or birch, to be used instead. As devastating as this is, it could provide the ideal wakeup call for professional baseball to protect their roots a little more.

Wet Troubles

With the increase in global temperatures, the melting of ice caps in our Artic and Antarctic, and more volatile precipitation patterns, sports near and around the water have needed to adjust and adapt for the future. One prominent, and somewhat bittersweet, example could be seen in the world's most financial sport, golf. With the increasing prevalence of warm weather, North American golf seasons might be extended by seven weeks, which comes as good news for enthusiasts. However, rising tides on coastlines have already begun to make large impacts

[14] http://www.bbc.com/news/magazine-31608062

[15] As opposed to the FIFA recommendation to move the world-famous tournament to the winter, which would decrease the temperatures to slightly more bearable 30ºC weather.

[16] The 2022 Qatar World Cup has had a very rocky existence: allegations of corruption to achieve the games, an early estimate that 4000 migrant workers will perish working on constructing the stadium, and camps of workers held without passports, visas, or outside contact have rocked the Qatar World Cup organizing committee.

[17] https://www.huffingtonpost.com/elliott-negin/major-league-baseball-cop_b_1785103.html

[18] https://www.npr.org/2016/08/03/488432537/a-beetle-may-soon-strike-out-baseballs-famous-ash-bats

on the courses around the world. In Scotland, the coastline on the Montrose Golf Course has moved in 70 meters, swallowing the third tee box in the process[19]. In the 2016, flooding in West Virginia forced the cancellation of the Greenbrier Tournament[20], with it stuffing a $1.7M purse for a rainy day.

Golf is not a sport that can easily handle the weather, with many players and fans traveling around the world to attend weekends away at the Master, PGA Championship, and US and British Opens. These events are planned years in advance, with the assumption that weather will be the last event to postpone or cancel events. Even with the purses of these events rising to over $11M[21], and the winner taking home a hair shy of $2M, golf can only be played in ideal conditions with limited or no rain. It makes one wonder if the golf industry could self-preserve itself with donating some of its large purses towards the preservation of their beloved greens and fairways. Especially if it means there are fewer bunkers of sand around the edges of their courses.

Surfing could also become the next world sport to radically alter their season or sport due to the effects of climate change. Ocean acidification impacts the world's surf spots, which are often dangerously close to collapse without the effects of a rising climate. Ocean temperatures are 90% of the increasing warmth taken in by climate change. As carbon-based pollution is spewed into the air, the ocean, a collector of over 25% of the world's carbon emissions, increases its acidity and increases the temperature of the earth. This in turn effects fishing, especially shellfish, but also coral reefs, which are extremely fragile ecosystems built upon steady temperatures. Even a 1°C increase could mean life or death for a coral reef, which provide the world's best surf breaks for surfers. The Great Barrier Reef, the largest reef in the world, also provides the Gold Coast of Australia the title of "world greatest surf" and is extremely vulnerable to climate change. Rising tides also affect surfers, as waves on higher sea levels break differently, and often, in unpredictable ways that make surfing too difficult.[22]

The community most adversely affected by the rising climate has been fishing. Two major effects of the climate crisis, rising ocean temperatures and increasing acidification of ocean waters, are affecting fishing communities around the world. In addition to rising sea levels, which affect coastal communities, fishing could see radical changes in the coming years unlike any other sporting industry. Tuna varieties, especially Asian skipjack tuna and bigeye tuna, are moving away from traditional fishing communities in southeast Asia and towards the open seas of the South Pacific, putting small islands in the South Pacific at extreme risk of starvation.[23] Atlantic Cod, a now-endangered fish, declined severely in the 1990's due to the overfishing, are severely at risk due to the lack of genetic diversity and young ages.[24] When combined with the oceanic changes, the Atlantic Cod, once harboring the world's largest population of cod, could be wiped out entirely. Low-lying countries, such as the Maldives and Tuvalu, are particularly at risk as their dependence lies entirely with fishing. Without their one source of income and sustenance, these countries may become the first climate refugees.[25]

Climate change is not only affecting the fishing community abroad, there are serious challenges faced within the United States. In Louisiana, rising seas have taken over 1,800 miles of coastline away from fisherman in the last seventy years.[26] Lobster harvest in New England fell from 9.4M pounds to just over 200,000, a fall of 97%[27]. This fall is not due to overfishing or a fall in the population; it's due to location. Lobsters are flowing away from their traditional home on the New England coastline to cooler water up in Maine, where the industry has surged 219% in the same timeframe as New England's 97% drop.

[19] http://www.telegraph.co.uk/sport/2018/02/07/climate-change-disrupting-british-sport/
[20] http://www.bbc.com/sport/golf/36630684
[21] http://www.telegraph.co.uk/golf/2017/04/09/masters-2017-prize-money-much-will-winner-earn/
[22] https://thinkprogress.org/endless-summer-how-climate-change-could-wipe-out-surfing-9209acd46fb9/
[23] https://think-asia.org/bitstream/handle/11540/531/spso-201003-fisheries-climate-change.pdf?sequence=1
[24] https://www.nytimes.com/2012/02/12/us/cod-fishermens-alarm-outlasts-reprieve-on-catch-limits.html
[25] https://www.youtube.com/watch?v=M4VRb6myfzU
[26] https://pubs.usgs.gov/sim/3164/
[27] https://www.climate.gov/news-features/climate-and/climate-lobsters

Rain has also been a problem, especially in one of the most rain-affected sports, cricket. The stick-and-ball sport, famed for its popularity amongst former British colonies, cannot be played if the grass is too wet. England, besides its reputation for its incessant rain, has had six of seven wettest years in the past decade. England's national cricket side has had to shorten 27% of their home One Day Internationals since 2000 due to rain.[28] Since 2011, this rate has doubled for shortened games, and around the British Isles, county championships have taken away "at least 175 days [of cricket] in five of the last ten years."[29] These effects have fallen down to the The England and Wales Cricket Board, who have seen a rise in emergency grants due to weather from £1M to £1.6M. While a sport of little importance in North America, the largest fanbase of cricket can be found in India, which also happens to be one the world's largest producers of greenhouse gases. Coming only second to religion, cricket can be a major motivator to limit and negate greenhouse gas production, especially when the threat of fewer games looms larger and larger each passing day.

What's Next

With the climate in crisis, and many aspects of our pop-culture, from sport to business to politics, seemingly devoid of any interest to right the ship, there are still some rays of hope. The sports industry is a $70B industry, with fans in hundreds of sports around the world, rooting for millions of teams. Sponsors invest more into football than any other single industry, and the American Super Bowl draws nearly $5M for a single 30-second advertisement to the world's largest audience. Surely, sport can utilize its influence to rally troops into fighting the climate in crisis.

Strides have been made, absolutely. The largest and most supported sporting club in the world, Manchester United, developed a nature reserve on their training property in Carrington, and their crosstown rival Manchester City developed 30 hectares of marshland and wetland from a toxic wasteland. The NHL has been the first league to develop a sustainability department, NHL Green, given that ice hockey is the only major professional sport that requires a cold climate to compete, train, and learn. This has been followed up by the other major professional sports, NBA Green, MLB Green, and NFL Green. These initiatives, such as the NHL's Green Week, can bring serious light to the climate change crisis, and ultimately help drive change in the social sphere that sport has begun to occupy.

While some might believe sport to be independent of the world's geo-social-political problems, the sports world has been inexplicably tied to the tides of people. In the 1936 Olympics, held in Hitler's Germany, Jesse Owens, a black man from Alabama, stormed the podium, winning gold medals and muffling the Aryan superiority held by the ruling Nazi Party. In the wake of the civil rights movement, John Carlos and Tommie Smith held up raised fists in black gloves without shoes after winning the gold and bronze medal in the 100m sprint at the 1960 Summer Games in Mexico City, cementing their place in history standing in solidarity with their brothers in arms fighting for their rights stateside.

Many have tried to stifle the voices of athletes and social causes throughout history; The St. Louis Cardinals threatened to strike if they were forced to play against Jackie Robinson[30]. Peter Ueberroth, President of the US Olympic Committee during John Carlos's and Tommie Smith's sign of solidarity in Mexico, revoked the medals of the athletes and banished them from representing the USA. When Jesse Owens returned from Berlin, he was not greeted with a message from the President or an invitation to the White House, an honor only given to white gold medalists. More recently, Fox News correspondent Laura Ingraham told NBA superstar Lebron James to "shut

---

[28] One Day Internationals (ODI's) are a form of cricket which takes place over the course of one day. Traditional "test" cricket takes places usually over four days of play, whereas ODI's are quicker, more fan-friendly events.

[29] http://www.telegraph.co.uk/sport/2018/02/07/climate-change-disrupting-british-sport/

[30] Jackie Robinson was the first African-American to play Major League Baseball after the institution of the color barrier, a gentlemen's agreement between the owners and managers of MLB teams to not sign any players of dark complexion after 1880.

up and dribble" rather than speak on political topics. Sports and the national topics of discussion have always been controversial but discussing the impact of climate change might pose the best use of athletes' united voices.

While hopeful, there is one serious hurdle for the sports industry as a whole must do to achieve the same level of success in developing, sustaining, and achieving public interest: industry interest. Frankly, the sports industry has little interest currently in raising awareness on climate change. Jonathan Toews, captain of the NHL's Chicago Blackhawks, has been outspoken on the effect of the warming Earth, but he is the only voice representing the millions who grew up plying their trade on outdoor rinks in Canada. These superstar caliber athletes already have the world's attention at their fingertips. Social movements, such as the Ferguson riots of 2016, were heightened due to athlete interest. Colin Kaepernick's choice to kneel during the National Anthem sparked debate on police brutality, amongst many other topics. The industry has made steps; forming these departments on sustainability, starting the conversations with the public and press, and changing the tide supplemented with facts and reason rather than feelings or trust. These are a good start, but more can always be done.

These are the issues athletes currently care about, and rightfully so. Many athletes come from poor and underprivileged backgrounds; to this group, highlighting the injustices of their past are natural and empathetic. However, these same communities where the athletes call home are also the very areas where climate change will have its largest effect. Global football, a sport played by billions, and whose elite athletes notably come from poor homes, are the most at risk of flooding, desertification, and monsoons. Unfortunately, football clubs in Chelsea, Manchester, and Paris, owned by billionaire tycoons from Russia or the Middle East, will largely be unaffected. It's a paradox seen before, the players, traditionally of a lower social standing, at ends with elite business leaders with little care or regard for the qualms of the lower class.

However, for this campaign to be successful, and turn elite athletes into the social movers they have embraced in other ways, it will take a combination of both the rich and poor. The situation of the climate crisis is too dire to leave into the hands of the common people. Ultimately, sport provides a unique platform no other subgroup can provide. While fights emerge daily about politics, sport is a fabric that unites the world together. No other social construct can put the rich and poor, black and white, gay and straight, liberal and conservative, together, united on a simple, objective goal. With it, the athletes and celebrities of the game holds immense power to educate, promote, and sustain the momentum of a cause they choose. Rather than picking sides, or starting arguments, there is a pressing, necessary issue that can be solved with awareness and education. The climate needs the attention of the world and its leaders. There is no better mouthpiece for the world to hear than the pastors, champions, leaders, and best of today, the sport industry.

Climate Crisis "Keep/Stop/Start" Actions

Below for your consideration, are several author-suggested action items. For a further explanation of the Keep/Stop/Start organizational tool, please see How to Read This Book.

KEEP:

There are many things that are going well in the sports world. Here's a few of my favorite and easiest ways to keep the ball rolling (pun intended).

1) *Keep holding hold athletes to a higher standard* than other public figures. There's a strange and odd fascination with holding our star athletes, whether it be Tom Brady or LeBron James, to a high moral

standard. Athletes are still put on teams, just like their sports, regarding their support or advocacy of certain topics. (See "What's Next" Section)

2) *Keep advocating for and encouraging* professional athletes to use their platform to motivate others to act on climate. Winter sports athletes are the most obvious candidates for this type of platform, as their livelihoods directly depend on the climate and subsequent snowfall and cold temperatures, but many other athletes are poised to make a difference as well. A perfect example of this was when long distance swimmer Lewis Pugh swam across the North Pole in 2007 to highlight melting Arctic sea ice.[31] Sports fans across the board took notice and many were inspired to look into the effects of climate change and see what they could do to help. As such, we should continue to encourage athletes of all sports to use their platforms to progress toward mitigating the climate crisis.

STOP:

1) *Stop supporting environmentally harmful sports-business practices.* This one will likely take a bit of research on your part, but it is also likely to make the biggest difference. Some things to start with are to first research where your sports apparel comes from and their supply chain. Companies like Nike and Reebok have already begun to cut carbon emissions from their supply chains and have made significant efforts to make environmentally-friendly, sustainable products. Research where you buy your workout clothes from and see if they are putting in the same efforts.

2) *Stop supporting mountain businesses that do not sustainably produce artificial snow.* Making snow is a catch-22: the less natural snowfall in a season, the more of a need to produce artificial snow to cover the slopes, which requires more water and energy to produce, and in turn burns more carbon emissions and removes water from the natural environment. As climate change worsens, more and more ski slopes are beginning to implement this practice, however very few are doing so in a sustainable manner. Research the mountain that you prefer to take your turns and see how they compare.

3) *Stop supporting sports-businesses that don't recycle.* It is now 2018 and there is no excuse for any private enterprise that doesn't provide recycling infrastructure. This is essentially the bare minimum that businesses can do to help combat the climate crisis and yet many still have not caught up with the times. Before buying a ticket to the big game, do a quick Google search about the recycling and waste stream infrastructure of the stadium and they don't support climate action, don't support them.

START:

1) *Start supporting the of athletes making a stand,* I would urge fans and supporters of sport to look toward the executive boxes as well, and support the owners, general managers, and coaches who also make stands towards positive growth in the environment. Frankly, while players have the most visibility of any of the characters in the sport industry, the real power lies in the *team behind the team*; the support staff and bankrollers of the sporting world. Scott Jenkins, the general manager of Mercedes-Benz Stadium in Atlanta, is also the chairman of the Green Sports Alliance of over 500 teams in 15 leagues over 14 countries[32]. Jenkins is right in saying that impactful business leaders "can't afford to live on the sidelines"

---

[31] https://www.sporttechie.com/4-eco-athlete-stars-fighting-climate-change/

[32] https://www.usatoday.com/story/sports/nhl/2018/03/28/save-ice-nhl-climate-change-sustainability-report-green-environment/463158002/

anymore. Athletes making a stand is the easiest and most visible way to impact change, but the real changes occur behind the scenes. Like most businesses, following the money is route towards success, and sports is no different.

2) *Start realizing that sports are at the intersection of society*, culture, athleticism, and politics. To keep politics "out of sports" is to keep sports out of sports. Sport has been political, since the dawn of men-only Olympics in Ancient Greece to NFL players kneeling to support the awareness of police brutality. Supporting politicians and leaders who understand this intersectionality is a large asset in the sports world, and leaders such as Adam Silver, commissioner of the NBA, is a great example of which leaders to believe in. It is assumed that great sport owners and leaders will be successful, financially or athletically. Understanding the externalities of their leadership, such as their support of athletes or political causes should be the next step in addressing the climate crisis.

3) In addition to keeping players, coaches, owners, and teams accountable for their positive actions, we need to *start holding them accountable* for their negative actions as well. In this day in age, the power of your vote with your dollar is arguably as significant as the power of your vote in the polls. Due to the absence of federal leadership for climate action, many consumers are turning to their pocketbooks to make political statements—and it is working. Whether it is Apple's decision to become 100% powered by renewables, Starbucks' quest to eliminate plastic, or Microsoft's company-wide carbon tax, private enterprises are seeing a demand for sustainable business practices and accountability from consumers, and they are providing it. We should extend this practice to sports teams and enterprises as well. Athletes that stand by the Trump Administration's bombardment on the climate and environment regulation need to be held accountable, and the best way to do so is through publicity and refusal of their brands and products. The sooner we make it known that professional athletes are held to a higher standard, the sooner we can begin to hold their negative actions accountable and encourage these leaders to act on the climate crisis.

~ ~ ~

Brij M. Singh is a student at the University of Michigan, Business Administration at the Stephen M Ross School of Business and Sport Management at the School of Kinesiology. Brij's experience began with baseball and basketball, but soon found a niche in field hockey. After playing for Northern California's Under-21 regional team at 13, he transitioned to sport business. Subsequent high school internships with USA Field Hockey and the US Olympic Committee's Diversity and Inclusion branch confirmed a career in sport. Singh's college student education in sport business education led to interning at the USA Hockey's National Team Development Program in Plymouth, Michigan, and at the National Hockey League (NHL) in New York City, where he was a Corporate Social Responsibility Intern for the head NHL office. In Manhattan, his work included developing tactics for hockey development, charity work, and the *Hockey Is For Everyone* department of the NHL, that has ties to NHL Green, the sport's sustainability department. Brij is currently living in London for a study abroad program.

Chapter 30

Climate Crisis & Civilization's Unacknowledged Energy Economic Constraints

by

Richard Nolthenius, Ph.D.

"A sum can be put right: but only by going back till you find the error and working it afresh from that point, never by simply going on." – **C.S. Lewis**

People make their economic decisions "at the margin", and therein lies the tragedy. By "at the margin", economists mean that past decisions are past. The question instead is – given the past, for the decision now in front of me, do I get more happiness from deciding X or instead deciding Y? Taking a Sunday drive into the country may add more CO_2 to the atmosphere versus staying at home with the lights and TV turned off, but the *marginal* cost to the environment, versus the immediate marginal value to your family's happiness, means the rational and nearly universal decision will be to take that Sunday drive. How decision-making happens, and what energy is involved, is crucial if we are to truly understand the economics of our climate tragedy. Decision-making, in the widest sense, is the subject of economics.

This chapter will focus not on how cheaper wind and solar power are going to save climate and bring prosperity to everyone—as you've no doubt seen proclaimed many times—but instead how looking beyond "the margin" gives a very different picture. I'll help you see beyond the "checkers thinking" that dominates most promotions in media, and instead see the "chess thinking" that both physical laws and the laws of human nature demand. (Note: Presentations with the links supporting all of what follows are listed in the "Start" section of this chapter.)

Energy Efficiency – The Path to Climate Solved?

We see the laments in the environmental press every day... if only we can eliminate "vampire" power appliances, replace lighting with LEDs, raise mileage standards for vehicles, etc... then life would be golden. The unstated assumption is that energy efficiency has so far been neglected as our climate savior. And yet, we've been doggedly increasing energy efficiency ever since the invention of the wheel. Since 1950, U.S. energy consumption per dollar of real GDP has steadily dropped an astounding 63%. And yet our total energy consumption rate has gone up an equally astounding 300% (EIA figures). This is despite the off-shoring of much of our energy-intensive manufacturing to Asia. Why? Understanding this is central to understanding the economics of climate change.

Civilization as a Thermodynamic System

Cloud physicist Tim Garrett pondered this connection. After all, civilization is a system that takes in raw materials and converts these into human values, powered by dissipating energy along its growing networks. This suggests it should be expected to obey thermodynamic laws. Thermodynamics is the branch of physics identifying how energy, heat, and entropy (disorder) flow within systems. It's vitally important to understanding behaviors as diverse as animal foraging, power systems, snowflakes, and the evolution of the Universe, and... the evolution of civilization. Let's see how Garrett's insights constrain the evolution of future climate. While I've borrowed heavily from Garrett's published work, I believe that seeing civilization in thermodynamic terms is easier for most to grasp in the language of order and disorder: entropy – the imposition of "order" and Nature's tendency to

273

evolve towards "disorder". Garrett's complimentary framing in terms of energy dissipation from higher to lower potential might be harder for the non-scientist to get a feel for.

"Civilization" is the process of satisfying innate human desires through networked connections between peoples, cities, and nations, and globally. "Civilizing" is the imposition of order onto a disordered array of raw materials by using accessible stores of energy, where every human act has the goal of bringing increased "order" (usefulness) to an otherwise less ordered or decaying state. Think of this as "civilization thermodynamics". All of it requires energy. In physical thermodynamics, your textbooks showed that the amount of "ordering" is directly proportional to the amount of energy consumed. In life, as in physics – **Energy is Everything!**

So how do we quantify in a unified way the vast array of civilization activities? Very naturally - money! It is the universal marker for, and store of, value. The *change* in disorder is, on average, proportional to the *cost* of the doing. And in a market economy, competition lowers this cost as efficiently as possible. Thus, we might expect that the total inflation-adjusted sum of all spending since society began, to be proportional to the net *civilization* (in the widest sense) created.

Conventional economics regards the fundamental elements determining economic production to be <u>capital</u> (money, land, resources...) and *labor* (look up the Cobbs Douglas equation). Amazingly, energy is entirely neglected, assumed taken for granted. But again, energy is everything! Human civilization is not a static collection of things. Human values exist only through *active* relationships, expressed along the networks connecting us. Life, and value, is *action*. Inaction... is death. <u>Action</u> *is* <u>value</u>: communicating with friends, electricity flowing through wires, fluids through pipes, people travelling roads and rails... All are action, requiring continuous energy consumption to accomplish, and then to support in an ongoing way against friction and decay. Take away the energy, then action ceases, and the value of civilization goes to zero.

All "order" must fight the *2nd Law of Thermodynamics*: Total entropy ("disorder") in a closed system always trends towards a maximum. And, since economic growth creates more order out of disorder, it encumbers new ongoing power to support that newly created order against the 2nd Law. What's more, a larger civilization has greater ability to access new energy reserves at a faster rate - and that is precisely what we do. Every dollar of spending increases the ordered state of civilization, even past spending on things long turned to dust. Even those old black and white TVs and Nokia brick phones sitting in your basement were a stair-step enhancing growth now reflected in today's world. Their ghosts live on in what they enabled, building on the initial value they created. What drives this? The biological imperative of *Evolution by Natural Selection* insures that the impulse to growth is bred into all life.

Clearly, like the red queen in "Alice in Wonderland", existing civilization needs constant power just to stay in place even with zero growth, and so Garrett realized that power consumption cannot simply be proportional to economic growth. He made the inference instead that the sum total of *all* past spending should be directly proportional to today's rate of primary energy consumption – power which maintains the value created, plus any new growth. Note the relevant energy here is *primary* energy - raw energy supplied by nature. After all, converting this into *usable* energy like electricity, entails losses which must remain in any fair accounting (beware of promotional graphs which don't). Now, the widest measure of civilization spending readily available is **Gross Domestic Product** (GDP) for each country. And so Garrett's formulation of the relation becomes:

The current rate of global primary energy consumption ("Power") should be directly proportional to the total inflation-adjusted Gross Domestic Product (GDP) summed over the entire past history of civilization ("Wealth"). Power/Wealth=Constant

Call it **"The Garrett Relation"**. It has profound consequences for projecting future climate's relation to economic growth. "Wealth" is Garrett's shorthand for total summed global spending over all time. I'll retain this definition here, and use a capital "W" as a reminder.

Much like a striking symmetry principle in physics it is revealed as a simple truth only for *global* civilization. It doesn't apply to individual cities, countries, or continents, since global trade sends large amounts of money, materials, and energy across borders. Claims which look at only one country or region will not be valid globally. For example, as previously mentioned, the Western nations have outsourced much of their CO_2-generating manufacturing to Asia, and the improving energy efficiency seen in the U.S. shows a perfect correlation with the percentage of our manufacturing sent to Asia. They spend the energy, generate the CO_2, and get our money in return, while the U.S. gets the goods, higher GDP, and bragging rights to happier energy efficiency and CO_2 trends. Don't be lulled - 76% of global GDP is generated outside the US. That the Garrett Relation applies only globally is not a problem, because we're going to relate this to climate. And CO_2 likewise is determined only globally. All atmospheric gases are "well mixed", diffusing globally from their source within weeks.

Implicit in the Garrett Relation (explicit, in the math) is what I call **"Generalized Jevons' Paradox"** (or in some moods "Jevons' Revenge"). William Stanley Jevons observed in 1865 that improving the efficiency of coal-fired steam engines would not <u>reduce</u>, but instead <u>increase</u> the rate of coal consumption – the classic **Jevons' Paradox** – as efficiency promotes growth, and raises civilization's need for - and also its ability to mine - more coal. Higher efficiency enables faster growth. Now, even with no improvements in energy efficiency, civilization can and does grow. But higher energy efficiency enables that growth to be even faster. Some economists wrongly dismiss **Jevons' Paradox** as leading to only a partial "rebound" in consumption so that net consumption still goes down. For example, if you get double the miles-per-gallon from your car, are you going to drive twice as many miles with those savings? Probably not! But this is "checkers thinking". They are failing to make the vital distinction between the original narrow coal-for-coal Jevons' Paradox and the more relevant Generalized Jevons' Paradox which I'll now explain. Generalized Jevons' Paradox recognizes savings can, and are, spent everywhere.

Generalized Jevons' Paradox

Any increase in energy efficiency will lead to savings. Those savings will not be destroyed but rather they will be spent, and the Garrett Relation shows that ALL spending requires the ongoing consumption of new energy to support the resulting civilizing against decay, while also expanding our ability to discover and exploit new energy at a faster rate. These combined effects more than offset the efficiency-gained reductions in power. Future global power

Historically, this is exactly what happens. The proof's in the data. Global primary energy efficiency (energy consumption required per dollar of global GDP generated) improved by a hefty 24% from 1990 to 2014, yet annual global GDP rose 94%, nearly 4 times higher, carrying net higher primary power consumption with it.

*If the **Garrett Relation** proves true, it not only shows how all spending encumbers higher future power consumption, it also confirms **Generalized Jevons' Paradox**. The implications for climate are sobering.*

Now, everyone appreciates that civilization growth must be accounted for in future projections, but realize that it is the very creation of new energy efficiencies which directly *cause* additional growth. We think that to solve the problems of growth, we need to improve energy efficiency, but fail to realize that it is the very creation of each new efficiency which both requires and enables corresponding new growth. The solution is part of the problem! It's like trying to solve a shopaholic's maxed-out credit card trouble with another credit card. It just digs the hole deeper. Claims that our climate salvation lies in improving energy efficiency implicitly assume those savings will never be spent. Worse, their writings lead the reader to believe that the idea of improving energy efficiency is something new. No! We've been improving energy efficiency since the invention of the wheel, and we've been doing it as fast as we can afford to, so don't expect a change in the trend of these improvements.

275

The Data: Is the Garrett Relation Really True?

Yes!

Figure 1 shows that accumulated global GDP and current power consumption track each other very closely, so that **Power/Wealth=Constant**. Garrett's original papers count just global GDP spending, but the thermodynamic arguments tell us that we should count ALL spending, including the "shadow economy" (barter, housework, black markets, etc, which total about 22% as large as official GDP). I've done this here, and find that Power then tracks Wealth even better (blue curve). Tracking is better still (not included here) when post-1994 low-biased official inflation is corrected using *e.g.* the M.I.T. "Billion Prices Project" or ShadowStats – but that issue would command a whole chapter in itself.

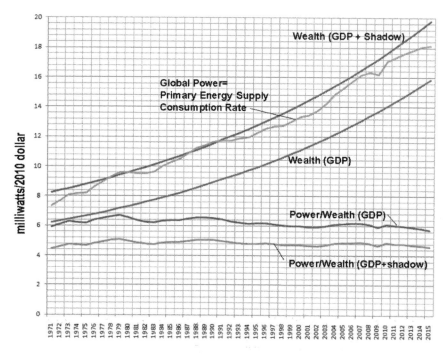

Figure 1: Global Primary Power / Wealth vs. Time.

You might worry that this figure only goes back to 1970, not the dawn of civilization. But earlier data does exist and is included in calculating Wealth in Figure 1. Most important, note that 58% of all global spending ever done, has been done just since 1970 - when high quality yearly records begin. Earlier data goes back at least two thousand years.

But doesn't the Garrett Relation fly in the face of improving energy efficiency? No. **Figure 2** below shows the global efficiency of primary energy to generate a dollar of GDP. Globally, energy efficiency improved fully 24% from 1990 to 2014, a period accounting for 40% of all accumulated global GDP spending ever done. Yet annual power consumption continues to grow, and the Garrett Relation continues to hold flat. But how? As long as GDP is rising faster than energy efficiency, the Garrett Relation can (and does) remain valid.

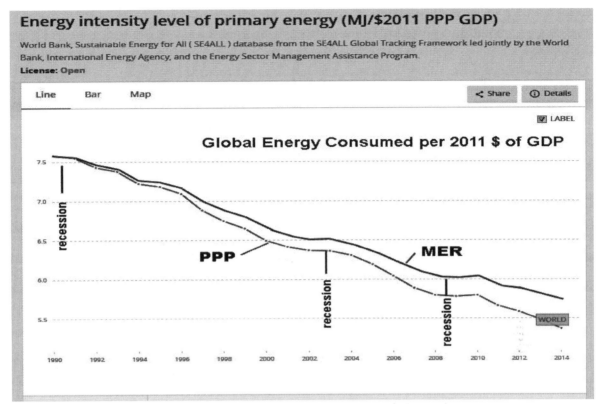

Figure 2: Energy efficiency improvements pause or reverse, during recessions.

Implications for Future Climate

Garrett ran calculations conservatively assuming that the growth rate of Wealth no longer rises, but instead holds constant at its recent 2.2% rate through 2100, even though exploiting new renewable energy sources argues the growth rate of Wealth should continue to rise. Further, he assumed we steadily decarbonize our energy, with the CO_2 emitted per calorie of energy consumed dropping exponentially with a halving time of only 50 years – a record-breaking pace compared to slow 180-year halving time seen in the 20th century (during rapid nuclear and hydroelectric construction). Despite the increasing number of solar panels, wind turbines, and Teslas you might see, the sad truth is that in the 21st century we've not de-carbonized at all. Rapid Asian growth continues to be mostly fueled by fossil fuels, dwarfing any decarbonizing in the developed world. Consider - China pledged recently to cut their CO_2 emission intensity (CO_2 per dollar of GDP) by fully 60% by 2030. That sounds planet-savingly dramatic... until you convolve with their economic growth rate. China's pledge actually results in annual CO_2 emissions continuing to rise and are 30% *higher* in year 2030 than today!

The sobering results for atmospheric CO_2 are in Figure 3, and explain my focus on the importance of the Garrett Relation. It fundamentally changes how the economy and CO_2 can evolve in the 21st century. We see that despite earnest decarbonizing, Generalized Jevons' Paradox leads to growing atmospheric CO_2, together with much slower economic growth than the unrealistic IPCC eco-friendly scenarios (see below); and this is true for almost all conceivable scenarios, even the most climate-crippled case, highlighted in the figure.

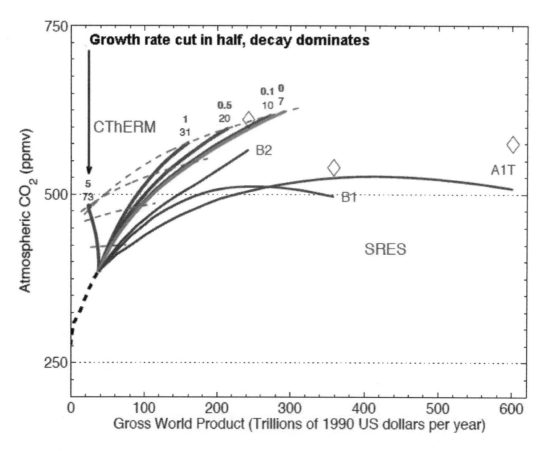

Figure 3: The Garrett Relation means atmospheric CO2 continues to rise even while we decarbonize our energy sources (red). Green dashed lines are years 2025, 2050, 2075, and 2100. Different red curves assume different resilience of civilization to the ravages of climate change. More resilience means faster GDP growth and higher CO2 rise.

Why? Global GDP spending is positive every year, adding to past accumulated GDP. Hence, by the Garrett Relation, power consumption rates must continually rise. Decarbonizing must happen much faster than 50% per 50 years to hope to overcome this. High resilience to climate change only means faster growth, meaning HIGHER atmospheric CO2. Perfect resilience results in atmospheric CO2 over 600 ppm and still rising. Only for the least resilient case (left-most curve, labelled decay rate 5% per year per CO2 doubling), with decay increasingly eclipsing growth, now cut in half from today, does atmospheric CO2 stabilize at just under 500 ppm (and this does not include non-CO2 GHG's).

The IPCC "Standard Representative Emission Scenarios" (SRES) curves are shown for comparison. The IPCC models were constructed to give a certain range of assumed (under UN political direction) representative outcomes, with population growth, global growth, and emissions reductions all independently tailored in order to arrive at the pre-desired outcomes. They do not take account of how civilization *actually* behaves, as revealed by the **Garrett Relation**. Is it physically possible to follow instead the eco-friendly SRES pathways? Even with strong government policy forcing very rapid decarbonization, it's extremely unlikely. To so severely cut CO2 emissions will require decommissioning perfectly working fossil fuel power plants and transforming the grid, as well as strong carbon capture and sequestration costing (at climate-significant scales) up to $400/ton of CO2– all of which add little to expanding civilization since this work merely replaces perfectly working energy systems with an entirely new energy system, all at great diversion away from the supposed civilization growth shown on the SRES curves. That would be painful – not something humans do voluntarily.

Even if accomplished, so that climate change is now reduced, this only permits higher resilience and hence faster growth in consumption of other finite resources. Assuming we have already dropped emissions to only 30% of today's, so that atmospheric CO_2 remains constant (but not including indirect human-caused emissions), then to further lower atmospheric CO_2 to 350 ppm would cost about $27,000 for every man, woman, and child on the planet. Globally, the median GDP per person is $2,700, so this cost represents a full 10 years of personal income.

Even if global Wealth grows at only 1.5% per year, and even if our goal were only to have direct human CO_2 emission rates stop rising and instead stay merely constant at 37 billion tons per year, it would require the equivalent of 11 square miles of solar panels deployed every single day. That's an area the size of Iceland every decade; removing it from other human uses, and removing it as home to millions of other species we share this planet with. Would that at least solve climate? No. Constant emission rates means atmospheric CO_2 would still rise, but now only along a straight line sloping upward, instead of accelerating upward as it has been.

But even that is too optimistic. The Figure 3 red curves, like the old IPCC models, do not include climate science discoveries after the last IPCC Assessment Report of 2013, showing manifest deficiencies in the old IPCC carbon budgets meant to keep the +2C Paris agreements temperature limit, further doctored in the final "Statement to Policy Makers" by U.N. political meddling. Yet every policy and techno-fix promotion I've read includes these old carbon budgets as if they are true. Let's look at the deficiencies...

The IPCC climate models do not include: the new pre-industrial temperature baseline from Michael Mann's team which is 0.2C below the conventional 1880-1910 "pre-industrial" baseline. They don't include any permafrost melt, which is triggered at 1.5C above true pre-industrial (which we hit briefly in 2016). They don't include the collapse of Arctic Ocean ice cover. They don't include dropping reflectivity of ice due to dark microbes, wildfire smoke, algae colonies, nor new ice sheet dynamics, nor the 2014 discovery of the ungrounding of the West Antarctic Ice Sheet dooming an eventual further 12 ft of sea level rise. They don't include the new satellite data of strong soil carbon emissions due to drought and deforestation, nor methane production rates in wetlands rising 44 times faster than temperature, nor the plummeting ability of coral and phytoplankton to convert CO_2 into calcium carbonate, nor how *Equilibrium Climate Sensitivity* of temperature to a CO_2 doubling (ECS) rises in hotter climate states. ECS for our future may be as high as 4.9C, not the 3C of the recent past and most modelling.

And, it does not include any non-CO_2 greenhouse gases. Nobel Prize winning physicist Steven Chu points out that the CO_2 equivalent GHG load today is actually 500 ppm, not the 410 ppm of CO_2. Under the politically motivated direction of the U.N., the scientists made the rosy assumption of no methane emissions, no nitrogen oxide emissions, no CFC's, HFC's... that all would simply end and decay away. *This is fantastically unrealistic.* Despite methane's rapid oxidation, concentrations in the atmosphere are going up 0.39% per year for the past decade. Even replacing all HFC's with their safest refrigerant equivalents (which are still GHG's) would only lower their corresponding global warming forcing by 30%, not 100%. The reduction in total greenhouse forcing by an HFC ban would only be 2%. All of these non-CO_2 GHG's worsen heating and thus worsen CO_2 emission from Earth's natural systems.

These are indirect human-caused sources, not answering to our political laws as direct emissions might. Remember that the "+2C limit" dates back to a 1975 op/ed by economist William Nordhaus - a compromise in favor of economic growth over climate stability. There's nothing safe about it! Climatologist James Hansen shows +2C is a "prescription for disaster", in his words.

Homo Economicus – Who ARE We?

The validity of the Garrett Relation shows that distant past spending counts just as much as yesterday's spending in determining the power consumption of today. I'll argue this shows civilization is ruthlessly efficient in executing its prime biological directive – ***growth*** - limited only by our ability to access new energy and use efficiently. ***Energy is Everything***. This argues against any easy behavioral changes to avert climate disaster. Yet on this finite planet - growth must, and will, end. The same biological drives which drove our successful dominance, become our Achilles heel when the limits to growth are reached.

What drives us to follow the Garrett Relation? Studies of happiness *vs.* income give a clue. In most realms, once the initial glow of an achievement fades, we need bigger and bigger rewards just to maintain the same glow. Figure 4 shows that to sustain a constant rate of improvement in life satisfaction, we must grow our income exponentially. Sounds just like drugs! Actually, it IS drugs – dopamine – a natural brain chemical, and key to addictive behavior. Perhaps it is only Zen masters who can find happiness in the pure stillness of being. But we'll all need to appreciate this mindset if we're ever to achieve a truly sustainable society. We must balance growth in life quality with de-growth elsewhere and forgo the biological drive for net growth — whether in terms of the economy, global population, or most other areas — because planet Earth isn't growing exponentially. It's still just 7,918 miles across (and there's no other planet out there we can live on. Don't get me started on Mars!). Worse, Earth's ability to regenerate and support life is actually shrinking by 1.1% per year through environmental degradation, brought on by our thirst for dominance.

You might think the tragedy is when a species loses in this ecological competition. No, the real tragedy is when a species as powerful as *Homo Economicus* **wins**. Because then **every** species loses. And then we lose too. After 6,666 generations, we have reached that point, and beyond. We must reprogram ourselves or face an existential threat to civilized society. We already commandeer ~37% of all primary energy falling on the planet. Humans and their livestock comprise 97% of the total biomass of all vertebrates on all continents.

Wild animals are now just scattered remnants. You may feel we can stabilize at today's civilization size. But that's an illusion from the comfort of your immediate surroundings. Beyond your view, we're losing 2 football fields worth of essential rainforest – the lungs of the Earth – every second. 1/3 of all edible ocean species have suffered population crashes by over 90%, and that was as of 2006. We're eating through our environmental seed corn. Like shopaholics overspending on credit, we yet cling to the illusion we all can continue to consume more, forever.

A perfect new example is cryptocurrencies. Bitcoin mining is increasingly energy intensive deliberately to insure hack-proof security By May 2018, a single addition to the blockchain, enabling *one* transaction, required as much electricity as an American home uses in a month, and total bitcoin electrical power consumption was greater than that in the entire nation of Chile. The future of cryptocurrencies is not clear. The Bitcoin Mining Energy Index, after rising spectacularly since inception, dipped by 30% in late '18, but started rising again as of this writing in early '19.

All of this testifies that *Homo Economicus* will access any and all energy available. Renewables - rather than replacing fossil fuels - are only being added on top of them, filling in niches where there is economic advantage. Reports of the death of fossil fuels are "greatly exaggerated", as Mark Twain would say. Until an expensive new grid is created, the marginal cost of additional solar photovoltaics rapidly becomes uneconomical once penetration reaches about 20%. Energy storage is essential, but in another nod to Generalized Jevons' Paradox, adding new energy storage today actually <u>increases</u> CO_2 emissions, since coal fired power plants running late at night can store low-demand power for use during high profit workday time, muscling out solar PV in the process. Improvements will happen, of course, but the growth enabled fights against our primary goal of restoring a healthy climate.

Garrett points out the double-bind we're in. We can't continue economic growth and successful atmospheric CO_2 reduction at the same time. We can't cut atmospheric CO_2 without extremely rapid decarbonization. But rapid decarbonization is expensive, and by the Garrett Relation, therefore requires high energy expense to accomplish. But this means high CO_2 emissions - because in 2018, just as in 1973, 87% of global power still comes from fossil fuels.

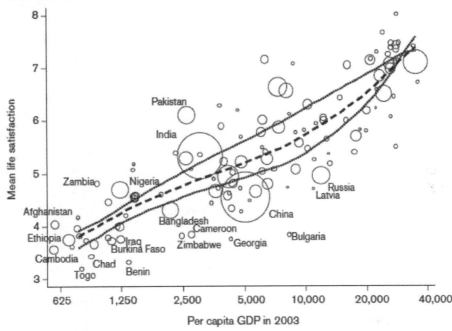

Each Doubling of GDP Is Associated With a Constant Increase in Life Satisfaction

Source: Penn World Table 6.2.
Note: Each circle is a country, with diameter proportional to population. The scale on the x-axis is logarithmic. The middle line shows average life satisfaction for each level of per capita GDP while the outer two lines show the same thing, but for two age groups, ages 15 to 25 — the upper line for most of the figure — and ages 60 and over — which is usually the lower line. GDP per capita in 2003 is measured in purchasing power parity chained dollars at 2000 prices.

Figure 4: Reported life satisfaction will rise at a constant rate only if income grows exponentially.

So, What Do We Do?

We make our decisions "on the margin", with harsh consequences. The atmosphere is one of the Earth's great commons: impossible to privatize, and shared by all. So *laissez faire* capitalism guarantees, as we've observed, that commons will be exploited for profit as rapidly as possible, before the next competitor can – this is "The Tragedy of the Commons". Can we trust business leaders? Alas, corporate CEO suites are as dense with psychopaths (21%) as hardened prisons (Brooks *et al.* 2016). The same criteria applied to the general population yields only 1%.

Counting on individual voluntary self-restraint is also a non-starter. Even if, by incredibly inspiring personal action, you convince 1 billion people to cut their carbon footprint in half, this only cuts global CO_2 emissions by 13%. The "low hanging fruit" of Progressive eco-friendlies who have the means and willingness for such sacrifice have already been picked; and indeed, were likely already enlisted after the environmental movement of the 1970's. Assuming the rest of the world will hop aboard has been a vain hope for these 50 intervening years. Live your values... but don't expect individual voluntary carbon footprint sacrifices to make a dent in the climate crisis.

Do we engineer a global long-term recession to rapidly bring CO2 emissions down? Yes and no. Do some math and you'll discover there's trouble there, too. The Garrett Relation, with a bit of calculus on Figure 2, implies that **when annual GDP is declining (that's a recession) then energy efficiency improvements actually reverse**. This makes rather grim common sense. During recessions our hands are full just supporting our past growth against decay, and the added expense of perfecting new energy efficiencies falls by the wayside. Indeed, note the flat points in Figure 2 occur precisely during global economic recessions. More unbiased data would likely show actual reversal in energy efficiency since Federal Reserve studies show that in recessions, China and other countries significantly overstate their GDP, so that energy efficiency indeed gets worse.

There are political reasons why the official GDP data reported to the World Bank and the U.N. have smoothed over these periods. In the U.S., even our *official* data, going back to the oil shock recessions of the early '70's, shows energy efficiency improvements went into reverse. I call this the **"Recession-GDP Bias"**, and its implications for climate policy are serious. As an old aphorism says... *"When you're up to your rear end in alligators, it's hard to remember the original intent was to drain the swamp"*. That's us, when economic times get hard. The Garrett Relation shows that what's actually needed is a strong global Spartan mindset, forcing us to tighten our belts for an engineered recession, and at the same time diverting maximum spending towards decarbonizing energy sources, rather than the growth we yearn for.

The physics and economics of human civilization make required climate policy much more severe than policy people are admitting. What is required is in strong conflict with the political/economic paradigm we've embraced, and indeed with human nature itself. *Solving climate will require a fundamentally transformed human nature, evolved to a level of maturity that only rare individuals have so far demonstrated. The recession-GDP bias is an essential constraint revealed by real-world data, and is a key take-away from this chapter.*

The changes that we as a society need to undergo will be dramatic, uncomfortable, and against the momentum of our everyday lives. We will need to empower governments to force us, against our individual willpower, to make the cuts necessary to return to the stable climate all existing species evolved in harmony with. Yet history shows we'd rather empower demagogues who blame and scapegoat phantoms, than face reality. Our societal natural inclinations only motivate us towards *"Climate Abandoned"*.

Shifting attention now – I need to make a few important points on economic opportunism warping the conversations on geo-engineering and carbon pricing.

As the reader probably knows by now, we have waited too late, and merely ending emissions will not save us. Climate geo-engineering is now essential to return climate to the stability we evolved in. As we seek to fund technological ideas, consider this framework: All *effective* strategies must either reflect additional sunlight back out to space, or enhance Earth's ability to re-radiate to space. This is already well understood by most. What is not well accepted, is a proper framework for safety. Here is the understanding I've come to - all *safe* strategies will have two important characteristics: **(1) They will guide our Earth systems approximately back along the trajectory that got us here. (2) They will leave the Earth's surface – where we and nearly all other species live - in as undisturbed and compatible a state with our ecologies as possible.**

Why? Because we're in a minefield - sanity says we do not stampede Earth systems into states that current ecologies have never experienced just because an idea seems cheaper or capitalizes on a patent you may own: Iron flakes across the open ocean, massive deployment of OTEC pipes, painting everything white, planting an India-sized area with weeds to repeatedly harvest, burn and capture their carbon, thereby depleting vast landscapes, sulfuric acid haze throughout the stratosphere significantly altering regional heat balances, spreading vast trillions of proprietary small white beads across the polar oceans as reflectors... are just some of the very dangerous ideas out there.

Many are motivated by carbon offset pricing laws (guess who motivates carbon laws? -see below) which fail to insure permanent sequestration (a few decades is considered "sequestered") and neglect ecological damage, let alone basic, fatal thermodynamics errors (OTEC pipes). Further, all safe solutions will seek to minimize changes to the Earth's **surface**, where we and virtually all life live. I know of only two geo-engineering proposals that clearly fit these criteria: (1) Direct Air Capture (DAC) of CO_2, then pumped underground in geological formations, and (2) Millions of wind-powered steel pumps on the Arctic Ocean, pulling ocean water to the cold winter ice surface to thicken it enough to survive the next summer (Desch 2017).

The estimated cost of thickening the Arctic Ocean winter ice by 1 m per year this way is $500 billion/yr for 10 years, which is only 0.64% of global GDP. A third idea – ships jetting seawater into the atmosphere generating salt spray to form clouds to reflect sunlight - might marginally qualify. However, at climate significant scales the damage to surface ocean ecologies is open to question, and the very regional reflectivity effect can cause global climate shifts hard to predict. DAC/sequestration at climate-significant scales may cost $400/ton CO_2. That's $1.5 trillion per year to capture and sequester our annual emissions, or 2% of global GDP.

Even if we end all human CO_2 emissions, DAC capture at such a rate is required to reasonably bring down CO_2 before climate chaos begins. Fossil fuel power plants can't compete on price with solar and wind once carbon capture is included (although solar and wind will require large scale energy storage). These costs are steep, but then - what's the future worth?

Climate Crisis "Keep/Stop/Start" Actions

Below for your consideration, are several author-suggested thought, behavior, and action items. Action nourishes inspiration! For a further explanation of the Keep/Stop/Start organizational tool, please see How to Read This Book.

KEEP:

Not an easy question! We need radical fundamental transformation. But the most obvious "keep" is one we're losing – our respect and financial support for basic scientific research. In the words of Nobel physics laureate Richard Feynman – *"Science is what we do to keep from fooling ourselves"*. We've instead come to think of science as mainly good for helping us get more and better gizmos. Yet I'm old enough to remember when science was respected even by Republicans. No, science is not about money-making. It's fundamentally an attitude of realism and honoring truth above all else. Dangerous attitudes, today's corporate and political powers realize.

STOP:

1) *Discourage population growth.* Aim for 1 child-per-family. Even this, if done globally and gradually brought to full by 2045, still results in 4 billion people in 2100, which is beyond what the Earth can sustainably support today. Politically impossible? Yet demographics and population projections which include no unintended births doesn't reduce population at all (Bradshaw and Brook 2014). It results in 7.5 billion through 2100, and more if we continue improvements in infant mortality.

2) *Stop oil industry subsidies*, which are as large as 3% of global GDP, even neglecting externalized environmental costs. Internalizing external costs would force all fossil fuel companies into new work, says a Cambridge Business School study.

3) *Stop Inflating Currencies*. Mandate a Federal Reserve goal of negative inflation, not +2% as we have today. Disinflation would encourage savings, discourage spending and hence energy consumption and CO2 emissions. We need a controlled glide path down to a smaller population and footprint on this planet, and that means de-growth.

4) *Stop Cap-and-Trade*. The promoted idea is that government sells credits to industry for the right to pollute - carbon in our case - and that the credits are capped so that emissions are capped. It's a failure, driven by fossil fuel industry lobbying which makes the appearance of limiting carbon emissions, but which in fact have always included far more carbon credits than they can use. This lets the big emitters sell those free credits for a profit to smaller players, while simultaneously guaranteeing that the allowances end up maximizing emissions. Worse, an emitter can either buy more credits if they need them, or they can invest in carbon offsets – equivalent on the carbon market. For example, you can strip mine a mountain of coal as long as you pay for the planting of palm oil trees on clear-cut rainforest land in Brazil. Or pay for spreading iron flakes across the open ocean (which has never been iron-rich), promoting algae blooms, and "red tide" and sending a only small part of the carbon from the algae under the ocean for at most a few decades (say studies) – but counting as big and cheap carbon sequestration the way the laws are written. But even if carbon offsets were not environmentally damaging, government severely limits the use of carbon offsets because they want the revenue that comes from industry buying more carbon credits. The results? No carbon reductions happen, it's a revenue-maker for the big carbon polluters and the government, further solidifying their industry dominance, and the whole cap-and-trade idea is greenwashed as earnest environmental concern by the players. And it's been going on for years and decades. Read or listen to economist Aldyen Donnelly's devastating critique.

START:

1) *Start advocating for a 28th amendment to the Constitution*:

"Congress shall pass no law which infringes on the right of all present and future citizens to stable environmental commons such as Nature provided - the great oceans, atmosphere, ice caps, and great forests, and our climate."

It is important to stipulate "commons", which cannot be privatized and therefore need special protection. South Korea is one of the few countries in the world with a Constitution that explicitly recognizes that "all citizens have the right to a healthy and pleasant environment." Other countries that have similar provisions include Ecuador, Peru, Hungary, the Philippines, and Portugal. It is long overdue that the American Constitution was amended to include this right.

2) *Start advocating for a carbon tax at the carbon source*: Levy a high tax where carbon enters the U.S.; at the wellhead, at the coal mine, and at our borders. Adjust import taxes if carbon is already taxed by the originating country. This is similar to what is advocated by climatologist James Hansen, and taken up by

Citizen's Climate Lobby. However, I advocate an important change – let an increasing proportion of the carbon tax go to funding carbon capture and sequestration technology advances and deployment.

Dividends to people may be politically popular, but dividends are spent by people in not necessarily climate-positive ways, and anything which encourages economic growth also raises energy consumption. We need maximum effort to pull CO_2 out of our atmosphere – immediately. Politics *vs.* climate realism are in conflict here. Still, regardless of where dividends go, any carbon tax at the source is a giant step in the right direction. Carbon taxes today are much closer to the consumer (think gas taxes), which preferentially hurt the poor and don't effectively motivate leaving carbon in the ground. This tax must be much higher than currently debated.

When British Columbia's carbon tax/dividend stopped rising after 3 years, halting at $30 (Canadian), carbon use resumed its rise. Even while in effect, it made an insignificant dent. In only one 18-month period did gasoline consumption dip, slightly. Claims that the BC carbon tax was a great success, are based on the promoters' unreasonable assumption that in the absence of this small carbon tax, and during the worst recession since the Great Depression, that BC fuel use would have somehow undergone the biggest 3 year increase in recorded history. It's an completely unrealistic counterfactual assumption. The 2018 IPCC report estimates to meet the 1.5C Paris target, carbon taxes would have to be between $135/ton to $5,500/ton of CO_2, depending on scenario.

3) *Start taxing consumption rather than income.* It is consumption that most burdens climate and the environment. Sales taxes already do this – a good start, but we need to keep going. Yes, the poor spend more of their income on consumption than the rich, so this is change is regressive, as many will be quick to point out. I'm talking climate; wealth inequality should be dealt with as a separate issue.

4) *Start demanding strong government policy:* It is Essential. How to bring it about? Alas, Gilens and Page (2014) showed there is zero correlation between what the average citizen wants and what laws are enacted, yet laws enacted show near perfect correlation (78%) with what the Economic Elites want. This was true over their entire 20-year study, carefully corrected for cross-correlations, and which included political control periods for both parties. The evidence is clear - despite Oscar-worthy posturing, that our lawmakers care far more about themselves than about your children's future. Congress makes the laws that Congress obeys, and we, the average citizens, don't count. Writing your congressman will continue to fail. We need to be more insistent.

I advocate **Occupy DC** with at least a hundred thousand peaceful but determined protesters exercising their First Amendment rights, until Congress and the President agree to the passage of the legislation above. A hundred thousand is enough to shut down Washington DC's "Business as Usual". It's too many to jail, and guarantees every media camera and smart phone will attend. It requires convincing only 0.03% of Americans of how dire climate really is, as I hope I've shown (see links below).

However, never corner a dangerous animal. Always leave an escape path. Occupiers must come with specific legislative demands and a promise to disperse when publicly agreed to by the leadership of Congress and the President. A limited march for a day or two, on the other hand, will not force change. I agree with James Hansen – the political revolution must be led by young people for maximum effect. It's their world at stake.

5) *Start accept that global economic growth is ultimately in conflict* with returning Earth to a sustainable, stable, healthy climate. We must commit to de-growth and de-carbonize – rapidly.

6) *Start educating yourself,* then others, about the straight science and economics. Check out the following presentation pdf's which contain the links supporting the many points I've made in this chapter:

- http://www.cabrillo.edu/~rnolthenius/Apowers/A7-K43-Garrett.pdf

- http://www.cabrillo.edu/~rnolthenius/Apowers/A7-K42-FutureClim.pdf

- http://www.cabrillo.edu/~rnolthenius/Apowers/A7-K44-Policy.pdf

- http://www.cabrillo.edu/~rnolthenius/Apowers/A7-K46-StrategiesGeoEng.pdf

~ ~ ~

Richard Nolthenius, Ph.D. is the head of the Astronomy Department at Cabrillo College in Aptos, California. He earned his degree in astronomy and astrophysics after doctoral work at Stanford University and UCLA. He was a member of the Thermodynamics Group at General Dynamics in their space program in San Diego and performed thermal analysis and design for the Atlas/Centaur rocket missions and space satellites, and was thermal systems designer on their proposal for what became the International Space Station. His post-doctoral work involved galaxy clustering algorithms and comparisons between numerical cosmological simulations and real observations. He was a visiting researcher and lecturer at UC Santa Cruz, in the city in which he now resides. Since 2009, Rick's focus has been on the science of climate change and its relation to political/economic systems. Read more at: www.climateabandoned.com.

Chapter 31

Climate Crisis & Solutions

by

Hari Krishna Nibanupudi, Ph.D.

(To small farmers, fishing, hill and forest tribes, indigenous communities, who with their simple sustainable lifestyles resiliently fight back the impacts of climate change and play a huge role in saving the future of life on earth.)

"We stand now where two roads diverge, but unlike the roads in Robert Frost's familiar poem, they are not equally fair. The road we have long been travelling is deceptively easy, a smooth superhighway on which we progress with great speed, but at its end lies disaster. The other fork of the road — the one less travelled — offers our last, only chance to reach a destination that assures the preservation of the Earth." **– Rachel Carson**

Climate Solutions Are many, and They Are Working

As many of us learned in grade-school biology, *homeostasis* is "the ability of the body or a cell to seek and maintain a condition of equilibrium or stability within its internal environment when dealing with external changes." The immune system valiantly fights viruses and bacteria through an arsenal of fevers, sneezing, coughing, and emesis, but it sometimes requires assistance. This is precisely the reason why doctors prescribe medicines, antibiotics, and additional fluids.

In essence, Planet Earth is undergoing a similar process. Anthropogenic activity, chiefly caused by humans, has polluted the planet with a surplus of carbon dioxide, methane, and other greenhouse gases. This has thrown Earth's equilibrium off balance, resulting in many of the detrimental climatic changes we are just beginning to witness. Though climate change is inherently natural, it is accelerating at a rate today that none of us can withstand. As such, we need to be doctors. We need to prescribe the additional, external solutions that will slow climatic warming and mitigate environmental changes, to maintain Earth's homeostasis; otherwise, we will continue on this trajectory, and the grim prognosis is that we will not survive.

Reversing climate change, mitigating the negative impacts, and restoring the Earth system to a hospitable, more productive state requires the implementation of e a wide range of solutions. As mentioned in Paris at the UNFCCC COP21, these include:

1) Halting sea ice retreat,

2) Protecting oceans and marine life from excess CO2, reducing extreme weather events,

3) Slowing sea level rise, reducing CO2 in the atmosphere, suppressing the methane released from melting permafrost,

4) Reducing droughts that result in fires and threaten food and water security,

5) Increasing wood biomass, increasing marine biomass, changing lifestyles and altering human behavior,

6) Preserving and managing existing forests, scaling existing renewable energy technology, and

7) Identifying and supporting early-stage innovation for the next wave of technological breakthroughs to help solve the climate crisis.

While many of these actions span the domains of policy, science, and technology, various actions can also be taken by individuals. Furthermore, citizens of the world must consider not only individual actions but also pressure national, state, and local governments, as well as the private sector, to prioritize climate mitigation policies and practices. Fortunately, the majority of the world has recognized the potential catastrophe before us. Governments, business leaders, academics, religious groups, scientists, farmers, and many others are already undertaking dedicated efforts to reverse the adverse impacts of climate change and preserve our Earth for future generations.

A Gallup World Poll, published in *Nature Climate Change Journal*, revealed that in North America, Europe, and Japan, more than 90 percent of the public is aware of climate change (Lee et al., 201[1]) . Furthermore, a 2017 national survey conducted by Yale Program on Climate Change Communication discovered a significant improvement in climate change awareness since 2008; according to the survey, 22 percent of Americans are very worried about global warming, and 63 percent see it as a significant threat. Moreover, over 50 percent of Americans anticipate personal harm and harm to the family from global warming, while 75 percent agree that global warming will undermine secure life for future generations (Leiserowitz, A. et al., 2017[2]).

Such awareness of the climate crisis among the American public is not new, and it is indeed not the result of "climate change brainwashing," as the Trump camp purports. This critical awareness is rooted in American concern about the environment and the willingness to do something about it. Way back in 1995, Jacquelyn A. Ottoman's *Green Marketing: Challenges and Opportunities for the New Marketing Age* revealed that more than one-quarter of U.S. voters (27 percent) pulled the lever for candidates based on their positive track records for environmental responsibility.

In the 1990s, about 10 percent of the American population contributed money to environmental groups and even shunned products made by companies that were not environmentally responsible. Furthermore, quoting a December 1994 nationwide study commissioned by the National Environmental Education Training Foundation, *Green Marketing* revealed that children held a healthy environment as a higher priority than adults did (Ottman, J.A, 1995[3]). Those children of the nineties are today's active voters, and they teach environmental awareness to the future generations. These are facts that no *sensible* political administration can ignore. Though the millennial generation may be ridiculed by baby boomers for killing the diamond industry, ruining the sanctity of marriage, or ditching religion, the fate of the climate and future of the world is mainly in their hands. Fortunately, they are ready to take action, especially at the polls.

Paris Climate Agreement Offers Hope (but America Abandons It)

The 2015 UN Framework Convention on Climate Change 21st Conference of the Parties succeeded in amassing global political will to attempt to limit global warming below 2°C through their committed nationally determined contributions. On November 4, 2016, the Paris Agreement was placed in force. As of August 2018, out of the 197 countries party to the convention, 179 have ratified it (UNFCC, 2018[4]).

---

[1] Lee, Ming Tien, Ezra M. Markowitz, , Peter D. Howe, , Chia-Ying Ko, & Anthony A. Leiserowitz, Predictors of public climate change awareness and risk perception around the world (2015), Nature Climate Change, 2015/07/27/online, Nature Publishing Group, http://dx.doi.org/10.1038/nclimate2728, https://www.nature.com/articles/nclimate2728#supplementary-information (Accessed on 26 February, 2018)

[2] Leiserowitz, A., Maibach, E., Roser-Renouf, C., Rosenthal, S., Cutler, M., & Kotcher, J. (2017). Climate change in the American mind: October 2017. Yale University and George Mason University. New Haven, CT: Yale Program on Climate Change Communication.

[3] Green Marketing: Challenges and Opportunities for the New Marketing Age, NTC Business Book, ISBN: 0844232904, 9780844232904, New York

[4] Paris Agreement - Status of Ratification, United Nations Framework Convention on Climate Change (UNFCC), 06, March, 2018, http://unfccc.int/paris_agreement/items/9444.php (Accessed on 19 August 2018)

However, in an unfortunate turn of events, U.S. President Donald Trump withdrew from the accord to the shock and dismay of the entire world. Even Syria, a nation amid unprecedented conflict and humanitarian tragedy, recently signed the accord, leaving the United States as the only country to oppose the globally renowned, common-sense treaty. The old adage seems to be making a comeback: Common sense isn't always that common.

The European Union (EU) took no time to reject Trump's proposal to renegotiate the Paris agreement, stating that the fight against climate change will continue with or without help from the United States. Sadly, the decision of the federal government is definitive proof that it is abandoning its moral and leadership role. Contrastingly, the EU is directly engaging with states, municipalities, and businesses to implement the Paris Accord collectively. Furthermore, Europe and China have vowed to raise $100 billion a year by 2020 to help poorer countries cut emissions and make up for the funding to the Green Climate Fund that was tragically ceased by the U.S. (Aniel Boffy et al., 2017[5]).

Edging out a large country like the United States from such a collective global effort is unusual, particularly in the recent history of international relations. For the last few decades, the U.S. has enjoyed global leadership in many fields, backed not only by its economy and military power but also thanks to its educated society, time-tested democracy, and progressive values. The U.S. was never previously dismissed from its position at the global leadership table, even during the peak of the recession in 2008 or even when it unfairly waged war on Iraq and destroyed it with false nuclear allegations.

As the *New York Times* rightly opined, by withdrawing from the Paris Accord, Trump ceded U.S. leadership (Lisa Friedman, 2017[6]), not only in climate affairs but also on many other fronts. As Simon Reich notes, "A global leader has to be willing to cast aside its short-term interests in favor of a longer-term outlook" (2017). Unfortunately, with Trump administration's America First platform seems to indicate that the nation has shed her eyeglasses and returned to poor-sighted myopia that only extends as far as her nearly walled borders.

Thankfully, this leadership vacuum on the global stage was not left open for long, as other world leaders have stepped up to fill the gap. Today, the world looks to the collective stewardship of other leaders like German Chancellor Angela Merkel, France President Emmanuel Marcon, China President Xi Jinping, Canadian Prime Minister Justin Trudeau, and Indian Prime Minister Narendra Modi, along with United Nations Secretary-General Antonio Guterres. All of these have repeatedly articulated their determination to ensure full implementation of the Paris Accord and achieve its targets with or without U.S. participation (Lisa Friedman, 2017). These remarkable leaders champion climate commitments in their respective countries, leading global efforts and even initiating other smaller, technical, thematic, and regional pacts in the collective pursuit of saving climate for posterity.

The Trump administration's politically suicidal frenzy to forsake the global climate, not to mention his nation's moral and political leadership on world affairs, is being led by a small minority of conservative-minded individuals, administrators, institutions, and special interest groups. Many states, cities, and significant private enterprises have decided not to be party to the federal administration's backward stance on the Paris climate agreement. According to news reports, Hawaii is the leader among twelve states that have formally defied Trump's withdrawal. The governor of that island state signed a law in 2017 that requires the state to honor the commitment to reduce carbon emissions targeted in the Paris Accord (Clark Mindrock, 2017[7]). Big businesses like Facebook, Apple, Ford, and Microsoft, along with hundreds of other corporations and companies, have publicly

---

[5] Aniel Boffey, Kate Connolly and Anushka Asthana (2017): EU to bypass Trump administration after Paris climate agreement pullout: The Guardian, 02 June 2017, https://www.theguardian.com/environment/2017/jun/02/european-leaders-vow-to-keep-fighting-global-warming-despite-us-withdrawal (Accessed on 28 February 2018)

[6] As U.S. Sheds Role as Climate Change Leader, Who Will Fill the Void?
The New York Times, 12 November, 2017, https://www.nytimes.com/2017/11/12/climate/bonn-climate-change.html (Accessed on 28 February, 2017)

[7] Hawaii defies Trump and becomes first state to legally support Paris Agreement on climate change, Independent, 07, June 2017, http://www.independent.co.uk/news/world/americas/us-politics/hawaii-trump-paris-agreement-climate-change-laws-first-us-state-a7778226.html (Accessed on 27 February, 2018)

disapproved of the administration's decision and have vowed to continue their support of environmental efforts, in spirit and action (Johannes Urpelainen 2017[8]).

Research, Technology, Social, Institutional, Business, and Governance Solutions

While some national governments do fluctuate in their climate commitments, there are many local governments in many parts of the world that are employing refreshing, impressive, concrete solutions to reverse the impacts of climate change. These include urban garden initiatives in Detroit, Michigan and Todmorden, England; recycling and composting programs in San Francisco, California; inter-cropping in Normandy, France; a decentralized democracy model in a village near Chennai, India that integrates ecosystem health and urban development; and the novel idea of prioritizing renewable in place of carbon-intensive energy and transportation practices in Copenhagen, Denmark.

For a cinematic illustration of many of these examples, check out the film Demain (*Tomorrow*) (Herrington, Nicole, 2017[9]). If the internet isn't your thing, just keep reading to see other specific examples of innovative climate solutions going on around the world.

Many people know it by the name of Silicon Valley, but the Pacific coast region of the United States is quickly earning a new nickname: Sustainability Valley. In an inspiring radio conversation, the former Palo Alto, California mayor outlined the initiatives the city took under his leadership to reduce greenhouse gases that simultaneously improved the economy. Specifically, Palo Alto formed a community of green warriors and, under their recommendation, adopted a series of green initiatives, such as:

1) Setting the goal of becoming powered by 100 percent renewable energy by 2030;

2) Reducing greenhouse gas emissions by an estimated 43 percent in 2017 from the 1990 baseline; and

3) Passing a 2018-20 Sustainability Implementation Plan, which outlines proposed measures in four areas that could reduce greenhouse gas emissions to about 54 percent below 1990 levels by 2020. It is highly recommended that you listen to this inspiring conversation!

Though fashion over function may have been true in the past, today, we can have both. A recent report published by the Macarthur Foundation exposed the scale of waste and pollution in the fashion industry, revealing that less than 1 percent of clothing is recycled, resulting in the release of a half-million tons of plastic microfibers pollution annually, equal to roughly 50 billion plastic bottles. As such, major fashion brands Nike, H&M, Burberry, and Gap have joined hands to curb this massive pollution, aiming to improve the industry record on sustainability by recycling raw materials and products (Gerretsen, 2018[10]).

Similarly, in Britain, more than forty companies—including Britain's biggest supermarkets, Coca-Cola, Nestle, and Procter & Gamble—signed the UK Plastics Pact, pledging to eliminate unnecessary single-use plastic packaging by 2025. A few big supermarkets in Britain have also launched plastic-free logos to allow shoppers to identify products with plastic packaging, as companies come under growing pressure to use eco-friendly alternatives (Taylor, 2018).

You can't spell Zero-carBon transItion without B. I. Z. The 2018 Responsible Business Trends Report shows that an impressive 75 percent of corporate CEOs are convinced by the value of sustainability, with a reasonable

[8] Trump's withdrawal from the Paris agreement means other countries will spend less to fight climate change, The Washington Post, 21 November, 2017, https://www.washingtonpost.com/news/monkey-cage/wp/2017/11/21/trumps-noncooperation-threatens-climate-finance-under-the-paris-agreement/?utm_term=.8ad-46ca989a5 (Accessed on 26 February 2018).

[9] Review: Worried About a Sustainable Tomorrow? There's Hope; New York Times, 19, April 2017, https://www.nytimes.com/2017/04/19/movies/tomorrow-review.html (Accesse don 28 May 2018)

[10] Nike, H&M and Burberry join forces for sustainable fashion, Thomson Reuters Foundation, May 16, 2018, https://www.reuters.com/article/us-britain-fashion-recy-cling/nike-hm-and-burberry-join-forces-for-sustainable-fashion-idUSKCN1IH2EU

level of awareness about Sustainable Development Goals (SDG). The survey revealed that 65 percent of businesses engage in SDG 13, on climate change mitigation and adaptation (Ethical Corporation, 2018[11]). Responsible business is not only ethical; it is also essential to surviving in the new age of conscious consumerism. According to a *New York Times* report, investors in control of endowments and portfolios worth more than $5 trillion have committed to divesting from fossil fuels in some capacity. Either they know something we don't, or else they are just reading the markets and operating their businesses accordingly.

Americans are sometimes negatively characterized by stubbornness, but this can be an asset as well. Many U.S. towns and cities are in no mood to compromise when it comes to their survival and sustainability. New York City is suing five of the world's largest oil companies for knowingly polluting the ocean. Meanwhile, other cities like San Francisco and Seattle insist that companies who reaped profits from fossil fuels should now pay for the disasters their business practices caused.

From masses to reducing greenhouse gases, in a 2015 effort to boost Paris climate negotiations, Pope Francis issued Laudato Si', his second encyclical, which emphasized the need to be responsible guardians of the environment, particularly in the face of climate change and carbon pollution. Following this call for saving the climate, Catholic groups in Italy, Germany, Canada, Australia, the U.S., Brazil, and some other nations announced a range of measures to distance themselves from fossil fuels (Page, Samantha 2016[12]).

Specifically, a dedicated campaign by Global Catholic Climate Movement has prompted thirty other Catholic institutions to announce divestment plans from the fossil fuel industry, including three German Catholic banks—Pax-Bank, Bank im Bistum Essen, and Steyler Ethik Bank—which hold more than $9.2 billion in wealth (Sadowski, Dennis, 2018[13]). Another is Caritas Internationalis, a confederation of 165 Catholic development and social service organizations operating in over 200 countries and an official institution of the Catholic Church. It is among the institutions who commit to divest USD 7.5 billion from fossil fuels (Fossil Free, 2018).

What is Zürich's best invention since the Swiss Army Knife? A Swiss company, Climeworks, has initiated another exciting, innovative initiative called Direct Air Capturing (DAC). "The first of its kind DAC plant is designed to pull 900 tons of carbon dioxide from the air per year and feed it to a neighboring greenhouse" (Jospe Christophe, 2017[14]). The DAC plant captures "atmospheric carbon with a filter. Air is drawn into the plant, and the CO2 within the air is chemically bound to the filter. Once the filter is saturated with CO2, it is heated (using mainly low-grade heat as an energy source) to around 100 °C (212 °F). The CO2 is then released from the filter and collected as concentrated CO2 gas to supply to customers or for negative emissions technologies" (Climate Works, 2018[15]). More information on DAC can be seen here: https://www.youtube.com/watch?v=jJwLJwDMBkM.

There is needed relief for the Great Barrier Reef. In Australia, scientists are innovating new technology to save vast expanses of the reef from climate change-induced extinction. Australia's Great Barrier faces enormous threats from the climate crisis, such as warmer and more acidic seawater and increased ultraviolet (UV) radiation from the sun. Fortunately, scientists have come up with a coral reef sunscreen, which they claim can protect the reef during heatwaves. This ultra-thin layer of calcium carbonate, the same material naturally found in coral skeletons, can be applied to the water surface above the reef. The biodegradable screen will form a layer just one molecule thick, 50,000 times thinner than a human hair. It is expected to give the coral time to adjust to the changing conditions of high temperature and doses of UV light so the coral can form different chemical structures that help it survive. Recent tests suggest that the thin film will reflect up to 30 percent of UV light that

[11] The Responsible Business Trends Report 2018, 24 May, 2018, http://www.ethicalcorp.com/just-published-responsible-business-trends-report-2018 (Accessed on 26 May 2018

[12] Catholic groups announce massive divestment from fossil fuels, Think Progress, 0 October 2016, https://thinkprogress.org/catholic-groups-announce-massive-divestment-from-fossil-fuels-f774e4346bc9/ (Accessed on 23 May 2018)

[13] Church teaching leads Catholic entities to divest from fossil fuels, National Catholic Reportes, 24 April 2018, https://www.ncronline.org/news/environment/church-teaching-leads-catholic-entities-divest-fossil-fuels

[14] What is Direct Air Capture (DAC), Part 1 and Part 2, Carbon A List, 24 May 2017 and 31 May 2017, http://carbonalist.com/2017/05/what-is-direct-air-capture-pt1/

[15] How Climate Works Plants capture CO2 from Air: http://www.climeworks.com/our-technology/ Accessed on 18 August 2018

hits the surface of the water, thus keeping the water at just the right temperature for the reef below (Campanaro, Amanda, 2018[16]).

Individual Actions Have the Power to Mitigate the Impacts of Climate Change

"The reality we now face implores us to act." – **Al Gore**

Climate change is at work so we cannot rest. Climate-induced environmental disasters scientists have long predicted are happening now, all around us, ever faster and even more detrimental. We still have a chance to stop the impending catastrophe, but the job can't be left to politicians alone. Solutions to the climate crisis must be developed and carried out by you and me, through lifestyle and behavioral changes, as well as intentional participation in the facilitation of technical solutions and forcing political action, both nationally and globally. There are three types of actions we all should take to save ourselves and future generations from climate change-induced catastrophes: things we should KEEP doing, things we should START doing, and things we should STOP doing. Below, for your consideration, are several suggested action items.

Conclusion

We have reestablished the fact that unprecedented, unsustainable interference of human actions has eroded the strength and resiliency of the natural environment and the climate. Reversal of the damage resulting from human-induced climate change and preventing future damage require many crucial, urgent actions by global powers, national governments, scientific institutions, local actors, and individuals. Concerned and responsible individuals should not only alter lifestyles and behaviors; we must also demand action from elected leaders, demanding that they support and implement a comprehensive set of climate solutions before it's too late.

Climate Crisis "Keep/Stop/Start" Actions

Below for your consideration, are several author-suggested thought, behavior, and action items. Action nourishes inspiration! For a further explanation of the Keep/Stop/Start organizational tool, please see How to Read This Book.

KEEP:

1) *Keep transitioning toward sustainability*: Many intelligent, motivated citizens worldwide are rapidly transitioning toward more sustainable lifestyles. Though this transition is excellent, it is not happening fast enough. Individuals should continue practicing climate-conscious behavior while imploring others to follow suit.

2) *Keep pressuring politicians*: Like it or not, a lot of the action on climate change must be taken by our elected officials. Fortunately, they work for us; thus, they are obligated to hear our concerns. A tactic that has proven effective is invoking rights enshrined in the Constitution to drag responsible parties, particularly within the government, to stop pollution. As an example: "Exercising my 'reasoned judgment,' I do not doubt that the right to a climate system capable of sustaining human life is fundamental to a free and ordered society." This remark was made by U.S. District Judge Anne Aiken, while hearing

---

[16] Can this ultra-thin 'sunscreen' save the world's largest coral reef? NBC News, 05 April, 2018, https://www.nbcnews.com/mach/science/can-ultra-thin-sunscreen-save-world-s-largest-coral-reef-ncna863001 (Accessed on 26 May 2018)

the landmark constitutional lawsuit *Juliana v. United States*. In this lawsuit, a group of young people, 10 to 21 years, under the umbrella of Earth Guardians organization (https://www.earthguardians.org/), asserted that "through the government's affirmative actions that cause climate change, it has violated the youngest generation's constitutional rights to life, liberty, and property, as well as failed to protect essential public trust resources." On March 7, 2018, the Ninth Circuit Court of Appeals rejected the Trump administration's "drastic and extraordinary" petition for writ of mandamus aimed at intimidating and silencing the youth and set October 29, 2018, as the trial date for *Juliana v. United States*. Keep tracking developments in this case and many such other cases at https://www.ourchildrenstrust.org/us/federal-lawsuit/ and extend your moral and vocal support.

Another exciting battle against the Trump administration's regressive policies that seek to allow climate destruction also came from young children, who have the right to grow up in a clean environment. The lawsuit, filed in federal court in Pennsylvania, alleges the U.S. is using "junk science" to roll back policies designed to limit the impact of climate change. Two child plaintiffs who were personally impacted by climate change in the form of health issues and by Hurricane Irene collaborated with Clean Air Council to file the lawsuit (https://www.youthvgov.org/#intro). They asked the court to stop the government from pushing "any rollbacks from climate commitment that increase the frequency and intensity of the life-threatening effects of climate change." KEEP giving this case your moral and vocal support, and seek replication of such legal battles to force the U.S. administration to take responsible policy measures to save the climate (Henry, Devin, 2017[17]).

STOP:

1) *Stopping climate change and saving our lives* will require reversing many misconceptions and changing lifestyle habits. It's high time we STOP treating nature as property, but as an entity that also has rights. We must end commodification, ownership, and exploitation of all ecosystems such as forests, grasslands, deltas, rivers, and wildlife preserves. Recently, the Whanganui River in New Zealand and the Ganges in India were granted human rights. Nature's intrinsic right was enshrined in the constitution of Ecuador (Tanasescu, Mihnea, 2017). It is time for other countries and societies to understand that the natural world has the right to equal care to all forms of life and not just to the human species. Individuals, governments, and institutions must end market-based mechanisms that promote the industrial and mono-cultural destruction of biodiverse regions and hotspots.

2) *Stop abusing the environment*: The misguided, arrogant belief that humans are the masters of all-natural resources on Earth is a powerful motive for our careless abuse of it. This has effectively resulted in us heaping catastrophe upon ourselves. In this backdrop, it is no surprise that in his encyclical Laudato Si', Pope Francis underlines the importance of social and cultural values in protecting the world climate. He wisely says that the sustainability of physical, moral, spiritual, and social ecology hinges on the culture and moral conditions created by it. Cautioning that climate change is a manifestation of violence in our hearts, Pope Francis noted that developing attitudes of peace, love, harmony, and coexistence are essential to stop climate change. He further elaborated that the key to preventing climate change and consequent environmental challenges lies in reversing recent trends in human lifestyles that indulge in consumerism

---

[17] Green group, children sue Trump over climate change policies, The Hill, 11 June 2017, http://thehill.com/policy/energy-environment/359015-green-group-children-sue-trump-over-climate-change-policies (Accessed on 23 May 2018)

and destruction of the ecosystem for luxuries and wastage. Whether you consider yourself religious or not, you have to admit he has a point.

3) *Stop permitting misinformation*: Like many political issues, climate change is a polarizing, hot-button topic. The inherent uncertainty about the degree to which humans are causing it and the best steps to mitigate the damage are frequent topics of debate. However, there are still many narratives that are being put forth that the climate is not changing at all, and if it is, it is not due to human activity, which is blatantly false. Instead of agreeing to disagree, try to have productive conversations with the people who say these things and present the facts in a polite, coherent, and persuasive manner. Though we cannot convince every human on Earth that the facts are the facts, we must still try. No one is born with an inherent knowledge of these things; we all had to be taught the basics of climate science by someone else. As such, we must continue to disseminate the information to the masses, but above all else, we must make a concerted effort to stomp out the false narratives, as difficult as that may be.

4) *Stop overheating and air conditioning*: We have only had modern-day heating and cooling systems for less than a century, yet we are already abusing these privileges. Heating and cooling our homes contribute to our exacerbated energy demands, as well as the excess emission of greenhouse gases in our atmosphere. By keeping the thermostat just two degrees warmer in the summer or cooler in the winter, and by turning off the air conditioning while we are asleep or away, we can individually save about 2,000 pounds of carbon dioxide each year (West, Larry, 2018[18]).

START:

"The future depends on what we do in the present" – **Mahatma Gandhi**

1) *Start building green infrastructure*: Buildings account for more than one-quarter of all greenhouse gas emissions (GHGs), according to the Global Alliance for Buildings and Construction. "A 2014 UC Berkeley study found that by building to the Leadership in Energy and Environmental Design (LEED) system, buildings contributed 50 percent fewer GHGs than conventionally constructed buildings due to water consumption, 48 percent fewer GHGs due to solid waste, and 5 percent fewer GHGs due to transportation. By building green, we can reduce the impact our buildings have on contributing to climate change while also building resilience into our homes and communities" (Huynh, Christina, 2017[19]).

2) *Start talking to your children about climate change*: Children are active learners. They grasp things faster than we think, and they seldom forget what they learn. Perhaps most importantly, they pass on what they learn. Therefore, it's vital and powerful to talk to children about climate change and the action we must take to inhibit further damage to the climate. In 2016, I visited Houston to attend a Climate Leadership Corps training conducted by Vice President Al Gore's Climate Reality Project. In the middle of the training, I was paired with another participant to exchange ideas for local actions we can pursue to save the climate back home in India. I was pleasantly surprised when I realized my partner was only 11 years old. That young lady had much brighter, more innovative, more energetic ideas, and she exhibited far greater commitment for climate action. She had plans to organize a collective of students in her school

[18] Things You Can Do to Reduce Global Warming, ThoughtCo, 22 February, 2018, https://www.thoughtco.com/how-to-reduce-global-warming-1203897 (Accessed on 26 May 2018)
[19] How green buildings can help fight climate change, LEED, 19 April 2017, https://www.usgbc.org/articles/how-green-buildings-can-help-fight-climate-change (Accessed on 23 May 2018)

and to create momentum in her town toward climate awareness and action. I learned so much from her and congratulated her father, who brought her to the training and has been guiding her all these years.

3) *Start buying locally.* Though many of us are enthralled with the ease and convenience of online shopping, it is one of the most carbon-intensive activities you can do. Ordering items from across the country or internationally leaves a massive carbon footprint, due to emissions from shipping, production of plastic and other materials needed to secure your items on route, and energy demands for massive warehouses and other production needs. Moreover, the rate at which you receive your online order has a large impact too: For instance, opting for slower ground transportation results in half the carbon emissions of overnight air delivery. Not only does buying locally support your local businesses but it also greatly reduces the carbon footprint of your purchases. Whether it is books, clothing, or kitchenware, resist the urge to hop on Amazon. First, see if you can find it somewhere in your city. Though smaller businesses sometimes charge a higher price, the planet and climate will benefit in invaluable ways.

4) *Start adding cow manure to your list of viable renewable energy sources*: The average cow produces over 100 pounds of manure a day, and it is a rich source of energy. Millions of low-income households in many developing countries have been using cow manure as an energy source for years. Cow dung is even dried and burned in earthen stoves for the cooking of meals, but it doesn't even stop there. In advanced methods, manure can be converted into concentrated gas in an anaerobic digester system and used for heat. In the U.S., there are currently 2,200 sites where cow manure is used to produce biogas, and there is huge potential for growth. "David Simakov, Assistant Professor of Chemical Engineering at the University of Waterloo in Ontario, is working to make the anaerobic process more efficient. He and his colleagues are researching ways to boost the energy content of this raw biogas with a refining process that uses hydrogen in a chemical reaction to convert carbon dioxide into methane. This conversation doubles the production rate of the older separation techniques, producing so much renewable natural gas that the volume could be stored and eventually used as a natural battery. As a bonus, the new method results in lower carbon dioxide emissions" (Efstathiou, Jim Jr, 2018[20]). Individuals and communities should come together to create collective bio-energy plants that can save energy costs and also reduce climate stress.

5) *Start being an active leader* on climate action in your community: The environment is sensitive to climate changes, and localized action can either reduce or aggravate disasters, depending on how current policies and practices treat the climate. This wisdom is known by simple farmers, nomads, traditional fisherfolk, forest dwellers, mountain inhabitants, small producers, and humble workers in rural and urban areas. They did not cause climate destruction, yet they suffer from it. Rather than complaining, many try to adapt to the changes, innovate new solutions, and move on with their commitment to family, society, and nature. There is a lot we can learn from the strength of their wisdom and adaptability in dealing with the climate crisis.

In the developed nations, innovative civil society leaders and institutions can help keep the country's leadership in knowledge, policy, and technology moving in a positive direction. For instance, former U.S. Vice President Al Gore's Climate Reality Project has trained tens of thousands from over a hundred countries about climate leadership, providing the necessary tools to spread climate awareness and climate actions around the globe. Such leadership initiatives from the Climate Reality Project and many other voluntary organizations in many parts of the world hold promise for climate-secure future.

[20] Add Cow Manure to Your List of Renewable Energy Sources, Bloomberg.com, 14 March 2018, https://www.bloomberg.com/news/articles/2018-03-14/add-cow-manure-to-your-list-of-renewable-energy-sources (Accessed on 23 May 2018)

There are a large number of good examples and climate heroes to be followed. The Climate Heroes Initiative honors individuals worldwide, those who are making significant, innovative contributions to reverse the negative impacts of the climate crisis. Illac Diaz from Manila, Philippines, has innovated an environmental-friendly lighting concept, made out of recycled plastic bottles through his nongovernmental organization, Liter of Light. This innovation helped illuminate the homes of over 300,000 low-income families, without burning any fossil fuels. The successful concept has also been replicated in many other countries, benefiting over 600,000 families worldwide.

Similarly, Bren Smith in Connecticut started an NGO, Green Wave, in 2013, with the aim of augmenting an open-source model of ocean farming he innovated with years of struggle. Through his NGO, Bren promotes a three-dimensional marine farm model that involves growing kelp, algae, oyster, clams, and mussels. This method, according to experts, helps to restore a healthy ocean ecosystem while reducing pressure on fish stock and capturing carbon and nitrogen from the atmosphere and from the water column. The inspiring stories of many such climate heroes can be followed on https://climateheroes.org/heroes/. Also visit https://climategamechangers.org/, which summarizes the main topics meaningful to people looking for ways to contribute effectively to solving climate change.

Here are some simple actions every individual should take to save our climate:

- Be energy efficient! Transition from CFL to LED bulbs and choose renewable power when possible.
- Always buy energy-efficient products. Look at labels and research the companies you buy from.
- Use less hot water.
- Always switch off your lights, fans, and the heating/cooling when these luxuries are not needed.
- Choose to eat organic, locally grown food.
- Reduce and reuse waste.
- Drive less, and use a bicycle or walk if possible.
- Plant a tree and contribute to greenery growth in your residential areas.
- Inform, educate, and encourage others to adopt all these actions.
- Read this inspiring, empowering article on *The Guardian*, 50 easy ways to save the planet

Hari Krishna Nibanupudi, Ph.D. is a Senior International Climate, Resilience, Humanitarian, Peace and Development Nexus Thinker, Programmer, and Policy Advisor. Currently, serving United Nations missions globally through the expert's rosters of the United Nations Development Program (UNDP), Swedish Civil Contingency Agency (MSB), Norwegian Capacity (NORCAP) and CANADEM International Civil Response Corps. Read more at: www.climateabandoned.com

Chapter 32

Climate Crisis & Editor's Tidbits and Video Clips

A smattering of news and videos to pique your interest for more information (by chapter)

Background Resources:

- Climate Science Special Report- *"Fourth National Climate Assessment (NCA4), Volume 1"*: https://science2017.globalchange.gov

- The Real News - "Michael Mann: We Are Even Closer To Climate Disaster Than IPCC Predicts": https://therealnews.com/stories/michael-mann-we-are-even-closer-to-climate-disaster-than-ipcc-predicts

- National Center for Climate Restoration – *"What Lies Beneath: The Understatement of Existential Climate Risk"*: https://www.breakthroughonline.org.au

- Common Dreams – December 6, 2018, *"How The Iconic 1968 Earthrise Photo Changed Our Relationship To The Planet"*: https://www.commondreams.org/views/2018/12/06/how-iconic-1968-earthrise-photo-changed-our-relationship-planet

- The Real News Network – *"Dr. Michael Mann Video Interviews"*: https://therealnews.com/?pum_form_popup_id=178694&s=Michael+Mann

- John Englander – *"Single Image Proves Human-Caused Global Warming"*: http://www.johnenglander.net/sea-level-rise-blog/single-image-proves-human-caused-global-warming/

- Yale Climate Connections: https://www.yaleclimateconnections.org

- Frontline – *"As UN Sounds Alarm on Global Warming, Revisit Frontline's Recent Climate Reporting"*: https://www.pbs.org/wgbh/frontline/article/as-un-sounds-alarm-on-global-warming-revisit-frontlines-recent-climate-reporting/

- Bulletin of the Atomic Scientists – *"Climate report understates threat"*: https://thebulletin.org/2018/10/climate-report-understates-threat/

- Vice - *"Extinction Rebellion Is Telling the Terrifying Truth About Climate Change."*: https://www.vice.com/en_uk/article/bjqdbd/xr-is-telling-the-terrifying-truth-about-climate-change

Chapter 1- Climate Crisis & Greenhouse Effect:

- The Real News – *"Climate Emergency: Greenhouse Gas Levels Surge to Historic Levels"*: https://therealnews.com/stories/ktrenberth1101co2

- Greenhouse Gases: *"Not just a bunch of hot air"*: https://www.youtube.com/watch?v=ZNPhn3TBOZE

- The Real News Network video -*"Global Carbon Emission Set to Hit Record High in 2018"* - December 7, 2018, From COP24 in Poland, Greenpeace USA's Naomi Ages says that carbon emissions are set to rise by 2.7% in 2018 due to more coal use in Asia, and that to combat this trend nations need to do more than what they agreed to in the 2015 Paris Accord: https://therealnews.com/stories/global-carbon-emissions-set-to-hit-record-high-in-2018

- Green World Rising (Leonardo DiCaprio/ Thom Hartmann) film short: *"Carbon"* - http://www.greenworldrising.org/carbon

Chapter 2 - Crisis & Biodiversity:

- Visual Capitalist – *"Animation: The Heartbeat of Nature's Productivity"*: https://www.visualcapitalist.com/animation-the-heartbeat-of-natures-productivity/

- Kalahari Lion Research – *"Global Terrestrial Mammal Biomass"*: http://www.kalaharilionresearch.org/2015/01/16/human-vs-livestock-vs-wild-mammal-biomass-earth/

- The Guardian - November 6, 2018: *"Stop biodiversity loss or we could face our own extinction, warns UN"* https://www.theguardian.com/environment/2018/nov/03/stop-biodiversity-loss-or-we-could-face-our-own-extinction-warns-un

- Bio4Climate: *"Biodiversity for a Livable Planet: Restoring Systems to Reverse Global Warming"*: https://bio4climate.org and *"Compendium of Scientific and Practical Findings Supporting Eco-Restoration to Address Global Warming"*: https://bio4climate.org/resources/compendium/

- The New York Times Magazine – November 27, 2018, *"The Insect Apocalypse Is Here: What does it mean for the rest of life on Earth?"* https://www.nytimes.com/2018/11/27/magazine/insect-apocalypse.html

- The Guardian – *"Climate change driving species out of habitats much faster than expected"*: https://www.theguardian.com/environment/2011/aug/18/climate-change-species-habitats

- Visual Capitalist Animation - December 1, 2018: *"Animation: The Heartbeat of Nature's Productivity"*: Today's unique cartogram animation comes from geographer Benjamin Hennig at Worldmapper, and it depicts ongoing cycles in the productivity of ecological systems around the world. https://www.visualcapitalist.com/animation-the-heartbeat-of-natures-productivity/

- The New York Times – April 4, 2018: *"5 Plants and Animals Utterly Confused by Climate Change"*: https://www.nytimes.com/2018/04/04/climate/animals-seasons-mismatch.html

- The Guardian – May 21, 2018: *"Humans just 0.01% of all life but have destroyed 83% of wild mammals – study"*: https://www.theguardian.com/environment/2018/may/21/human-race-just-001-of-all-life-but-has-destroyed-over-80-of-wild-mammals-study

Chapter 3 - Climate Crisis & Ice:

- America Progress: *"The Big Melt: Curbing Arctic Climate Change Aligns with U.S. Economic and National Security Goals"*: https://www.americanprogress.org/issues/green/reports/2017/05/08/432016/big-melt-curbing-arctic-climate-change-aligns-u-s-economic-national-security-goals/.

- The Guardian: *"We've never seen this': massive Canadian glaciers shrinking rapidly"*: https://www.theguardian.com/world/2018/oct/30/canada-glaciers-yukon-shrinking.

- Live Science – *"The Artic"*: https://www.livescience.com/topics/arctic.

- Live Science – *"Every Year, the Swiss Cover Their Melting Glaciers in White Blankets"*: https://www.livescience.com/61951-swiss-glacier-blanket.html.

- Artic News: http://arctic-news.blogspot.com.

- BBC News – *"Can painting a mountain restore a glacier?"*: https://www.bbc.com/news/10333304.

- This Week (June 2, 2017) – *"Spitsbergen, Norway: Doomsday Threat"*: Norwegian official, Hege Njaa Aschim: *"Unusually high arctic temperatures caused permafrost to melt and seep into the 'Doomsday seed vault' – a fail-safe trove intended to protect food supplies in case of a global calamity – it was revealed last week. The Svalbard Global Seed Vault, which is buried in a frozen mountain on a Norwegian island, stores some 500 million seeds from around the world. But late last year [2016] temperatures soared on Svalbard, pushing the permafrost around the vault above melting point. Water seeped into the entrance tunnel, but didn't reach the seeds, "It was not in our plans to think that the permafrost would not be there."*

- BBC Earth - *"There are diseases in ice, and they are waking up"*: http://www.bbc.com/earth/story/20170504-there-are-diseases-hidden-in-ice-and-they-are-waking-up

- Think Progress – "Earth's thawing permafrost threatens to unleash a dangerous climate feedback loop": https://thinkprogress.org/global-warming-permafrost-thaw-97436404e353/

Chapter 4 – Climate Crisis & Oceans Part I –

- Join the Monterey Bay Aquarium: https://www.montereybayaquarium.org. Conservation & Science Blog: https://futureoftheocean.wordpress.com

Chapter 5 - Climate Crisis & Oceans Part II –

- Join Oceana: *"10 Things You Can Do to Save the Oceans"*: https://oceana.org/living-blue/10-things-you-can-do.

- CBS News – *"Salt-Water Fish Extinction Seen By 2048"*: https://www.cbsnews.com/news/salt-water-fish-extinction-seen-by-2048/

- The Guardian – *"Heatwaves sweeping oceans like wildfires, scientists reveal"*: https://www.theguardian.com/environment/2019/mar/04/heatwaves-sweeping-oceans-like-wildfires-scientists-reveal

- Heriot Watt University – *"Deep sea mining zone hosts Co2-consuming bacteria, scientists discover"*: https://www.hw.ac.uk/about/news/2018/deep-sea-mining-zone-hosts-co2-consuming.htm.

- Zillow – *"Ocean the Door"*: https://www.zillow.com/research/ocean-at-the-door-21931/.

- Earther – *"Deep Reefs Might Not Be the Climate Refuge We Hoped"*: https://earther.gizmodo.com/deep-reefs-might-not-be-the-climate-refuge-we-hoped-1827714088

- NPR – *"Massive Starfish Die-Off is Tied to Global Warming"*: https://www.npr.org/2019/01/30/690003678/massive-starfish-die-off-is-tied-to-global-warming

- Yale Environment 360 – *"How Long Can Oceans Continue to Absorb Earth's Excess Heat?"*:

- https://e360.yale.edu/features/how_long_can_oceans_continue_to_absorb_earths_excess_heat

- The Guardian - *"Scientists study ocean absorption of human carbon pollution"*: https://www.theguardian.com/environment/climate-consensus-97-per-cent/2017/feb/16/scientists-study-ocean-absorption-of-human-carbon-pollution

- Frontline Documentary – "The Last Generation": http://apps.frontline.org/the-last-generation/?utm_source=newsletter&utm_medium=email&utm_campaign=last_generation

Chapter 6 - Climate Crisis & Putting a Price on Carbon

- Carbon-Prince.com- *"Global Carbon Pricing: The Path to Climate Cooperation"*: https://carbon-price.com

- Bloomberg News: *"Yellen Touts Carbon Tax as 'Textbook Solution' to Climate Change"*: https://www.bloomberg.com/news/articles/2018-09-10/yellen-touts-carbon-tax-as-textbook-solution-to-climate-change.

- Green World Rising – *"Carbon"* 8 min. video: http://www.greenworldrising.org/carbon.

- Axios - *"The Age of America Oil"*: https://www.axios.com/us-oil-exporter--1b94a4b8-4492-4019-b542-4d0767e50bc4.html

- The Real News Network – *"Humanity's 'Carbon Budget' is Smaller Than Generally Believed"* video: https://therealnews.com/stories/mmann0906reportpt2.

- Scientific American – *"NASA satellite reveals source of el niño-fueled carbon dioxide spike"* video: https://www.youtube.com/watch?v=4JDfEb5MiJE.

- Stockholm Environment Institute – *"Principle for aligning U.S. fossil fuel extraction with climate limits"*: https://www.sei.org/publications/principles-for-aligning-fossil-fuel-extraction-with-climate-limits/

Chapter 7 - Climate Crisis & Extreme Weather

- Business Insider – *"How extreme summer heat affects your body and brain"*: https://www.businessinsider.com/weather-in-summer-heat-impact-on-health-2018-7

- Tech Times – *"Heat Waves Can Make People Dumber By 13 Percent"*: https://www.techtimes.com/articles/232143/20180712/heat-waves-can-make-people-dumber-by-13-percent-says-harvard-study.htm

- Clean Choice Energy: The Latest – How Climate Change is Connected to Extreme Weather": https://cleanchoiceenergy.com/news/how_climate_change_is_related_to_extreme_weather/

- The Guardian - *"Met Office: global warming could exceed 1.5C within five years"*: https://www.theguardian.com/environment/2019/feb/06/met-office-global-warming-could-exceed-1-point-5-c-in-five-years.

- NOAA Climate.gov – *"2017 U.S. billion-dollar weather and climate disasters: historic year in context"*: https://www.climate.gov/news-features/blogs/beyond-data/2017-us-billion-dollar-weather-and-climate-disasters-historic-year

- Live Science – January 8, 2018: *"Hundreds of 'Boiled' Bats Fall from Sky in Australian Heat Wave"*: https://www.livescience.com/61372-boiled-bats-australia.html

- Rolling Stone - *"What Happens When a Superstorm Hits D.C.?"*: https://www.rollingstone.com/politics/politics-news/what-happens-when-a-superstorm-hits-d-c-196856/

Chapter 8 - Climate Crisis & Archeology:

- Pacific Standard – *"How Climate Change Could Destroy Thousands of Archeological Sites"*: https://psmag.com/environment/rising-tides-swallow-all-sites.

Chapter 9 - Climate Crisis and Eco-Restoration

- EcoWatch – January 19, 2018: *"Monsanto's Roundup Destroys Healthy Microbes in Humans and in Soils"*: https://www.ecowatch.com/monsanto-roundup-microbes-2526878891.html

- BBC News – October 25, 2018

- San Jose Mercury news – *"Carbon-rich soil may be the ticket to sustainability"*: https://www.mercurynews.com/2019/03/03/protecting-the-soil-stemple-creek-ranch-is-pushing-for-a-more-sustainable-farming-practice/

Chapter 10 – Climate Crisis & Selling Doubt Part I: (*See extensive footnotes in chapter*)

- The Years Project video– *"Big Oil Knew: The Confusion Memo"*: https://www.youtube.com/watch?v=uMd6hE6Rr6M.

Chapter 11 – Climate Crisis & Selling Doubt Part II: (*See extensive footnotes in chapter*)

- • Coal Rollers! – Videos: https://www.youtube.com/watch?v=09BZ6ikBZ54 and https://www.youtube.com/watch?v=_1PbgwowkeQ

Chapter 12 – Climate Crisis & Selling Doubt Part III: What can we do?

- Quartz – *"Those 3% of scientific papers that denial climate change? A review found them all flawed"*: https://qz.com/1069298/the-3-of-scientific-papers-that-deny-climate-change-are-all-flawed/

- Global Climate Summit – *"Call to Global Climate Action"* tab: https://www.globalclimateactionsummit.org/call-to-action/.

Chapter 13- Climate Crisis & the Media

- Ad Fontes Media – *"Media Bias Chart: Version 4.0"* - www.mediabiaschart.com

- Climate Crocks – *"How Fox News Works, Spoiler, it's a Drug"*: https://climatecrocks.com/2018/02/20/how-fox-news-works-spoiler-its-a-drug/.

- Forbes – A Rigorous Scientific Look Into the 'Fox News Effect'": https://www.forbes.com/sites/quora/2016/07/21/a-rigorous-scientific-look-into-the-fox-news-effect/#1fb82cd712ab

- Business Insider –*" STUDY: Watching Only Fox News Makes You Less Informed Than Watching No News At All"*: https://www.businessinsider.com/study-watching-fox-news-makes-you-less-informed-than-watching-no-news-at-all-2012-5

- Inside Climate News – *"Survey: More Fox News, More Climate Doubts"*: https://insideclimatenews.org/content/survey-more-fox-news-more-climate-doubts

- Fact Finding Websites – www.mediamatters.org), www.snopes.com, www.mediabiasfactcheck.com, and www.politifact.com (Pulitzer Prize winner).

- Politico – *"The Media Bubble is Worse Than You Think"*: https://www.politico.com/magazine/story/2017/04/25/media-bubble-real-journalism-jobs-east-coast-215048

- CNN Business (re: Deepfakes) *"When seeing is no longer believing: Inside the Pentagon's race against deepfake videos"*: https://www.cnn.com/interactive/2019/01/business/pentagons-race-against-deepfakes/

- The New Yorker – *"The Making of the Fox News White House"*: https://www.newyorker.com/magazine/2019/03/11/the-making-of-the-fox-news-white-house

- Axios – *"4 Ways to Fix 'Fake News'"*: https://www.countable.us/articles/12926-4-ways-fix-fake-news

- Vanity Fair - *"Staggering, Marvelous Our Planet Is the Nature Show We've Been Waiting For"*: https://www.vanityfair.com/hollywood/2019/04/our-planet-netflix-nature-show-review-climate-change

Chapter 14 - Climate Crisis & Climate Stress

- American Psychological Association – *"Environment"* tab: https://www.apa.org/topics/environment/index.html.

- Medium – *"Where can we find hope after the 1.5° UN climate report?"*: https://medium.com/@felixkramer/where-can-we-find-hope-after-the-1-5-un-climate-report-880d6a617ebe.

- Medium - *"Where can we find hope after the 1.5° UN climate report?"*: https://medium.com/@felixkramer/where-can-we-find-hope-after-the-1-5-un-climate-report-880d6a617ebe

Chapter 15 – Climate Crisis & Zombie Myths:

- Skeptical Science – *"Global Warming & Climate Change Myths"*: https://www.skepticalscience.com/argument.php

- Skeptical Science – *"Debunking Handbook"* pdf: https://skepticalscience.com/docs/Debunking_Handbook.pdf

- Mother Jones – *"A Scientist Just Spent 2 Hours Debating the Biggest Global Warming Deniers in Congress"*: https://www.motherjones.com/environment/2017/03/michael-mann-lamar-smith-house-science-committee/

Chapter 16 - Climate Crisis & Ideology v. Science:

- Quartz – *Those 3% of scientific papers that denial climate change? A Review found them all flawed"*: https://qz.com/1069298/the-3-of-scientific-papers-that-deny-climate-change-are-all-flawed/.

- Skeptical Science (excellent worldwide team resource): Website excerpt: Scientific skepticism is healthy. Scientists should always challenge themselves to improve their understanding. Yet this isn't what happens with climate change denial. Skeptics vigorously criticise any evidence that supports man-made global warming and yet embrace any argument, op-ed, blog or study that purports to refute global warming. This website gets skeptical about global warming skepticism. Do their arguments have any scientific basis? What does the peer reviewed scientific literature say? https://www.skepticalscience.com

- Cabrillo College: Astro 7 PowerPoint Deck: Planetary Climate Science: *"Chapter 0: Principles of Clear Thinking and Science"*: http://www.cabrillo.edu/~rnolthenius/astro7/A7PowerIndex.html.

- America Meteorological Society – *"21 June 2017 Letter to Energy Secretary Rick Perry"*: https://www.ametsoc.org/ams/index.cfm/about-ams/ams-position-letters/letter-to-doe-secretary-perry-on-climate-change/

- In 2018, climate science was put on trial. Several California cities sued Chevron, ExxonMobil, Shell, and BP arguing that these world dominant, powerful corporations were responsible for the planetary damages related to global warming. Chevron's position was the climate change is a product of the way people live their lives and not the fault of the corporations. The oil companies did not present any science but only contrarians who call climate scientists as "a glassy-eyed cult" ... this is a perfect example of ideology versus science. Even the judge, William Alsup, recognized how vital science is and convened a tutorial answering key questions.

- Mother Jones – *"The Worst Anti-Science BS of 2018"*: https://www.motherjones.com/environment/2018/12/the-worst-anti-science-bs-of-2018/

Chapter 17 - Climate Crisis & the Failure of Political Leadership

- Think Progress – *"Scientists say Trump's first 2 years have been fatal for a livable climate"*: https://thinkprogress.org/trump-presidency-two-years-climate-f23d266c15ac/.

- The Washington Post – *"I'm a scientist who has gotten death threats. I fear what may happen under Trump"*: https://www.washingtonpost.com/opinions/this-is-what-the-coming-attack-on-climate-science-could-look-like/2016/12/16/e015cc24-bd8c-11e6-94ac-3d324840106c_story.html?noredirect=on&utm_term=.2d089d83c060.

- The New Yorker - *"The Hard Lessons of Dianne Feinstein's Encounter with the Young Green New Deal Activists"*: https://www.newyorker.com/news/daily-comment/the-hard-lessons-of-dianne-feinsteins-encounter-with-the-young-green-new-deal-activists-video.

Chapter 18 - Climate Crisis & Energy

- 2019 Sustainable Energy in America *"Factbook"*: https://www.bcse.org/factbook/?fbclid=IwAR2n8IYNIM0F4Sn6YWAf8JovAetvJhjVj1gloAiD38MjpKTryAOFWzNF6kc

- Navigant Research – *"To Meet the Paris Climate Goals, Focus on Deep Energy Efficiency"*:

- https://www.navigantresearch.com/news-and-views/to-meet-the-paris-climate-goals-focus-on-deep-energy-efficiency?utm_campaign=NR_INT_10%2F12%2F18_Cleantech_Update&utm_medium=email&utm_source=Eloqua. Climate Citizen Lobby email (by revlen): *Some ask "what do you mean 'deep' efficiency". Here's an example: A commercial building with a 27-year-old cooling system could reduce their cooling energy consumption by 33% simply by replacing the old chiller with a new one. But if they also replace their windows with high-performance glass with spectrally-selective coatings tuned to solar exposure, upgrade all their lighting to LED and add automatic lighting occupancy-based controls with dimming, re-configure the chilled water pumping system, add reflective coatings to the roof, control ventilation based on interior CO_2 levels, and tune up the temperature controls for heating, cooling, and air supply, they could install a smaller chiller that would satisfy the reduced cooling load, permanently reduce pumping volumes and fan air volumes, and overall reduce the entire energy use of the building, including electricity and heating fuel, by over 40%. That's "deep" efficiency, and requires a comprehensive look at how energy use of different systems and components in a building or industrial process interact.*

- CBS News video – *"It's now cheaper to build a new wind farm than to keep a coal plant running"*: https://www.cbsnews.com/news/its-now-cheaper-to-build-a-new-wind-farm-than-to-keep-a-coal-plant-running/.

- In These Times – *"The unimaginable is now possible: 100%renewable energy. We can't settle for less."*: http://inthesetimes.com/features/bill_mckibben_renewable_energy_100_percent_solution.html

Chapter 19 - Climate Crisis & Corporations

- Yes! *"9 Strategies to End Corporate Rule"*: https://www.yesmagazine.org/issues/9-strategies-to-end-corporate-rule/9-strategies-to-end-corporate-rule.

- ActionPAorg - *"How to Overthrow Corporate Rule in 5 Not-so-easy Steps"*: http://www.corporations.org/solutions/

- The World Post video – German Lignite Coalmining: https://www.facebook.com/theworldpost/videos/529785177522090/.

- Join Corporate Accountability: www.corporateaccountability.org.

- Citizens Climate Lobby/ Santa Cruz email (by RN): *"As long as Big Business installs and controls our politicians, don't expect Big Business to be motivated beyond the window-dressing, towards actually creating a healthy planet - not if instead they can manipulate the rules (carbon offsets, where pollution taxes actually fall, etc.) to their primary goal - maximize their profits. Don't be taken in by the greenwashing that they're concerned about saving the planet. They're PRIMARILY concerned about their profits, and planet saving is only a convenient ad point. That's the ONLY CONCLUSION I can draw by watching the disgraceful antics of the OTEC pipe advocates and proprietary trillions of white beads to replace lost ice, iron flakes across the open ocean, and all the rest. Despite the science against these "fast, cheap, but out of control" schemes, despite the attention I have attempted to bring to how dangerous and flawed they have been shown to be, in peer-reviewed journal studies - they carry on in exactly the same way as the financial media does when it comes to climate change -* **they simply ignore it**. *It's an effective strategy alas - spinning you towards the view that the valid scientific criticisms are too unworthy to even merit a response."*

Chapter 20 - Climate Crisis & the Future of Democracy

- Center for Humans & Nature – *"Democracy Itself is the Solution to the Climate Crisis"*: https://www.humansandnature.org/democracy-maude-barlow.

- The Washington Post – *"Democracy may fatally slow climate action"*: https://www.washingtonpost.com/news/theworldpost/wp/2018/09/13/saving-the-planet/?utm_term=.7d34cfda18ce.

Chapter 21 - Climate Crisis & National Security

- Council on Foreign Relations – *"Climate Change and National Security"*: https://www.cfr.org/report/climate-change-and-national-security

- United Nations Climate Change – *"Climate Change Threatens National Security Says Pentagon"*: https://unfccc.int/news/climate-change-threatens-national-security-says-pentagon

- Vimeo – *"Film: The Burden Trailer"*: https://vimeo.com/109036109

- The Real News – *"Col. Larry Wilkerson: Pentagon Plans for Climate Change Despite White House"*: https://www.youtube.com/watch?v=la3V_V6OfyI

Chapter 22 - Climate Crisis & Europeans' Viewpoint

- European Social Survey – *"European Attitudes to Climate Change and Energy"*: https://www.europeansocialsurvey.org/docs/findings/ESS8_toplines_issue_9_climatechange.pdf

- Cardiff University – *"European Perceptions of Climate Change"*: https://orca.cf.ac.uk/98660/7/EPCC.pdf

Chapter 23 - Climate Crisis & Children

- No Ordinary Lawsuit Podcast – *"Juliana v. United States: Behind the Scenes Access to the Trial of Our Century"*: https://www.noordinarylawsuit.org

- Medium - "Where can we find hope after the 1.5° UN climate report?": https://medium.com/@felixkramer/where-can-we-find-hope-after-the-1-5-un-climate-report-880d6a617ebe

- Our Children's Trust – *"Securing the Legal Right to a Safe Climate":* https://www.ourchildrenstrust.org

Chapter 24 - Climate Crisis & Education

- Brookings Institute – *"Combating Climate Change through Quality Education"*: https://www.brookings.edu/research/combating-climate-change-through-quality-education/

- National Education Association – *"Climate Change Education: Essential Information for Educators"*: http://www.nea.org/climatechange

- The Commonwealth Education Hub – *"The role of education in propelling climate action":*

- https://www.thecommonwealth-educationhub.net/the-role-of-education-in-propelling-climate-action/

- Frontline – *"Mailings to Teachers Highlight a Political Fight Over Climate Change in the Classroom"*: https://www.pbs.org/wgbh/frontline/article/mailings-to-teachers-highlight-a-political-fight-over-climate-change-in-the-classroom/

Chapter 25 - Climate Crisis & Millennials

- Harvard Kennedy School Belfer Center – *"Climate Change and Millennials"*: https://www.belfercenter.org/publication/climate-change-and-millennials

Chapter 26 - Climate Crisis & Health – A Physician's View

- World Health Organization – *"Climate Change and Health"*: https://www.who.int/news-room/fact-sheets/detail/climate-change-and-health

- United Nations Climate Change – *"Climate Change Impacts Human Health"*: https://unfccc.int/news/climate-change-impacts-human-health

Chapter 27 - Climate Crisis & Health – A Nurse's View

- Global Change.gov – *"The Impacts of Climate Change on Human Health in the United States: A Scientific Assessment"*: https://health2016.globalchange.gov

- CBS News – *"New government report reveals staggering economic and health toll of climate change"*: https://www.cbsnews.com/news/climate-change-report-national-climate-assessment-released-today-reveals-economic-health-toll-climate-change-2018-11-23/

Chapter 28 - Climate Crisis & Religion

- CBS This Morning/Katherine Hayhoe Interview: https://www.cbsnews.com/news/katharine-hayhoe-to-fight-climate-change-talk-about-it/

- BBC – *"Climate Change: Pope urges action on clean energy"*: https://www.bbc.com/news/world-europe-44424572

- Public Religion Research Institute – *"Most Believe Science in Conflict with Religion—Except Their Own Religious Beliefs"*: https://www.prri.org/spotlight/conflict-science-religion-debate/

Chapter 29 - Climate Crisis & Sports

- The News-Review – *"Where will the Iditarod go when snow and ice are gone?"*: https://www.nrtoday.com/where-will-the-iditarod-go-when-the-snow-and-ice/article_1b87904a-62c7-59c2-ac8a-d7c037ab226b.html

Chapter 30 - Climate Crisis & Civilization's Unacknowledged Energy Economic Constraint

- Information for Climate Action – *"The Economic Case for Climate Action in the United States"*: https://feu-us.org/case-for-climate-action-us/

- Axios – *"Wall Street is starting to care about climate change"*: https://www.axios.com/wall-street-is-starting-to-care-about-climate-change-1513303205-f97cf14c-c921-4ad0-b37a-12832acea4fb.html

Chapter 31 - Climate Crisis & Solutions

- The Guardian – *"Overwhelmed by Climate Change? Here's What you Can Do"*: https://www.theguardian.com/environment/2018/oct/08/climate-change-what-you-can-do-campaigning-installing-insulation-solar-panels.

- New Consensus - *"Green New Deal in Summary"*: https://newconsensus.com/green-new-deal/

- The Guardian - *"Don't know how to save the planet? Here's what you can do."*: https://www.theguardian.com/environment/2019/mar/25/dont-know-how-to-save-planet-this-is-what-you-can-do

- Axios – *"What tackling climate change means, illustrated"*: https://www.axios.com/what-tackling-climate-change-means-illustrated-5102232a-524b-46c7-8d40-9314535ac56e.html.

- Climate Changes Everything – *"Hopeful ways to think and act on climate"*: https://climatechangeseverything.org/hopeful-ways-to-think-and-act-on-climate/

- Yale Environment 360 – *"Climate Solutions: Is it Feasible to Remove Enough CO2 from the Air?"*: https://e360.yale.edu/features/negative-emissions-is-it-feasible-to-remove-co2-from-the-air

- Fight Global Warming Now, Bill McKibben: http://billmckibben.com/fightglobalwarmingnow/index.html

- The Guardian - "The destruction of the Earth is a crime. It should be prosecuted.": https://www.theguardian.com/commentisfree/2019/mar/28/destruction-earth-crime-polly-higgins-ecocide-george-monbiot

Plus:

- Climate Science Special Report- *"Fourth National Climate Assessment (NCA4), Volume 1"*: https://science2017.globalchange.gov

- Skywatch News video: *"Dire Climate Warning to Americans, Buried on Black Friday"*: https://www.youtube.com/watch?v=8MvsvfRlF8M.

- Frontline – *"As UN Sounds Alarm on Global Warming, Revisit Frontline's Recent Climate Reporting"*: https://www.pbs.org/wgbh/frontline/article/as-un-sounds-alarm-on-global-warming-revisit-frontlines-recent-climate-reporting/

- Truth Dig *"Extinction Rebellion"*: https://www.truthdig.com/articles/extinction-rebellion/

- CBS News/ Climate Change: https://www.cbsnews.com/climate-change/.

- College Astro 7 Power Deck: https://www.cabrillo.edu/~rnolthenius/astro7/A7PowerIndex.html.

- Green World Rising – Short films by Leonardo DiCaprio and Thom Hartmann – *"Carbon, Last Hours, Green World Rising, Restoration"*: https://www.greenworldrising.org.

- Common Dreams – *"We Are Climbing Rapidly Out of Humankind's Safe Zone': New Report Warns Dire Climate Warnings Are Not Enough"*: https://www.commondreams.org/news/2018/08/20/we-are-climbing-rapidly-out-humankinds-safe-zone-new-report-warns-dire-climate.

- Rolling Stone - *"This is How Human Extinction Could Play Out - Food system collapse, sea level rise, disease"*: https://www.rollingstone.com/politics/politics-features/bill-mckibben-falter-climate-change-817310/

- Climate Crocks - *"As America Wakes to Climate Impacts, Trump Sets Up "Gravity Denier" Panel"*: https://climatecrocks.com/2019/02/25/as-america-wakes-to-climate-impacts-trump-sets-up-gravity-denier-panel/

- Union of Concern Scientists - *"World Scientists Warning to Humanity"*, 1992, pdf: https://www.ucsusa.org/sites/default/files/attach/2017/11/World%20Scientists%27%20Warning%20to%20Humanity%201992.pdf

- BioScience - *"World Scientists' Warning to Humanity: A Second Notice"*, 2017: https://academic.oup.com/bioscience/article/67/12/1026/4605229

- Extinction Rebellion Video - *"Why we are heading for extinction and what to do about it"*: https://extinctionrebellion.nz/2019/01/15/xr-christchurch-why-we-are-heading-for-extinction-and-what-to-do-about-it-video/

- Extinction Rebellion US Demands and Principles: https://xrebellion.org/xr-us/demands

Acknowledgements

The idea and desire to create this book would still be rattling around in my head, and a goal left unaccomplished if it wasn't for my *hero hubby*, Jerry Grimes. All the demands around our home

and of life would have blocked me from finding the time if he had not supported my dream. Thank you to Kevin Quinn, who began as a chapter author, and then stepped up to be the team's Creative Editor. Thank you to Patch Mackenzie for brainstorming subtitles for *Climate Abandoned* and coming up with the one selected. Thank you to Martha Bullen, whom I met in New York at a Quantum Leap Publicity Summit two years ago, and has been my touchstone as an author coach ever since. Thank you to Deana Riddle for assuring a print version of the interior design for this book would happen. Thank you to Joe Jordon for brainstorming energy saving ideas for the Climate Crisis & Energy chapter. Thank you to Dr. Richard Gammon, Ph.D., for reviewing the book's climate science when requested.

A very special and sincere thank you to the chapter authors who responded to my "Call for Authors" on the Climate Reality Project Hub and for volunteering to write a chapter of their choice, plus another thank you goes to those additional chapter authors who joined the team. Thank you, everyone, for supporting my dream over this two-and-a-half- year journey while providing your time, expertise, passion, and offering your *urgent voice* by contributing to this anthology:

Cathy Bowen Becker, Ohio	Luis Camargo, Columbia
Kasper Eplov, Sweden	Solange Marquez Espinoza, Mexico
Bob Hallahan, Washington	Peter Hess, California
G. Elizabeth Kretchmer, Washington	Annamaria Lehoczky, Spain
Cindy Martinez, Missouri	William McPherson, Washington
Robert Mullins, California	Mike Newland, California
Hari Krishna Nibanupudi, India	Richard Nolthenius, California
Julie Packard, California	Rituraj Phukan, India
Kevin Quinn, Colorado	Lois Robin, California
Betsy Rosenberg, Texas	Maria Santiago Valentin, New Jersey
Jigar Shah, California	Brij Singh, Michigan

Mathieu Thuillier, France

About the Editor

Jill Cody, M.P.A.

As a citizen advocate, award-winning author, and KSQD 90.7FM radio host, Jill's 31-year career in public service, was and still is infused with a life-long passion for knowledge and a will to inspire change in politics, the environment, higher education, and organizational development.

Jill was personally trained and authorized by vice president Al Gore (2007 - 4th class) to share his astonishing *"An Inconvenient Truth"* presentation with schools and community groups across Silicon Valley. This revelatory information caused Jill to realize America's abandonment, not only of the planet but in many other important aspects of society. *America Abandoned – The Secret Velvet Coup That Cost Us Our Democracy*, explores the myriad ways abandonment is unknowingly permeating the lives of the American public and what "We, the People" can do to reverse this course. Now, with *Climate Abandoned - We're on the Endangered Species List*, she explores, with 22 additional chapter authors, how our planet has been abandoned and what we can do about it.

Jill was honored when asked to lead and facilitate national and international strategic planning meetings on Information Literacy in Washington D.C., Prague, and Egypt, that created and built teams from around the world to bring Information Literacy skills to their respective countries.

Also, Jill was delighted when asked to host a radio show on KSQD 90.7FM, www.ksqd.org, loosely based on her book, *America Abandoned*. She now is producer and host of "Be Bold America!" a bi-weekly, hour-long radio program for those who are motivated to step-out with *bold,* spirited actions necessary to reunite this country and save our democracy.

Jill earned a master's degree in public administration from San Jose State University's Political Science Department and was named distinguished alumni in the College of Applied Arts and Sciences. Jill is proud to currently serve on the university's Emeritus and Retired Faculty Association Executive Board.

Jill lives in Carmel, California with her husband, Jerry, and their two dogs, Sassy and Scooter.

www.jillcodyauthor.com
www.climateabandoned.com
www.americaabandoend.com
bebold@ksqd.org

Coming 2021

Thought Abandoned: How Greedy and Unsavory Characters
Gaslight Minds for Money and Power

Thought Abandoned is a quick reference guide to the subtle and subliminal techniques foisted on the American mind, minute by minute. As you are listening and watching cable news, opinion radio, product advertising flip through *Thought Abandoned* and identify which gaslight techniques their well-paid, professionally hired *gaslighters* inflict on your thinking. News flash! You don't own your mind. They do.

Gaslighting: *A form of intimidation or psychological abuse where false information is presented to the victim, making them doubt their own memory, perception and quite often, their sanity.*

A form of intimidation or psychological abuse where false information is presented to the victim, making them doubt their own memory, perception and quite often, their sanity. **–Urban Dictionary**

The end goal of gaslighting your thoughts is to buy a product you don't need or believe as reality something false or manipulate your perception to act or vote against your own interests. The *gaslighter* wants something from you, and it's either the power you can handover to them or the money in your wallet.

Byant Welch Ph.D., in his book *State of Confusion: Political Manipulation and the Assault on the American Mind*, explains how our mind creates a sense of reality. We think the reality is external, but it is a mental function where we select bits and fragments and weave them into our reality. What if those bits and pieces are meant to gradually transform our thoughts to make us anxious, confused, less trusting, and dependent on the *gaslighter's* needs and perceptions? As explained by Dr. Welch *"I became increasingly concerned with the growing instability of the American mind itself as it struggled under trauma and political manipulation to bear the psychological burdens to maintain our democracy."*

Thought Abandoned will be a reference guide listing numerous gaslighting strategies such as the use of dopamine, creating doubt, future pacing, puffing, inoculation theory, scapegoating, and emotional, crowd & data manipulations. Today, we suffer an onslaught of bits and fragments of gaslighting ever minute. Win back your mind with:

Thought Abandoned: How Greedy and Unsavory Characters
Gaslight Minds for Money and Power

Made in the USA
San Bernardino, CA
23 April 2019